NUCLEAR CHEMISTRY: NEW RESEARCH

NUCLEAR CHEMISTRY: NEW RESEARCH

AXEL N. KOSKINEN
EDITOR

Nova Science Publishers, Inc.
New York

For permission to use material from this book please contact us:
Telephone 631-231-7269; Fax 631-231-8175
Web Site: http://www.novapublishers.com

LIBRARY OF CONGRESS CATALOGING-IN-PUBLICATION DATA

Koskinen, Axel N.
Nuclear chemistry : new research / Axel N. Koskinen.
p. cm.
ISBN 978-1-60456-957-5 (hardcover)
istry. I. Title.
008

2008032305

ublished by Nova Science Publishers, Inc. ✦ New York

CONTENTS

PREFACE

Nuclear chemistry is a subfield of chemistry dealing with radioactivity, nuclear processes and nuclear properties. It is the chemistry of radioactive elements such as the actinides, radium and radon together with the chemistry associated with equipment (such as nuclear reactors) which are designed to perform nuclear processes. This includes the corrosion of surfaccs and the behaviour under conditions of both normal and abnormal operation (such as during an accident). An important area is the behaviour of objects and materials after being placed into a waste store or otherwise composed of:

- the study of the chemical effects resulting from the absorption of radiation within living animals, plants, and other materials. The radiation chemistry controls much of radiation biology as radiation has an effect on living things at the molecular scale, to explain it another way the radiation alters the biochemicals within an organism, the alteration of the biomolecules then changes the chemistry which occurs within the organism, this change in biochemistry then can lead to a biological outcome. As a result nuclear chemistry greatly assists the understanding of medical treatments (such as cancer radiotherapy) and has enabled these treatments to improve.
- the study of the production and use of radioactive sources for a range of processes. These include radiotherapy in medical applications; the use of radioactive tracers within industry, science and the environment; and the use of radiation to modify materials such as polymers.
- the study and use of nuclear processes in non-radioactive areas of human activity. For instance, nuclear magnetic resonance (NMR) spectroscopy is commonly used in synthetic organic chemistry and physical chemistry and for structural analysis in macromolecular chemistry.

In the Short Communication, results on elemental concentrations in bulk deposition samples collected in the city of Megalopolis showed that the As, Se, Mo, Br, Sb, U and Zn were enriched compared to the crustal abundances, due to the fly ash escaping form the power station, since the same elements were also enriched in fly ash samples. Natural radionuclide measurements in surface soil samples collected from areas around the city showed that the specific activities of ^{238}U and ^{226}Ra were higher compared with the average values for Greek soil, indicating slightly enhanced natural radionuclide concentrations, due to the fly ash deposition on the surface environment. Indoor radon measurements were also carried out in

50 dwelling using SSNTD. The average annual arithmetic and geometric means were found to be 53 ± 34 Bq m^{-3} and 43 ± 2 Bq m^{-3}, respectively, which are comparable with radon values found in other Greek cities. The resulted total effective dose (1.4 mSv y^{-1}) for the local population due to outdoor external irradiation of terrestrial origin and to indoor internal irradiation from the short-lived decay products of ^{222}Rn was lower than the global effective dose due to natural sources (2.4 mSv y^{-1}).

Cesium-137 (^{137}Cs) is an artificial radionuclide which did not exist in the environment until man began atmospheric testing for nuclear weapons development. The phrase "cloud forest" and its` characteristics are very ambiguous and can be misleading, thus care needs to be taken into account when this term is encountered. The activity of ^{137}Cs was determined mainly in surface soils, as well as mosses, lichens and other types of vegetation. The ^{137}Cs activities (Bq kg^{-1}) were measured by high energy resolution gamma-ray spectroscopy employing reference materials as comparators.

In Chapter 1, the spatial distribution of ^{137}Cs in various tropical cloud forests, foggy areas and their surrounding areas in Venezuela will be presented, in order to show the effect that cloud forests and foggy areas had on the deposition of radioactive fallout. The relationship between the ^{137}Cs activities in the surface soils and the elevations of their sampling sites not only clearly showed the effect that the cloud forests had on the enhancement of ^{137}Cs but also can be used to estimate the base of the cloud cover. Cloud forests were usually considered to be in mountainous areas at high altitudes, but the peaks of some of the tropical cloud forests discussed herein were located at relatively low elevations. Finally, the ^{137}Cs activities in soils, mosses and lichens collected along the Caruay River will be presented with respect to its surrounding area. The area along the river had similar characteristics to a tropical cloud forest, but this was due to the fog which was produced daily.

In conclusion, it is very important to consider the effect of cloud forests and other similar pathways, such as foggy areas, in which ^{137}Cs can be directly deposited from the atmosphere at much higher concentrations than by rainfall, or one could report anomalously high activity values, which could suggest that more ^{137}Cs has been recently introduced into the environment.

The application of the instrumental neutron activation analysis (INAA) in support of provenance research of archeological ceramics has been largely used over the past few decades. INAA is a sensitive technique useful for quantitative multi-element analysis of major, minor, and trace elements. Among these, trace elements can be particularly useful to determine provenance of the mineral components in the paste of archaeological ceramics or to distinguish different ceramic groups, better differentiating diverse raw materials. The chemical characterization of the regional potential raw materials is particularly important in the case of the lack of kilns structures in archaeological excavations, and if the ceramic artifacts study points to local productions. This is the case of Bracara Augusta (nowadays Braga), a Roman town located in the north of Portugal, recognized as a center of production and distribution of pottery. In Chapter 2 chemical data (major and specially trace elements) of the different types of clay materials of Braga region obtained by instrumental neutron activation analysis are used to establish the geochemical profiles of raw materials, which can be used as indicators of provenance of several ceramic types found in Bracara Augusta. Among the chemical elements studied rare earth elements play an important role differentiating resources and their correlation with ceramics. The results obtained suggest that during Roman times two main resources were used to produce different types of ceramics: (1)

the sedimentary clays of the Prado basin (located near Braga) to produce common wares, amphorae, as well as imitations of "Pompejanish-roten slip", and "Terra Sigillata Hispanica" and "African Red Slip Ware" forms; and (2) sedimentary kaolin deposits located 30 km west of the town, near the coast, for the production of the "Bracarense" ceramics, a fine yellow-pale paste covered with a brownish yellow slip. Until the finding of archaeological evidence of pottery workshops in the area, one may assume that potters, already settled in Bracara Augusta, may have went further west to look for especial clays to answer specific needs, functional and aesthetic. The chemical characterization of the ceramics pastes and their comparison with ceramics of well established origin also point to imports of several types of ceramics from other locations of the Roman Empire, such as Italic "Pompejanish-roten" production (Campania) as well as imitations of "Pompejanish-roten slip" produced in Lucus Augusti (NE Braga).

The characteristics of the ROS, such as hydroxyl radical, formation in depleted uranium (DU)-hydrogen peroxide (H_2O_2) systems and the effect of biosubstances on the radical formation were studied in detail. The hydroxyl radical formation in the UO_2^{2+}-H_2O_2 system can be described as a stepwise reaction process including the reduction of UO_2^{2+} to UO_2^{+} by H_2O_2 and the Fenton-like reaction of UO_2^{+} with H_2O_2. A part of UO_2^{+} will be reduced to UO^{2+} through disproportionation under appropriate pH conditions. The UO^{2+} ion is oxidized back to UO_2^{2+} through a couple of Fenton-like reactions under the coexistence of H_2O_2, which suggests that twice the amount of the hydroxyl radical will be formed in the UO^{2+}-H_2O_2 system than in the UO_2^{2+}-H_2O_2 one. The behaviors of biosubstances, such as saccharides, amino acids, and microbial cells, in the UO_2^{2+}-H_2O_2 system could be explained by the direct elimination of the hydroxyl radical and the inhibition of the hydroxyl radical formation through the coupling of the UO_2^{2+} ion with these compounds. Catechins, especially epigallocatechin, enhanced the hydroxyl radical formation by the reduction of UO_2^{2+} to UO^{2+}. Mimosa and quebracho tannins enhanced the hydroxyl radical formation by the reduction of UO_2^{2+} to UO^{2+}, while Chinese gallo, mylaboram, and chestnut tannins depressed the hydroxyl radical formation by strongly coupling with the UO_2^{2+} ion. The behaviors of vitamins, such as C and E, eliminated spin adducts in addition to the hydroxyl radical, which caused complicated behaviors in the spin-trapping procedure in the UO_2^{2+}-H_2O_2 systems. In summary, Chapter 3 explains that some biosubstances enhance the ROS formation in the DU-H_2O_2 system, and some of them reduce it, which suggests positive and negative effects, respectively, on the chemical toxicity of DU.

The possibility of a radiological/nuclear event and the consequences of its aftermath have led the scientific community to rethink its analytical approach toward environmental and biological radiological monitoring. Rapid procedures leading to precise and unequivocal results are required for timely response, whether dealing with an alpha, beta, or gamma-emitter. Chapter 4 will highlight recent technological developments within the Radiation Protection Bureau, Health Canada (RPB/HC) to cope with the unique requirements associated with an emergency situation. Three types of radioanalyses will be covered in this chapter, ranging from long-lived alpha actinides to radiostrontium and gamma emitters. First, automated methodologies for the rapid detection of $^{89,90}Sr$ in milk and water will be described. This automated system is developed around a computer-controlled radiochemical procedure followed by on-line counting using stop-flow liquid scintillation. For radiostrontium ($^{89,90}Sr$), a detection limit of 200 mBq (800 mL sample with a counting time of 50 minutes) is achieved for each nuclide in the absence of a complementary radionuclide. The two stage radiochemical separation of radiostrontium can be performed on eight samples simultaneously and enables the separation of other radionuclides such as ^{226}Ra. Second, a rapid methodology for the on-line detection of plutonium in urine using an inductively coupled plasma mass spectrometer (ICP-MS) will be described. For ^{239,240}Pu urinalysis, a detection limit of 0.58 mBq L^{-1} (^{239}Pu) at an analytical rate of one sample every 15 minutes is possible. The analytical performances of these methodologies have been demonstrated for both reference and inter-comparison samples.

Finally, an automated analysis platform for rapid gamma-spectral analysis used at RPB will be described. It consists of a database server, with storage schema named Linux System for Spectral Information (LINSSI), a high purity germanium gamma spectra analysis software (UniSampo/Shaman), and a communication scripts layer. This platform is used to process daily global Comprehensive Nuclear Test Ban Treaty (CTBT) and weekly Canadian radiological monitoring air filters for airborne radioactivity analysis. Raw gamma-spectral data is emailed to our server from field laboratories for analysis and assessment with results being available in near real-time. This analytical engine can be connected to any radiation spectrometry software, either commercial or in-house. Use of database storage allows the raw data, sample parameters, and final results to serve as a comprehensive radiological information archive. The nuclide detection limit of the system is suitable for both environmental radiation monitoring and emergency preparedness purposes. The performance of this platform including its capabilities and limitations for rapid gamma-peak search, nuclide identification, and radionuclide activity concentration evaluation will be described.

As an important waveguiding medium, optical fiber plays significant roles in optical communications, optoelectronics, and sensors. A new type of microstructure inscribed in the optical fibers, i.e., fiber Bragg gratings (FBGs), has received considerable attention in recent years. A FBG is a type of distributed Bragg reflector constructed in a short segment of optical fiber that reflects specific wavelengths of light and transmits all the other components. In Chapter 5, optical properties of FBGs will be reviewed first with the underlying physical mechanisms. Different techniques to fabricate FBGs will be illustrated with the comparison of their advantages and drawbacks. For their important sensing applications, FBGs as temperature and humidity sensors will be discussed. The FBG sensors exceed other conventional electric sensors in many aspects, for instance, immunity to electromagnetic interference, compact size, light weight, flexibility, stability, high temperature tolerance, and resistive to harsh environment. A novel approach to realize individual temperature or humidity measurement, and their simultaneous measurement will be demonstrated by the use of FBGs coated with different polymers. The polymer-coated FBGs indicate linear shifts in the Bragg resonance wavelengths of the gratings with the temperature changes. A polyimide-coated FBG is sensitive to humidity due to the unique hygroscopic properties of polyimide while an acrylate-coated FBG shows insensitivity to humidity. The experimental results are in good agreement with the theoretical analysis.

N,N-Dialkyl-3-Oxa-Glutaramic Acid ($RR'N(COCH_2)_2OHO$) (DAOGA) and N,N,N'N'-tetra-alkyl-3-oxa-pentanediamide ($(RR'NCOCH_2)_2O$) (TAOPDA) are two serials of powerful ligands to lanthanides (III) and actinides(III and IV) . Their solubility in organic phase or aqueous phase and other properties vary with their different substituted alkyl groups, R and R'. Due to these merits, they are able to be used as extractants or stripping agents in nuclear chemistry, specifically treatment of nuclear spent fuel. In Chapter 6, the progress of lately researches on organic-soluble oxa-pentanediamide as extractants, which is world widely studied including in our lab, is reviewed. Secondly, the synthesis, properties of aqueous-soluble DAOGA and TAOPDA, and their applications are summarized, of which the following work were initiated in our lab: synthesis and identification of six ligands of these two series; structure identification with IR and EXAFS of the extracted complexes of Ln(III) with N, N'-dimethyl-3-oxa-glutaramic Acid (DMOGA); determination of stability constants of Np(IV) and Pu(IV) complexes with DMOGA; and measurements of the distribution ratios of actinides and some fission products between 30% trialkyl phosphine oxide (TRPO) in kerosene and HNO_3-DMOGA solutions. Based on these results, DMOGA is proposed to be used as an effective stripping agent for An(III, IV) and therefore the TRPO process, well-known as one of the promising processes for treatment of high level liquid waste(HLLW) , is simplified.

High level waste (HLW) produced from reprocessing of the spent nuclear fuel still contains actinides, ^{90}Sr, ^{137}Cs, ^{99}Tc, ^{129}I, etc., which engender permanent hazard to environment. The safe treatment and disposal of HLW are crucial to sustainable development of nuclear energy. The partitioning and transmutation (P&T) strategy for converting long-life nuclides from HLW into short-life radioactive or stable species has increasingly attracted interest. In recent years, a total partitioning process, including removal of actinides by the TRPO process, removal of Sr by the crown ether strontium extraction (CESE) process, removal of Cs by ion exchange using potassium titanium hexacyanoferrate (II) or by solvent extraction with 25,27-Bis(2-propyloxy) calix[4]-26,28-crown-6 (iPr-C[4]C-6), and separation of lanthanides and actinides with Cyanex 301, has been developed to treat Chinese HLW at Institute of Nuclear and New Energy Technology (INET), Tsinghua University, China. A series of cold and hot tests of both the separate processes and the total partitioning process were carried out with centrifugal contactors, and satisfactory results were obtained from these tests. In particular, a hot laboratory-scale test of the total partitioning process was carried out with the genuine HLW at INET in 1996. In this test, 50 stages ϕ10 mm centrifugal contactors were applied successfully for removal of actinides and Sr. The test results showed that the decontamination factors (DFs) for total α activity, ^{99}Tc, ^{90}Sr and ^{137}Cs were 588, 125, >2500, and >200 respectively. In addition, a cold pilot-scale test of the total partitioning process was carried out with the simulated HLW at INET in 2005. In this test, 24 stages ϕ70 mm centrifugal contactors were applied successfully in the stripping section of the TRPO process for removal of actinides. The test results showed that 99.9% of Nd, Zr and U was stripped respectively. In order to perform successfully these tests, a series of new centrifugal contactors have been developed at INET. In Chapter 7, developments of centrifugal contactors for partitioning of HLW are introduced, and applications of centrifugal contactors in tests for partitioning of HLW in China are reviewed in detail.

In the frame of a European project (Eureka Σ!2210 "Eurocare-Bronzart"), the effectiveness of several corrosion inhibitors for bronze surfaces: benzotriazole derivatives and 3-mercapto-propyl-trimethoxy-silane, was comparatively evaluated by the Thin Layer Activation method. Bronze surfaces were irradiated with a 11.5 MeV proton beam, generated by a MC-40 energy variable cyclotron. The tracer radio nuclide ^{65}Zn ($t_{1/2}$ = 244 days) was produced by the following nuclear reaction: ^{65}Cu(p,n)^{65}Zn. Activated surfaces of different bronze alloys, whose composition is suitable for production of outdoor sculptures, were coated with the inhibitors and exposed to natural and artificial corrosive environments. After determination of the activity versus depth functions, the thickness loss of the activated samples, due to the effects of corrosion, was calculated by measurements of the ^{65}Zn γ-activity at 1115 KeV ($I_\gamma = 0.50$), carried out with a multi channel analyser equipped with a Ge detector. In evaluating the sample thickness loss, two different procedures were applied in Chapter 8. The first one consisted in the quantification of the activity of ^{65}Zn on the solid sample, prior to exposure and after stripping. The second one consisted in a step by step electrolytic dissolution of the solid sample, after stripping the corrosion layer. The ^{65}Zn activity was successively measured in the pickling solution and in each solution arising from the electrolytic dissolution of the solid sample. The last method, because of its high sensitivity, was applied to evaluate the thickness loss at sub micrometer scale, as in the case of a mild corrosion.

We study the binding energies and radii of spin saturated light nuclei with nucleon number in the range $8 \leq A \leq 30$ in the framework of semiclassical Monte Carlo many-body simulations. Using a model Hamiltonian consisting of the usual kinetic term, hadronic nucleon nucleon (NN) interaction potential, Coulomb and an effective Pauli potential, which depends on density, results are obtained by the simulation of a set of isotopic chains for selected nuclei. In particular, especial attention is devoted to nuclei in the vicinity of proton and neutron drip lines. In this latter case proton and neutron skins develop. Chapter 9 suggests a pragmatic procedure for modelling a set of non-isosymmetric nuclei focusing on the description of the empirical binding energies and different (matter, neutron, charge) radii, deduced from experimentally available data, starting from a given effective nuclear interaction potential picture and statistical treatment.

Short Communication

In: Nuclear Chemistry: New Research
Editor: Axel N. Koskinen, pp. 3-11
ISBN: 978-1-60456-957-5

Environmental Impact of a Lignite-Fired Power Plant Operation: The Case of Megalopolis, Southern Greece

H. Papaefthymiou *

Division of Physical, Inorganic and Nuclear Chemistry, Department of Chemistry, University of Patras, Patras 25600, Greece

Abstract

Results on elemental concentrations in bulk deposition samples collected in the city of Megalopolis showed that the As, Se, Mo, Br, Sb, U and Zn were enriched compared to the crustal abundances, due to the fly ash escaping form the power station, since the same elements were also enriched in fly ash samples. Natural radionuclide measurements in surface soil samples collected from areas around the city showed that the specific activities of ^{238}U and ^{226}Ra were higher compared with the average values for Greek soil, indicating slightly enhanced natural radionuclide concentrations, due to the fly ash deposition on the surface environment. Indoor radon measurements were also carried out in 50 dwelling using SSNTD. The average annual arithmetic and geometric means were found to be 53 ± 34 Bq m^{-3} and 43 ± 2 Bq m^{-3}, respectively, which are comparable with radon values found in other Greek cities. The resulted total effective dose (1.4 mSv y^{-1}) for the local population due to outdoor external irradiation of terrestrial origin and to indoor internal irradiation from the short-lived decay products of ^{222}Rn was lower than the global effective dose due to natural sources (2.4 mSv y^{-1}).

Introduction

Combustion of coal for heat or electric energy production is one of the major anthropogenic sources of air pollution (Kampa and Castanas, 2008). Coal, as a sedimentary rock, contains chemically toxic trace elements and also naturally occurring radioactive elements (^{238}U, ^{235}U, ^{232}Th decay series and ^{40}K) in concentrations, in some cases higher than those in sedimentary

* E-mail address: epap@chemistry.upatras.gr

rocks (Swaine, 1990; Clarke and Sloss, 1992). During combustion, a considerable amount of toxic and radioactive elements are released into the atmosphere associated with the fine-sized fly ash particles, while large amounts of these remain as bottom ash (Clarke and Sloss, 1992; Seams and Wendt, 2000; Arditsoglou et al., 2004). Although the majority of fly ash is retained by particulate control devices, the smaller and respirable particles are not efficiently collected (Clarke and Sloss, 1992). Thus, they are transferred from their source over a wide range of distances and deposited on the soil around power plants or inhaled from the population and might cause environmental and health implications. The amounts of toxic elements and radionuclides released into the atmosphere mainly depend on the initial concentrations in the coal, but also on combustion engineering and pollution control devises at the plant. The large amounts of coal that must be burned make the impact of coal-fired power plants significant in some cases.

The lignite mining operations classify Greece in the second position between lignite burning countries in European Commission and fifth in the world, since more than 65% of the installed capacity of electrical energy is produced by lignite burning in two main power stations (Ptolemais, 3020 MW and Megalopolis 850 MW). Greek lignites especially that of Megalopolis basin are characterized by high moisture content, low calorific value and very high ash content. They also appear to have high concentrations of certain toxic elements and natural radionuclides, especially that of ^{238}U and ^{226}Ra (Simopoulos and Angelopoulos, 1987; Papaefthymiou et al., 2007).

Megalopolis, a small city in the centre of the Peloponnese peninsula with 10,000 inhabitants, is located about 2.5 and 4 km NE to the lignite power plants A (550 MW) and B(300 MW). In order to evaluate the impact of the lignite power plants operation on the environment of the Megalopolis city, a number of studies were performed using nuclear analytical techniques: 1) Determination of dust and elemental atmospheric inputs on surface environment using Instrumental Neutron Activation Analysis combined with high resolution γ-ray spectrometry, 2) Natural radioactivity measurements in surface soil samples using direct high resolution γ-ray spectrometry and 3) Indoor radon measurements in 50 dwellings in the city of Megalopolis by means of Solid State Nuclear Track Detector (SSNTD) technique.

Experimental

Collection of Bulk Deposition Samples and Elemental Analysis

Bulk (dry and wet) deposition samples were collected at four sites (M1-M4) in Megalopolis city and one (M5) at the power plant A area, during 1997, on a monthly basis. The sampling procedure is described in detail elsewhere (Papaefthymiou et al., 2005). Briefly, the samples were collected in high-walled PVC pots containing a layer of 2 cm in height ultra pure water, which were located on the flat floor of two-storey buildings (EML, 1992). At the end of the month, all the content of the pot was transferred to a porcelain dish and was concentrated till dryness, by evaporation at $75^{0}C$, in a water bath.

The elemental analysis of the collected samples has been performed by Instrumental Neutron Activation Analysis (INAA) combined with a high-resolution γ-ray spectrometry. INAA is a multielemental non-destructive nuclear technique, which is widely used in pollution studies. The samples were irradiated at the research reactor of the National Research

Centre for Physical Sciences "Demokritos", NCRP (Athens, Greece) for 30 min. After irradiation, the samples were counted using a γ-spectrometry setup consisting of a HPGe detector of 20% efficiency and 1.69 keV FWHM for the 1332 keV ^{60}Co γ-ray, connected to a Canberra 35+, 4K-MCA. The accuracy of the analytical results was verified by analyzing, together with the samples, the GRX-1 standard from the United States Governmental Geochemical Standard (Papaefthymiou and Anousis, 2006). The countings were performed at the Nuclear Technology Laboratory of the NCRP.

Natural Radioactivity Measurements in Surface Soil Samples

Ten surface soil samples (0-5 cm deep) were carefully collected from open areas around the city, which were thought to be undisturbed. All samples were air-dried. After removing the stones and grass, the samples were crushed to fine powder (200 mesh), homogenized and oven dried at a temperature of 105 ± 1°C for 24 h. Then, the samples, ranging in mass from 60-80 g, were packed into cylindrical containers weighted, sealed hermetically and stored for a minimum period of 4 weeks prior to counting to establish secular radioactive equilibrium between ^{226}Ra and its short-lived daughter products.

The activity concentrations of the natural radionuclides ^{238}U, ^{226}Ra, ^{232}Th and ^{40}K in the soil samples were determined by direct high resolution γ-ray spectrometry, using a Canberra HPGe detector with a relative efficiency of 25% and FWHM of 1.9 keV for ^{60} Co γ-ray energy line at 1332 keV. The detector was coupled to a PC-based 8K multichannel analyzer and the appropriate electronics. Energy and efficiency calibration procedures of the system are described in detail elsewhere (Papaefthymiou et al., 2007). The counting time of each sample was 24 h in order to achieve statistically meaningful results.

The ^{238}U specific activity was derived from the weighted mean of the photopeaks of ^{234}Th (63.3 and 92.8 keV). The ^{226}Ra specific activity was derived from the weighted mean of the activities of the photopeak of ^{214}Bi (609.3 keV) and of two photopeaks of ^{214}Pb (295.2 and 352.0 keV), while for ^{232}Th the photopeak of ^{228}Ac (911.1 keV) and the photopeaks of ^{212}Pb (583.1 keV) and ^{208}Tl (238.6 keV) were used. ^{40}K was directly determined using the 1460.8 keV photopeak. The accuracy of the method was checked by measuring under the same conditions as the samples, the reference materials IAEA-312 and IAEA-375. All countings for specific activity measurements were performed at the Nuclear Chemistry Laboratory of the Department of Chemistry, University of Patras.

Indoor Radon Measurements

Measurements of the indoor radon concentrations in the city of Megalopolis were performed by means of the passive track etching method with the LR-115 SSNTD film, type II (Kodak-Pathé, France) in "closed can" geometry (Kritidis et al., 1994; Papaefthymiou et al., 2003). The survey lasted one year and obtained as four three-month surveys during 2006. Fifty detectors for every season were prepared and distributed in 50 dwellings in different floor levels. After exposure, the films were etched in a 2.5 N analytical grade NaOH solution at a constant temperature of 60.0 ± 0.5 °C for 145 minutes. After etching, the detectors were washed for 30 min with running cool water and then with distilled water followed with a 50%

water/alcohol solution. After drying in the air, the films were subjected to track counting. Using the calculated calibration factor (2.3 x 10^{-3} tracks cm^{-2} per Bq h m^{-3}), the measured numbers of tracks were converted to radon concentration in Bq m^{-3} (Papaefthymiou et al., 2003). The combined uncertainty (statistical and calibration) of a single three month-long measurement at the average level of 40 Bq m^{-3} was about 15% (1σ).

Calculation of Radiological Effects

To evaluate the external exposure from naturally occurring radionuclides, the total air absorbed dose rate, D due to the mean activity concentrations of ^{238}U, ^{232}Th and ^{40}K (Bq kg^{-1}) in the surface soil samples was calculated using the formula of Beck et al. (1972):

$$D=0.462A_U+0.604A_{Th} + 0.042\ A_K \qquad (1)$$

where D is in nGy h^{-1} and A_U, A_{Th} and A_K the activity concentrations of ^{238}U, ^{232}Th and ^{40}K, respectively in Bq kg^{-1}. For the conversion from absorbed dose rate in air to annual effective dose to adults, the coefficient 0.7 Sv Gy^{-1} and an outdoor occupancy factor of 0.2 proposed by UNSCEAR (2000) have been used.

To evaluate the annual effective dose to the population of the city of Megalopolis in the indoor environment due to radon a conversion dose factor of 9 nSv Bq m^{-3}, a mean equilibrium value of 0.4 and an occupancy factor of 80% were adopted (UNSCEAR, 2000).

Results and Discussion

Elemental Concentrations

A total of 23 major, minor and trace elements were detected in all the collected samples using INAA. In this work, 16 of them, which are considered as most important from an environmental point of view, are presented (the elements which are not presented are of Eu, Hf, La, Rb, Ta, Tb, Yb). Since the differences in mean values of dust amount and mean elemental concentrations between the four sites located in Megalopolis city (M1-M4) were not statistically significant, the average values during the one-year sampling period were calculated. These values, along with the average values of dust and elemental concentrations deposited at site M5 during the same period, are presented in Table 1. Statistical analysis revealed that the differences in mean dust amount between sites M1-M4 and M5 were statistically significant, whereas the differences in mean elemental concentrations between sites M1-M4 and M5 were not statistically significant for most of the analyzed elements, except for As, Ca, Na, Mo, Se and U. The location of site M5 in the area of power plant A explains this difference (Papaefthymiou, 2008).

To evaluate if the source of an element is anthropogenic or of natural origin, the values of the Enrichment Factor, *EF,* were calculated according to the following equation:

$$EF = \frac{(X/C)_{sample}}{(X/C)_{reference}} \tag{2}$$

where X is the concentration of the element of interest and C is the concentration of a the reference element Sc. The concept of *EF* is based on the fact that some elements originating from well-defined sources can be distinguished from other elements derived from different sources. Thus, value of *EF* greater than ten is indicative that the source of the element is anthropogenic, while value close to one is indicative for natural sources of the element. The calculated *EF* values of the detected elements are also presented in Table 1.

Table 1. Average dust and elemental concentrations in bulk deposition samples from sites M1-M4 and average dust and elemental concentrations at site M5 (in mg kg^{-1}, *g kg^{-1}) during February 1997-January 1998. The calculated Enrichment Factor (EF) values are also included.

	Average value (M1-M4)		M5	
Element	**AM±SD**	**EF**	**AM ± SD**	**EF**
As	11 ± 1	11 ± 1	18 ± 7	16 ± 4
Br	62 ± 3	47 ± 5	76 ± 51	50 ± 34
Ca*	83 ± 6	3.7 ± 0.6	64 ± 10	2.5 ± 0.4
Ce	60 ± 3	1.7 ± 0	66 ± 22	1.7 ± 0.4
Co	22 ± 1	1.5 ± 0	24 ± 5	1.5 ± 0.1
Cr	202 ± 10	3.6 ± 0.2	200 ± 57	3.2 ± 0.9
Cs	4.7 ± 0.3	2.7 ± 0.1	5.1 ± 1.3	2.7 ± 0.1
Fe*	45 ± 3	1.4 ± 0.2	53 ± 12	1.5 ± 0.1
Mo	43 ± 5	67 ± 3	62 ± 18	67 ± 22
Na*	10 ± 2	0.8 ± 0.2	6.1 ± 2.6	0.4 ± 0.2
Sb	6.5 ± 1.9	72 ± 29	4.1 ± 2.4	35 ± 24
Sc	13 ± 1	-	14 ± 3	-
Se	12 ± 1	428 ± 17	20 ± 10	607 ± 233
Th	8.8 ± 0.5	1.6 ± 0	9.6 ± 2.4	1.6 ± 0.1
U	35 ± 3	23 ± 1	51 ± 14	30 ± 3
Zn	333 ± 41	10 ± 1	330 ± 130	7.5 ± 3.1
Dust , g	0.8 ± 0.1	-	1.6 ± 0.7	-

As shown in Table 1, the elements As, Br, Mo, Sb, Se and U present EF values higher than ten at all sites, which indicates that the main sources of these elements are anthropogenic. Zinc behaves as enriched at sites M1-M4, but not at site M5. The *EF* values of U, As, Mo and Se are higher for site M5, compared with the sites M1-M4. The opposite behavior stands for Sb. The elements As, Mo, Sb, Se and U were also found to be enriched in fly ash samples collected from the Megalopolis power plant A (Papaefthymiou et al., 2007). It should be noticed that As, Se, Sb and U are between the 189 substances cited as potentially hazardous air pollutants (Finkelman, 1994).

Natural Radionuclides in Surface Soil Samples

Table 2 presents the activity concentration values of the ^{238}U, ^{226}Ra, ^{232}Th and ^{40}K in each soil sample along with the world and Greek soil average values. This table also includes the specific activities of the aforementioned radionuclides in fly ash samples from the Megalopolis station (Papaefthymiou et al., 2007). The absorbed dose rates in air, calculated according to Eq. 1 are also given in this table. The ^{238}U activity concentrations ranged from 36 to 208 Bq kg^{-1} (average:73 Bq kg^{-1}), the ^{226}Ra concentrations ranged from 33 to 126 Bq kg^{-1} (average: 51 Bq kg^{-1}), that of ^{232}Th ranged from 25 to 39 Bq kg^{-1} (average: 33 Bq kg^{-1}), while that of ^{40}K ranged from 228 to 451 Bq kg^{-1} (average: 360 Bq kg^{-1}). The samples S7 and S8 present the higher ^{238}U and ^{226}Ra activity values, compared with the values of the other samples. It is important to notice that, in most of the examined surface soil samples, the ^{238}U and ^{226}Ra specific activities were higher, compared with the average values for Greek and world soil. Moreover, the values obtained for the activity concentrations of ^{238}U and ^{226}Ra show a rather high coefficient of variation, indicating that the dispersion of the results was relatively high. On the other hand, since the dispersion of the ^{232}Th activity values was small, the thorium values are considered to be the same for all the examined samples. It is interesting to notice that, fly ash samples (Table 2) were found to be enriched in ^{238}U and ^{226}Ra, but not in ^{232}Th. This indicates that the increased concentrations of ^{238}U and ^{226}Ra in a number of the examined soil samples could be attributed to the fly ash deposition on the ground of Megalopolis city, depending on the prevailing wind direction.

Table 2. Specific activities of ^{238}U, ^{226}Ra, ^{232}Th and ^{40}K in undisturbed surface soil samples. The calculated absorbed dose rates (D) are also included.

	^{238}U	^{226}Ra	^{232}Th	^{40}K	D (nGyh^{-1})
S1	56 ± 6	37 ± 1	34 ± 1	359 ± 27	61
S2	46 ± 6	47 ± 1	39 ± 1	387 ± 29	61
S3	45 ± 5	40 ± 2	35 ± 2	380 ± 28	58
S4	36 ± 4	36 ± 4	34 ± 1	320 ± 24	51
S5	84 ± 9	46 ± 1	36 ± 2	392 ± 34	77
S6	42 ± 7	51 ± 1	25 ± 2	228 ± 30	44
S7	208 ± 20	126 ± 2	39 ± 2	404 ± 33	137
S8	110 ± 8	51 ± 1	25 ± 1	382 ± 30	82
S9	38 ± 12	46 ± 1	30 ± 1	293 ± 23	48
S10	66 ± 7	33 ± 1	30 ± 1	451 ± 33	68
Average	73 ± 58	51 ± 27	33 ± 5	360 ± 64	69 ± 27
Fly ash	1007 ± 64	1042 ± 21	59 ± 1	446 ± 6	
World Soil[a]	35 (16-110)	35 (17-60)	30 (11-64)	400 (140-850)	57 (24-160)
Greek Soil[a]	25 (1-240)	25 (1-240)	21 (1-190)	360 (12-1570)	56 (30-109)

[a]UNSCEAR, 2000

The absorbed dose rate in air ranges from 44 to 137 nGyh^{-1} with an average of 69 ± 27 nGyh^{-1}. This value is slightly higher, compared with the estimated world average external exposure rate from terrestrial gamma radiation of 57 nGy h^{-1} and the Greek average of 56 nGy h^{-1}, but fall within the world and Greek ranges as shown in Table 2. Only in one case (S7) the absorbed dose rate falls out of the Greek ranges. The estimated annual effective dose for adults is 0.09 ± 0.02 mSv.

Indoor Radon Measurements

Since the distribution of the measured values of the indoor radon concentrations was log-normal (Kolmogorov-Smirnov normality test: Z=1,828; P=0.003), geometric mean values will be used for data description and statistical comparisons. The radon concentration levels observed in the surveyed dwellings (50) during the four seasons and the annual average values are shown in Table 3. As shown in this table, the values of radon concentration levels were highly variable from dwelling to dwelling. The highest value observed in a particular house was 241 Bq m^{-3} during winter, while the lowest was 10 Bq m^{-3} during spring and summer. Statistical analysis (One-way analysis of variance) applied to log-transformed concentrations showed that the differences in radon concentrations between the four seasons were statistically significant (F=8.420, on 3 and 196 *df*, P<0.001). Multiple comparison tests showed that the values in winter were significantly higher than those observed in spring and summer. A pressure difference caused by the difference between the inside and outside temperature forces the radon-rich air in soil to flow into the dwelling. This phenomenon increases the winter concentrations of radon. The average annual arithmetic and geometric means were found to be 53 ± 34 Bq m^{-3} and 43 ± 2 Bq m^{-3}, respectively.

Table 3. The average annual indoor radon concentration (Bq m^{-3}) and the concentrations during the 4 seasons of 2006.

	Indoor radon concentration levels (Bq m^{-3})			
	GM[a] ± GSD[b]	Min[c]	Max[d]	N[e]
Winter	62 ± 2	11	241	50
Spring	33 ± 2	10	143	50
Summer	38 ± 2	10	146	50
Fall	45 ± 2	13	144	50
Annual	43 ± 2	10	241	200

[a]GM: Geometric mean; [b]GSD:Geometric standard deviation; [c]Min:Minimum value; [d]Max:Maximum value; [e]N:Number of measurements

These values are comparable with radon values found in other Greek cities (Nikolopoulos et al., 2002, Papaefthymiou et al., 2003; Psichoudaki and Papaefthymiou, 2008) and the values given in UNSCEAR (2000) for Greece. It should be noticed that the reference level of 400 Bq m^{-3}, recommended by the European Commission (EC, Council Directive, 1990) for old houses is never exceeded.

As mentioned above, the radon concentrations ranged from 10 to 241 Bq m^{-3}. These values correspond to an annual effective dose ranging from 0.3 to 6.1 mSv. The resulting annual effective dose calculated from the geometric mean was estimated to be 1.3 ± 0.9 mSv, that is consistent with the world average annual effective dose of about 1 mSv for radon and its decay products (UNSCEAR, 2000), but is slightly higher than the annual average value of 0.8 mSv found by Nikolopoulos et al. (2002) in a relative larger radon survey, which has been carried out in Greece from 1995 to 1998.

The total annual effective dose, due to outdoor external irradiation of terrestrial origin and to indoor internal irradiation from the short-lived decay products of ^{222}Rn, is estimated to be 1.4 mSv. The worldwide average annual effective dose due to natural sources is 2.4 mSv for adults (UNSCEAR, 2000).

Conclusions

Results on elemental characterization of bulk deposition samples collected in Megalopolis city located in the vicinity of the lignite power station, showed that the concentrations of As, Se, Sb, U, Mo, Sb and Zn, which are considered as toxic elements, were enriched in all samples examined. This is possibly attributed to the fly emissions of the power station, since the same elements were also enriched in fly ash samples. Natural radioactivity measurements showed slightly enhanced specific activities of the ^{238}U and ^{226}Ra in a number of surface soil samples, indicating possible contamination of the soil around the city, that have been caused by deposition of fly ash escaping from the power station. On the other hand, the indoor radon concentration levels measured in the city of Megalopolis show no significant influence of the effluents of the plant on radon concentration, although fly ash particles escaping from the stacks of the power station are enriched in ^{226}Ra, precursor of ^{222}Rn. This may be attributed to the lower emanation of radon from fly ash compared with other materials. Further studies, especially on $PM_{2.5}$ concentrations in Megalopolis atmosphere are needed for assessing the potential hazard of stack emissions to the local population.

References

Arditsoglou, A.; Petaloti, Ch.; Terzi, E.; Sofoniou, M.; Samara, C. *Sci. Total Environ.* 2004, 323, 153-167.

Beck, H.L.; De Campo, J.; Cogolak, C. In situ Ge(Li) and Na(Tl) gamma ray spectrometry. *USAEK Report HASL*-**258**, NY, 1972.

Clark, L.B.; Sloss, L.L. Trace elements: Emissions from Coal Combustion and Gasification. In: IEACR, *IEA Coal Research*, London, 1992, Vol.49.

EML, *Environmental Measurements Laboratory, Procedures Manual*, U.S. Department of Energy, 27th ed., 1992, Vol. 1, p. 59.

European Commission (1990). Commission recommendations on the protection of the public against indoor exposure to radon, *Official Journal of the EC* **L80/26**; 90/143/Euratom, 1990.

Finkelman, R.B. *Fuel Process. Technol.* 1994, 39, 21- 34.

Kampa, M.; Castanas, E. *Environ. Pollut.* 2008, 151, 363-367.

Kritidis, P.; Kamenopoulou, V.; Kallithrakas-Kontos. *Radiat. Protect. Dos.* 1994, 55, 149-152.

Nikolopoulos, D.; Louizi, A.; Koukouliou, V.; Serefoglou, A.; Georgiou, E.; Ntalles K.; Proukakis, C. *J. Environ. Radioactiv.* 2002, 63, 173-186.

Papaefthymiou, H.; Mavroudis, A.; Kritidis, P. *J. Environ. Radioactiv.* 2003, 66, 247-260.

Papaefthymiou, H.; Kritidis, P.; Anousis, J.; Sarafidou, J. *J. Environ. Radioactiv.* 2005, 78, 249-265.

Papaefthymiou, H.; Anousis, J. *J. Radioanal. Nucl. Ch.* 2006, 399-405.

Papaefthemiou, H.; Symeopoulos, B.D.; Soupioni, M. *J. Radioanal. Nucl. Ch.* 2007, 274, 13-130.

Papaefthymiou, H. *J. Radioanal. Nucl. Ch.* 2008, 275, 433-439.

Psichoudaki, M.; Papaefthymiou, H. *J. Environ. Radioactiv.* 2008, in press.

Seams, W.S.; Wendt, J.O.L. *Adv. Environ. Res.* 2000, 4, 45-58.

Simopoulos, S.E.; Angelopoulos, M.G. J. *Environ. Radioactiv.* 1987, 5, 379-389.

Swaine, D.J. In: *Trace elements in Coal*, Butterworths, London, 1990.

UNSCEAR (2000). In: Sources and effects of ionizing radiation". United Nations Scientific Committee on the Effects of Atomic Radiation. Report to the General Assembly with Scientific Annexes, United Nations, NY, 2000, Vol. 1: Sources.

RESEARCH AND REVIEW STUDIES

In: Nuclear Chemistry: New Research
Editor: Axel N. Koskinen, pp. 15-49
ISBN: 978-1-60456-957-5

Chapter 1

THE ENHANCED DEPOSITION OF ^{137}CS FALLOUT IN TROPICAL CLOUD FORESTS

John J. LaBrecque
Instituto Venezolano de Investigaciones Científicas-IVIC
Apartado 20632, Caracas, 1020A, Venezuela

Abstract

Cesium-137 (^{137}Cs) is an artificial radionuclide which did not exist in the environment until man began atmospheric testing for nuclear weapons development. The phrase "cloud forest" and its` characteristics are very ambiguous and can be misleading, thus care needs to be taken into account when this term is encountered. The activity of ^{137}Cs was determined mainly in surface soils, as well as mosses, lichens and other types of vegetation. The ^{137}Cs activities (Bq kg^{-1}) were measured by high energy resolution gamma-ray spectroscopy employing reference materials as comparators.

In this chapter, the spatial distribution of ^{137}Cs in various tropical cloud forests, foggy areas and their surrounding areas in Venezuela will be presented, in order to show the effect that cloud forests and foggy areas had on the deposition of radioactive fallout. The relationship between the ^{137}Cs activities in the surface soils and the elevations of their sampling sites not only clearly showed the effect that the cloud forests had on the enhancement of ^{137}Cs but also can be used to estimate the base of the cloud cover. Cloud forests were usually considered to be in mountainous areas at high altitudes, but the peaks of some of the tropical cloud forests discussed herein were located at relatively low elevations. Finally, the ^{137}Cs activities in soils, mosses and lichens collected along the Caruay River will be presented with respect to its surrounding area. The area along the river had similar characteristics to a tropical cloud forest, but this was due to the fog which was produced daily.

In conclusion, it is very important to consider the effect of cloud forests and other similar pathways, such as foggy areas, in which ^{137}Cs can be directly deposited from the atmosphere at much higher concentrations than by rainfall, or one could report anomalously high activity values, which could suggest that more ^{137}Cs has been recently introduced into the environment.

Introduction

A tropical cloud forest is a specific type of rainforest that is located usually in coastal mountains at relatively high altitudes, where the air temperatures are much cooler than the surrounding areas at lower elevations. The forest is submerged by clouds very frequently, on a daily basis, which deposits water directly from them and/or mist (fog) onto the vegetation and land surface. As a result, cloud forests have a unique ecosystem, that renders them difficult for man to live in, but they are habitable for many other animals and plants, even some that are not known to populate other areas of this planet. Some of the cloud forest areas have been included in national parks which directly help to protect them from man, but similar to other high altitude forests in northeastern North America and Europe, they can be destroyed indirectly by the deposition of acidic and trace metal compounds [1-3]. The importance of preserving cloud forests has been presented in a synopsis: Mountain Cloud Forest Conservation and Research [4].

Cloud Forest

The usage of the term “cloud forest” can be very confusing and ambiguous. According to the American Heritage Dictionary of the English Language [5], a cloud forest is “a tropical forest, often near peaks of coastal mountains, that usually has constant cloud cover throughout the year”. While, the Wikipedia free encyclopedia [6], describes a cloud forest (also called a fog forest) as a generally tropical or subtropical evergreen montane moist forest characterized by a high incidence of low-level cloud cover, usually at the canopy level. Cloud forests often exhibit an abundance of mosses covering the ground and vegetation, in which case they are also referred to as mossy forests. Several forests in the temperate zones have similar characteristics as cloud forests in the tropical and sub tropical area, such as in Argentina and Spain, but they are not considered as true cloud forests. And sometimes in tropical countries, true cloud forests are labeled as temperate due to their cooler climate [6]. Thus, special care needs to be taken when doing literature searches and when comparing different cloud forests.

A literature search [7] performed on the second day of April 2008 using a Google search engine resulted in about 1,320,000 hits for cloud forest. When the search was refined by searching within the results for Cs-137, only about 610 hits were reported and when the search was further limited by the words “soils” and “soils and vegetation”, the hits were reduced to about 513 and 357 respectively. In comparison, a literature search using all the data bases of the Web of Science [8] for all years, since 1982 on the same day, resulted in 1,141 hits for “cloud forest” and only nine when the search was refined by including the term “Cs-137”. Therefore, it can be concluded that there has been very little interest from the scientific community with respect to radioactive pollution in cloud forest. The possible reasons being, that cloud forests are mostly located in remote high altitude mountainous areas, in developing countries and places where man does not want to inhabit.

According to reference [6], there is a relatively small band of altitude in which the atmospheric environment is suitable for cloud forest development. Cloud forests are characterized by persistent mist or clouds at the vegetation level, resulting in the reduction of direct sunlight. Trees in these regions are generally shorter and more heavily stemmed than those located in the near by lower altitude forests and the high amount of moisture promotes the development of vascular epiphytes. As a result, mosses and ferns are usually abundant and frequently orchids may be encountered. The soils are rich in organic material, mainly peats and humus. Within cloud forests, much of the water input is in the form of fog drip, where fog condenses on the leaves of the vegetation and then drips onto the ground below. The areas we considered as cloud forests in Venezuela fits this description very well, but they started at the base of the cloud cover to peak of the land formation. At the Altos de Pipe location, the elevation difference was more than 250 meters between the base of the cloud cover and the peak of the mountain, where the cloud cover was denser except when there were strong winds.

Venezuelan Tropical Clouds Forests

Before the studies of fog events at two tropical cloud forests near Caracas, Venezuela [9], little was known about the fog (cloud) frequency, duration and the importance of the fog (cloud) as a hydrological (water) input into these ecosystems. A second study [10], which was performed concurrently, investigated the chemistry of the fog. The principal objective of these studies was to investigate the hydrological input affect of fog on the deposition of atmospheric pollution. It was concluded that fog was a very important hydrological input in these tropical cloud forests, especially in the dry seasons. Preliminary data estimated that fog related deposition accounted for 35 and 71% of the annual water input respectively at the Altos de Pipe and El Avila cloud forests. Ion concentrations in the fog (condensed water) were determined to be up to 20 times higher than those collected in the rainwater at the same sampling sites. Thus, the pollution of cloud forests from wet fallout would be much greater from the cloud (fog) hydrological input than precipitation from rain itself. This could be the

Figure 1. Photographs of the peak of La Sierra Study area on the Island of Margarita (G): The left photo showing the peak deeply immersed in the cloud cover; the right photo showing the area where the dense cloud cover was in the above photo.

explanation for our previously found anomalously high ^{137}Cs activities in surface soils [11, 12] with respect to the baseline activities from a general country-wide survey and other similar close by areas that were not in cloud forests. Upon realizing this, we decided to further investigate this effect of fog on the distribution of ^{137}Cs activities in surface soils located in pristine areas of cloud forests. In Figure 1 photos of the peak of the La Sierra study area on Margarita Island (G) when there was dense cloud cover and when the same area was clear.

Cesium-137 (^{137}Cs) and Soils

Cesium-137 (^{137}Cs), sometimes also spelled as caesium-137, is a radioactive isotope of cesium. It is a product of nuclear fission and considered an artificial radionuclide (man-made). It has a half-life of about 30.2 years and decays by beta radiation to ^{137m}Ba, which in turn decays to the stable ground state of ^{137}Ba emitting only one gamma-ray with 661 keV of energy. Because of its long half-life and decay properties, man has received the largest tissue doses of nuclear radiation from ^{137}Cs, as the result of the deposition of world-wide radioactive fallout.

In general, ^{137}Cs did not exist in surface soils until after the beginning of atmospheric nuclear weapons testing in the 1950`s. But, since then it can be found almost everywhere on the earth's land surface and in the oceans. Only a small amount (activity) has been released into the environment since the beginning of the world treaty on non-proliferation of nuclear weapons in 1970, mostly from minor nuclear accidents. The Chernobyl nuclear power plant accident in May of 1986 was the most notable one, which revived world wide the importance of the monitoring of environmental radioactivity in the environment.

Shortly after the Chernobyl nuclear power plant accident, many countries that were not even actively involved in nuclear activities began to monitor radioactivity in the environment and foodstuffs. These monitoring programs were a result of the awareness that radioactive pollution knows no boundaries. Even countries far away from nuclear activities can be affected directly by atmospheric deposition or indirectly by importation of foodstuffs. But still, there is only a little information on the distribution, migration and the pathways of ^{137}Cs in the environment in many parts of the world, that do not have major nuclear activities, such as South America, Africa and Antarctica.

In general, the regions close to the global poles both North and South, as well as those near the equator were considered as zones with relatively low deposition of ^{137}Cs fallout from the nuclear weapon testing era with respect to the mid-latitudes [13]. But, recently it has been reported that high activities of ^{137}Cs have been determined in some organic rich soils near the Arctic Circle with factors as high as 20 times with respect to the other reference soils from the same area [14]. Similarly, some areas with anomalously high ^{137}Cs activities in surface soils have also been previously reported near the equator in Venezuela [12, 15-18]; most of these areas were located in very humid tropical cloud forests. Therefore, it was difficult at first to explain the anomalous high values of ^{137}Cs found in soils in mountainous regions near the Venezuelan coast [12, 16, 17, 19] with respect to other areas of Venezuela [11] and even areas in the same mountainous regions at lower elevations.

When ^{137}Cs was deposited on mineral soils which contain appreciable amounts of clay minerals, it was stabilized relative rapidly. Thus, the uptake of ^{137}Cs from mineral soils by the vegetation and its vertical migration in the soil profile was limited. In contrast, ^{137}Cs was taken up by vegetation much more readily from organic and tropical lateritic soils. Further information on the transference of ^{137}Cs from different types of sub-tropical and tropical lateritic soils can be found elsewhere [20]. It is also well known [21-23] that organic soils retain and accumulate radionuclides. These types of soils can act as a sink for radionuclides causing localized areas of elevated radioactivity [14].

Environmental Radioactivity Studies in Venezuela

In Venezuela, environmental radioactivity studies to determine the activity (Bq kg^{-1}) of ^{137}Cs in surface soils country-wide began in 1989 as part of an oriented research project (PC-075) and later continued as part of a multi-institution research project (RPVII 7600026), both funded in part by the Venezuelan National Science and Technology Foundation (CONICIT). It was determined that from more than 1 000 surface soils, that were collected country-wide, that more than 60% had ^{137}Cs activities less than 2 Bq kg^{-1} and only about 5% of the soils had values greater than 10 Bq kg^{-1}. Thus, when activity values greater than 10 Bq kg^{-1} of ^{137}Cs were found in surface soils, a more intensive follow up study of the area was performed. As a result of these follow up studies, we have been able to located many anomalously high areas of ^{137}Cs activities in surface soils, many of them have been related to rich organic soils in very humid tropical cloud forests [12, 15-18]. In Figure 2, the results of one of these more in depth studies [18] can be seen for surface soils, as well as a schematic of the cloud cover over the mountain range.

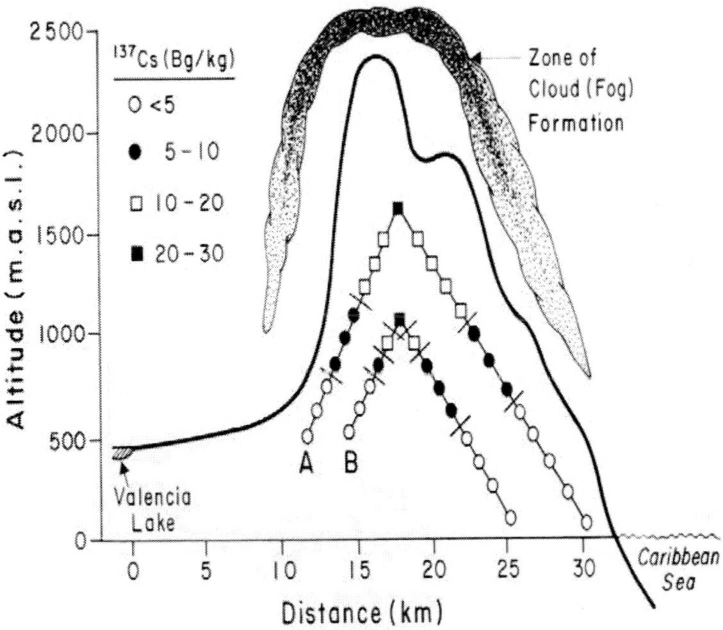

Figure 2. A schematic of the cloud cover over the mountain range in the Henri Pittier National Park (C) showing the determined ^{137}Cs activities ranges versus altitude in surface soils collected from Maracay to Choroni (A) and from El Limon to Ocumare (B).

In areas of relative high ^{137}Cs content in surface soils, we have also investigated the activity of ^{137}Cs in some of the available plant materials, such as grasses and cactus plants [24-25], fern (*Pteridium caudatum*) leaves [25], the leaves and fruit from the *Clusia grandiflora* tree [15,26], the leaves and trunks of the *Brochinina micrantha* palm plants [15], as well as mosses and lichen [15,26,27], which are well known bio monitors that concentrate and accumulate radionuclides [28].

Bio Monitors

The use of naturally growing mosses as bio monitors for studies of atmospheric deposition of is a well established technique [29]. Mosses are a carpet-like plant species obtaining most of their nutrients from both wet and dry fallout of airborne particles. They accumulate and retain several trace elements, as well as radionuclides. At first, they were employed to investigate toxic metals [30, 31] and more recently radionuclides [32, 33]. Mosses do not have root systems that interact with the top soil layer and basically depends on surface absorption for their nutrient input. Since, they concentrate particulate and dissolved chemical species from both dry and wet deposition they have been employed as bio monitors for atmospheric studies.

Mosses have two principal advantages over the conventional sampling method (rain water) for atmospheric deposition studies:

1) The pollutants are strongly concentrated in the mosses relative to their very low levels in rain water, thus the pre concentration methods are not required.

2) The mosses offer a simple and economic sampling procedure, which in turn allows for a large number of sample sites. But, employing mosses as bio monitors also has some limitations such as, other sources and mechanisms other than atmospheric deposition of pollutants may contribute significantly to the measured values. For survey studies, they still need to be calibrated against the amount of rainfall. Only areas where mosses grow naturally can be monitored. Other sources of the pollutants in mosses can be due to the marine effect, the vegetable effect and the soil factor, as well as the transport of soluble compounds from the soil or it's under cover into the moss during periods of extensive contact [34].

In general, lichens have a world wide distribution and can attach themselves to most any stationary material. They can survive in very harsh environments: in cold Polar Regions, in hot and dry deserts, and even in high altitude mountain cloud forests with 100% humidity and little sunlight. Thus, there have been many publications on the use and applications of lichens as bio-monitors for atmospheric pollution, most of which can be found in a recent review [28]. Some studies mention correlation between the radionuclide concentration in lichen and the environment, but there is no accepted relationship between the concentrations in lichens and their respective atmosphere [35]. Lichens obtain nutrients for growth and metabolism from both wet and dry precipitation [36]. Accumulation of dust or soil particles which settle in the intercellular space can also occur. Pollutants can be deposited onto lichens directly or indirectly from fog, dew or gas absorption [37].

Thus, it can be concluded that, both lichens and mosses can be employed to monitor the quality of the environment, remembering that they not only reflect the wet and dry precipitations from the atmosphere but also other inputs. It was shown that ^{137}Cs was successfully monitored and investigated for many years in both mosses and lichens on the Yamal Peninsula in Russia [32].

In this chapter, we shall present the spatial distribution of the ^{137}Cs activities in surface soils, as well as in vertical soil profiles from our investigations in tropical cloud forests, foggy areas and in their corresponding nearby areas. The ^{137}Cs activities from the respective nearby area can be employed as baseline (reference) values. It will be clearly seen that the water input directly from the clouds (fog) condensation and their respective mist increased the amount of the ^{137}Cs deposition in the tropical cloud forests with respect to the nearby areas at lower elevations. The relationship of the density of the cloud (fog) immersion in the cloud forests versus the ^{137}Cs activity will be shown and discussed. It will also be shown that the base of the cloud cover can be estimated from the relative amount of ^{137}Cs determined in surface soils at different elevations. The results of the ^{137}Cs activity ratios for mosses and lichens versus their respective soil samples will be presented and discussed. Finally, the results of the ^{137}Cs activities from various other types of vegetation that were collected in the tropical cloud forests and foggy areas will be presented and discussed.

Experimental

A map of Venezuela showing the locations of the different study areas in general is presented in Figure 3. The study areas were labeled as letters on the map, A to H: from the West to the East and H to M: from the North to the South and will de discussed in that order in this section.

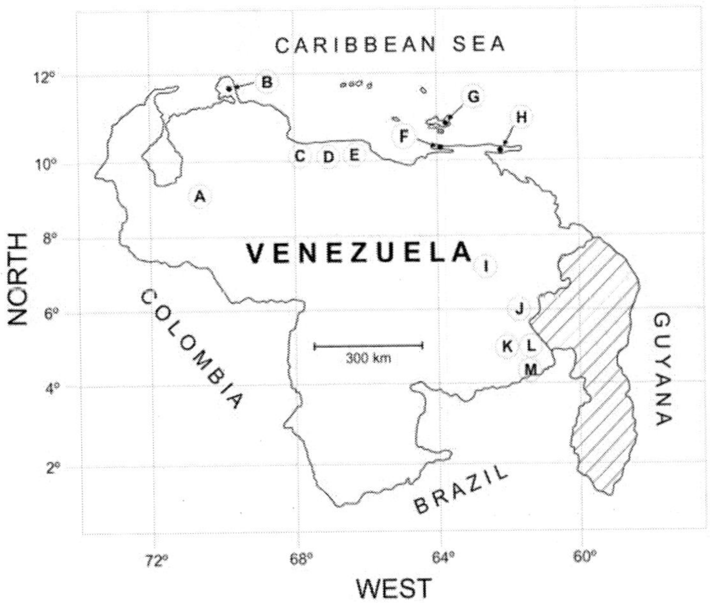

Figure 3. A geographic map of Venezuela, showing the general locations of the different study areas; they are coded with letters A to M, which will be used in the tables and throughout the text.

While, the type of areas, the geographic coordinates and elevation, meters above sea level (m.a.s.l.) ranges are given in Table 1. More detailed information can be found in the references for the different study areas, as well as other data both related and not related to the distribution of ^{137}Cs in cloud forests.

Table 1. The general locations and description of the study areas.

Area (location)[a]	Types of areas	Elevation[b] (m.a.s.l)	Geographic Coordinates North	Geographic Coordinates West	References
Pico El Aguila (A)	Cloud forest	4000-4100	8°50'	70°50'	[27]
Cerro Santa Ana (B)	Cloud forest	100-500	11°79'	69°56'	[19]
Henri Pittier N.P. (C)	Cloud forest	50-1600	10°25'	67°40'	[18]
Altos de Pipe (D)	Cloud forest	1300-1700	10°24'	66°59'	[16]
Central Coast (E)	Cloud forest Foggy areas	25-1000	10°28'	66°40'	[17]
Araya Peninsula (F)	Coastal (BKG) areas	25-250	10°35'	64°00'	[38]
Margarita Island (G)	Cloud forest Foggy areas	350-850 50-200	11°00'	63°50'	[12]
Paria Peninsula (H)	Coastal areas	25-350	10°40'	62°30'	[38]
Guri Reservoir (I)	Inland (BKG) areas	150-600	7°30'	63°30'	[39]
Sierra de Lema (J)	Cloud forest Foggy areas	475-525	6°04'	61°24'	[15]
Caruay River (K)	Foggy areas	900-1050	5°50'	61°55'	[26]
La Gran Sabana (L)	Inland (BKG) areas	900-1450	5°20'	61°00'	[40]
Sierra Pacarima (M)	Foggy areas	900-1100	4°35'	61°25'	[40]

[a] The general locations are shown on the map in figure 3.

[b] More specific elevations and geographic coordinates are given for each of the sampling sites in the original publications (see references).

A Short Description of the Different Study Areas

Pico El Aguila (A): this area was in the Los Andes mountain range, the state of Merida about 150 km South of the Caribbean coast where the elevation was about 4000 m.a.s.l.; the area was frequently foggy and covered by clouds. Thus, the surface was very wet most of the time due to the abrupt temperature changes daily. It can be considered a cloud forest even though most of the vegetation was grasses.

Cerro Santa Ana (B): In general, the Paraguana peninsula was flat, with altitudes less than 100 m.a.s.l and one small mountain (Cerro Santa Ana), which rose sharply to about 830 m.a.s.l. and was covered like the rest of the peninsula with thorny vegetation and small bushes and trees, but much more densely. The geological formation of Cerro Santa Ana was basically sand and clay sediments with a rocky component. The soils were Aridisols in general, but also included some Inceptisols too. The climate was classified as semi-arid with temperatures between 26 and 29°C and the annual precipitation was only between 200 and 600 mm [41].

Henri Pittier National Park (C): The area investigated was located partially in a tropical cloud forest classified as very humid premontane [42] in the state of Aragua, Venezuela. The park was situated in a coastal mountain range running East and West with the geographic coordinates from 10°15' to 10°30' North and from 67°41' to 67°29' West. The elevation of the summit along the roadway to Ocumare del Mar was about 1105 meters above sea level (m.a.s.l.) and along the roadway to Choroni about 1625 m.a.s.l. The atmospheric temperature ranges from 18° to 24°C and the annual rainfall was between 2000 mm and 4000 mm. More detailed information on the physical environment, vegetation and soil properties can be found in the following book [43].

Altos de Pipe-IVIC (D): All the sampling sites in the two study areas considered in this chapter were located in a tropical cloud forest, from the base of the cloud cover to the top of the mountain. The first study area was in a remote area, where only a few foot paths existed and the other was located near the roadways of Instituto Venezolano de Investigaciones Cientificas-IVIC.

Figure 4. A photograph of the summit of the tropical cloud forest at Altos de Pipe (D) showing the cloud cover at noon time; the remote study area was located behind the trees that were immersed by the cloud cover in the upper right corner of the photo.

The vegetation was dense at all the sampling sites and denser at the higher elevations. The vegetation consisted of large trees, bushes, grasses and weeds, as well as mosses and lichens. The surface features very inclined with rich organic soils at the higher elevations where the vegetation was dense with respect to soils at the lower elevations which had larger sand, clay and lateritic components. A photograph of the summit, where the roadway of IVIC ends and the remote study area started can be seen in Figure 4.

Central Coast (E): This study area was in a remote area near the town of Osma. It was only one of the fifthteen areas that were investigated in the original publication [17]. It had two distinct climates, a tropical semi-arid zone at low elevations which was frequently foggy and a tropical mountainous climate at higher elevations which included a tropical cloud forest. This study area consisted of metamorphic rocks that sharply rose from the coast to the top of the mountain range. The soils were generally sandy at the lower elevations near the coast with their origins from sedimentary and alluvial processes.

Araya Peninsula (F): As previously mentioned in Table 1, this study area was not in a foggy area or a cloud forest. But, it will be used as a reference for baseline values of ^{137}Cs activities along the Northern Venezuelan coast. In general, the Araya Peninsula had small hills and terraces. It was located between the geographic coordinates of about 10°29′ and 10°43′ North and 64°18′and 63°39′ West. In general, the temperature was between 24°C and 28°C and the annual rainfall was about 300-600 mm, which partially accounted for the dry desert areas with little vegetation. The soils were mainly Aridisols.

Margarita Island (G): The Island has the maximum insolation for Venezuela, with temperatures between 26-29°C most of the year and only a small amount of rainfall, in the range of 200-260 mm a year [41]. There were various study areas of interest on the Island of Margarita, but only three will be discussed in this chapter:

Figure 5. Photographs of the El Maco hill study area on the Island of Margarita (G).

G1) La Sierra was a tropical cloud forest area which rose to more than 850 m.a.s.l. The vegetation in general consisted of large trees and grasses at the lower elevations and bushes and grasses at the higher elevations. While, the surface features were rocky, covered with organic litter (mat) and very inclined.

G2) El Maco hill was an area affected by frequent fog as a result of the large differences in the surface and air temperatures with elevations less than 125 m.a.s.l. The vegetation consisted of small trees and bushes, grasses and weeds, as well as cactuses. The surface features were slightly inclined rocky with lateritic materials and many bare areas. Photographs of this area can be seen in Figure 5 and can be compared with those shown in Figure 1 of the peak of the La Sierra study area in a tropical cloud forest.

G3) El Mirador, an area that also was affected by daily fog caused by the Eastward winds from the Caribbean Sea. The vegetation was dense and included large and small trees, bushes grasses and weeds, as well as cactuses. While the surface features included lateritic materials and very inclined.

Paria Peninsula (H): Similar to the Araya Peninsula study area, the Paria Peninsula study area was not a foggy area nor located in a cloud forest, but cloud forests did exist at much higher elevations (about 1000 m.a.s.l.) which were inaccessible for sampling. Thus, this study area was employed to obtain reference values of ^{137}Cs activities for areas not affect by fog or cloud forest along the coast. In general, the Paria Peninsula had small hills and very inclined mountains and located between 10°32′ and 10°45′ North and 62°52′and 61°51′ West. In general, the temperature was between 24° and 28° C. The annual rainfall was about 600-1200 mm on the Paria Peninsula, which partially accounted for the lush savannas and semi-humid forest and the soils were mainly Inceptisols.

Guri Reservoir (I): This area was relative flat and was affected by fog and mist that was produced by the large body of water as the result of the daily temperature changes and winds. Thus, the ^{137}Cs activities determined here were employed as baseline data for inland areas. The Guri reservoir area was located between the following geographic coordinates: 6°00′ - 8°15′ North and 61°15′ - 63°35′ West with elevations between 25-550 m.a.s.l. The areas surrounding the reservoir had similar environmental parameters [44], except for the area in the Southeastern corner near the base of Sierra de Lema mountain range [45].

Sierra de Lema (J): Only a small area on the surface, around the edges and near a large diabase (rock) outcrop was investigated. It was called El Mirador (The Outlook) [45] at the base of the Sierra de Lema mountain range, Bolivar state (Venezuela). It was located between the following geographic coordinates: 06°03.760′ North and 061°24.243′ West at 478 m.a.s.l. and 06°03.930′ North and 061°24.232′ West at 518 m.a.s.l. This area was surrounded by a very humid forest located along the top of the mountain range in a dense cloud forest. Further information on the description of the area can be found elsewhere [45].

Caruay River (K): The Caruay River was located in the Gran Sabana, west of the town of Kavanayen. In this study area, the river runs south about 20 km to Untay, where it meets the Mauak River. The river had dense forests on its banks that were about 20-100 m wide that contained lichens and mosses, typical of cloud forests, but this area was not!. Along this section of the river, there were five water falls, which produces fog and mist; a photograph of the last one at the junction of two rivers is shown in figure 6. Fog was also produced daily from the changing temperature along the river. The soils were very rich in organic matter.

Figure 6. A photograph of the water falls along the Caruay River near Utay showing the production of mist (fog).

La Gran Sabana (L): This study area was in the Canaima National Park (about 30,000 km^2); it was sparsely populated with less than 3 persons per km^2 and mostly indigenous [41]. The geographic coordinates of this area were: 4°35′- 5°55′ North and 60°55′ - 61°45′ West with elevations ranging from 863 to 1426 m.a.s.l.; even though the elevations of the sample sites varied greatly in the Gran Sabana, they did not fluctuate much from one site to the next, but rather they declined gradually from the North to the South. The Gran Sabana was a large plain with coarse grasses and a few scattered patches of trees and is unique for its tepueyes. The most famous being the Auyan-tepuey, which has a surface area of about 700 km^2 and exhibits one of the world's highest (972 m) waterfalls [46]. The geological materials were predominantly acidic and basic igneous and sedimentary rocks. The soils of the Gran Sabana were generally acidic and lateritic, as a result of the large amount of rainfall and the high horizontal sides of the tepueyes, thus they were very infertile [41]. This study area was used for reference baseline values of ^{137}Cs.

Sierra Pacarima (M): This study area was mountainous and consisted generally of dense forest and fertile soils, except for zones that have been damaged by mining operations in the past. It was located between Santa Elena de Uairen and El Pauji at the base of a mountain overlook, which was along the southern border with Brazil; the geographic coordinates were: 4°30′ - 4°40′ North and 61°15′ - 61°35′ West with elevations between 942-1086 m.a.s.l.. The soils had been classified as Entisols. The rainy season in general consists of seven months from May to November and the average annual temperature was in the range of 18° and 24°C [47].

Sampling Procedures

A list of the different types of specimens analyzed for ^{137}Cs and their respective collection dates for the different study areas are presented in Table 2. More detailed information on the collected samples and the sampling procedures can be found in the references of the original investigations.

Surface soils were collected and analyzed for ^{137}Cs from 1992-2004 in many very different environments, thus the sampling procedures described below are general and the specific sampling procedures can be found in the corresponding original publications. In general, surface soil specimens were collected in pristine undisturbed areas. The soils should have not been tampered with by man, animal or nature. Of course, selecting such areas was very subjective and when possible duplicate samples were taken. The areas were usually covered with grasses, weeds or other low profile vegetation. After selecting an area, about a half to one square meter was first cleared of vegetation and mat (organic litter), which was some cases was more than 10 cm thick, such as in the dense rain and cloud forests. Then, an additional 2 cm was cleared before the soils were sampled from depths between 2-5 cm as recommended by the International Atomic Energy Agency's guidebook on the measurement of radionuclides in food and the environment [48], in some cases the surface soils were taken from different depth intervals between 2 and 20 cm. The soils were collected employing a small hand pick to cut and crush the chucks into smaller pieces so that they could be passed through sieves between 4 and 8 Mesh (4.25 to 2.36 mm) at the sites to eliminate stones and foreign materials, including plant roots, before transferring to labeled plastic bags. In some cases, the soils samples were too damp, wet or difficult to sieve in the field, thus much larger sample sizes had to be taken and transported to the laboratory. The geographic coordinates and elevation (m.a.s.l.) data were recorded, as well as other related information, depending on the specific study, such as surface features, types of vegetation, types of mineral material in the sampled depth interval, etc.

Vertical profile soil samples were collected from between 5-100 cm depths, after the surface was cleared of vegetation and mat, as described above for the surface soils. The first sampling interval was 5 cm deep, followed by 10 cm vertical intervals. The sampling pits were about one square meter horizontally, but in some cases much less, when the soils were very hard or contained large rocks. The upper layers were processed as the surface soils samples employing a small hand pick and collected with a plastic scoop, while the lower layers were processed with a 1 or 2 m bar. Again, the soil specimens were sieved in the field similar to the surface soil samples. Care was taken not to mix the soils from one layer with another.

Grasses were collected from areas, where previous surface soils samples were determined to contain relatively high ^{137}Cs activities. When the grasses, which included a variety of weeds were sparsely populated or lying on the surface rather than vertically, small handfuls were pulled out of the ground with their roots. The roots and an additional 2 cm of material that was above the surface were eliminated. When the grasses were densely populated, they were cut about 2 cm above the surface with a hand hose. It should be noted here, that very large volumes of vegetal material were required, about 1 cubic meter for each sample to ensure that 500 grams of clean dried material would result. Cactuses (pineapple-type) were also collected from areas, where previous surface soils samples were determined to contain relatively high ^{137}Cs activities. They were sampled by simply passing a machete between them and the surface to cut their roots.

Table 2. The types of samples and their respective collection dates of the study areas.

Study area (location)[a]	Types of areas	Type of samples	Collection dates	References
Pico El Aguila (A)	Cloud forest	Soils	1996	[27]
		Mosses	1996	
Cerro Santa Ana (B)	Cloud forest	Soils	1997	[19]
Henri Pittier N.P. (C)	Cloud forest	Soils	1994-1996	[18]
Altos de Pipe (D)	Cloud forest	Soils	1994-1996	[16]
		Mosses	1996	
		Grasses	1994	
Central Coast (E)	Cloud forest Foggy areas	Soils	1995	[17]
Araya Peninsula (F)	Coastal (BKG) areas	Soils	2001	[38]
Margarita Island (G)	Cloud forest Foggy areas	Soils	1992-1996	[12]
		Mosses	1996	
		Grasses	1993-1995	
		Cactus	1995	
Paria Peninsula (H)	Coastal areas	Soils	2001	[38]
Guri Reservoir (I)	Inland (BKG) areas	Soils	2002-2003	[39]
Sierra de Lema (J)	Cloud forest Foggy areas	Soils	2002-2003	[15]
		Black Mat	2003	
		Palm leaves	2003	
		Palm trunks	2003	
		Fruit bush leaves	2003	
		Fern leaves	2003	
		Fruit	2003	
Caruay River (K)	Foggy areas	Soils	2004	[26]
		Mosses	2004	
		Lichens	2004	
		Fruit	2004	
La Gran Sabana (L)	Inland (BKG) areas	Soils	2002-2004	[40]
		Lichens	2004	
Sierra Pacarima (M)	Foggy areas	Soils	2002	[40]

[a] The general locations are shown on the map in Figure 3.

Mosses and lichens were collected in the study areas, when they were encountered, principally in cloud forests and foggy areas. They were sampled by hand, shaking off any of the soil or other materials that they were attached to, as well as any other foreign material that may have been on their surface, for example leaves from bushes or trees. Surface soils, where collected either below them or close by when they were attached to other matrices. Figure 7 shows specimens of ferns, grasses, mosses and lichens at the Altos de Pipe (D) study area near the summit, where they were frequently immersed in the clouds.

Figure 7. Ferns, grasses, mosses and lichens at the Altos de Pipe (D) study area.

Black mat was collected on the surface of a diabase outcrop from areas in which there were little or no dry fallout deposits on its surface, when there was some dust or other fine dried material it was swept off by a small hand brush. About one cm of the bottom of the specimens that were attached to the diabase outcrop was discarded, before they broken into small pieces and pass through a 20-mesh (85 μm) sieve to eliminate small stones and other foreign materials, similarly as the surface soils that were collected on and nearby the outcrop.

Palm leaves and their respective trunks were cut from the *Brocchina micrantha* plants using a large machete. The leaves were cut completely and compactly folded into large plastic bags; again as was the case with the grasses, a large volume was required to obtain 500 grams of dried material for analysis. The trunks ranged between 20 and 30 cm in diameter and were cut into sections between 30 and 40 cm long and placed in separate plastic bags. Fern branches from the *Pteridium caudatum* specie were completed cut and processed in the field similarly as the palm leaves.

The leaves and fruit of the *Clusia grandiflora* were collected directly from the trees. Since the amount of the leaves required to obtain about 500 grams of dried material for analysis was a much larger volume than the fruit, the leaves were collected as branches about 30 cm in size and placed in separate plastic bags. Branches were selected that seem to be healthy and without any vision signs of dry deposits. Photographs of the *Brocchina micrantha* palms, the *Clusia grandiflora* fruit and leaves and the black mat can be seen in reference [15].

Sample Preparation for Analysis

In general, the soil and black mat samples that were not previously sieved with a coarse screen in the field were first air dried in large trays for a few days in the laboratory. If, they still were too damp to crush into a coarse powder to pass an 8- mesh sieve, they were oven dried at 25°C until they could be. After the coarse sieving to remove small stones and other foreign materials, they were transferred to large trays in layers less than 5 cm deep and placed in an oven at about 110°C until dried. When cooled they were hand ground with a pestle in a large ceramic mortar to a fine powder and passed through a 20 mesh (85 µm) sieve. A portion of this fine powder was transferred to a 1-l plastic measuring container (10 cm diameter) to occupy the standard volume of about 500 cubic centimeters. Finally, the measuring container was weigh and the weight (grams) of the soil sample was calculated by difference. Usually, there was sufficient dried material to prepared duplicate samples. The measuring containers with the soil samples were stored in a dry place both before after the gamma-ray measurements.

In the laboratory, the mosses were spread out on a large plastic sheet to air dry for a week or two. Then were broken into small pieces, approximately 2 cm x 2cm and placed in 4-liter glass beakers that were placed in an oven at 60°C for about 48 hours, followed by another two days at 60°C. The dried material was hand ground to pass a 8 mesh (2.36 mm) sieve to ensure that the material was thoroughly dried; and then to pass a 20 mesh (85 µm) sieve. Similarly, as the soils, a portion of this fine powder was transferred to a 1-l plastic measuring container (10 cm diameter) to occupy the standard volume of about 500 cubic centimeters. The measuring container was weigh and the weight (grams) of the moss specimen was calculated by difference. Depending on the amount of dried material obtained two or three aliquots were obtained, which were used to check the homogeneity of the sample preparation. The measuring containers with the moss samples were stored in a dry place both before after the gamma-ray measurements. Finally, it should be mention that a small (about 10 g) portion of each of the aliquots was saved for humidity measurements and future studies.

In general, the lichens were prepared similar as the mosses, except that the constant volume employed in the measuring container was reduced to about 250 cubic centimeters because of the limited amount of dried material obtained.

The palm and fruit leaves were first air dried for a week or two before they were cut into small pieces. The palm trunks had to be cut first into small pieces for air drying. The fern leaves were easily stripped from their branches after air drying. These materials were then placed in 4-l glass breakers and processed similar as the mosses and lichens.

Finally, the fruit and cactuses were the most problematical to process! Both the whole fruit and the edible parts were difficult because the edible part formed a paste and could not be even be completed dried at 110°C. Therefore, it was a problem to prepare the samples homogeneously and in a standard geometrical form and volume size. While, the cactuses similar to the palm trunks needed to be cut into small pieces for air drying, which was difficult because of the thorns and their hard exterior shell. Thus, the determinations of ^{137}Cs in these samples were not as accurate or precise as the other materials.

Instrumentation

All the ^{137}Cs activity measurements were made with conventional high energy resolution gamma–ray spectroscopic instruments. All the measurements presented herein were performed in the Atomic and Nuclear Spectroscopy Laboratory (ANSL) at the Venezuelan National Science Institute (Instituto Venezolano de Investigaciones Cientificas), except for those for the surface soils collected at the Osma site near the central coast of Venezuela [17], which were performed at the Litoral campus of Universidad Simon Bolivar (USB). Both groups used similar instrumentation. The gamma-rays from the samples, reference materials and blanks were collected by hyper-pure germanium (HpGe) semi-conductors that were shielded by lead housings. The measured energy resolution, full width at half maximum (FWHM) was about 1.8 keV for the 1.33 MeV peak from a ^{60}Co point source for the IVIC system and about 2.0 keV for the USB system. The counting efficiency for the IVIC system was greater than 25% over the many years that it was employed, while the counting efficiency for the USB system was more than 20%. Thus, both of the detection systems were very similar and of high quality.

The semi-conductors were cooled with liquid nitrogen in 31 l dewars and were equipped with preamplifiers to complete the spectrometers component. Detector bias supplies (0-5000 volts) were connected to the pre amplifiers and the signal output to spectroscopy amplifiers, which in turn were connected to the data acquisition systems.

The data acquisition was performed employing IBM-compatible PC`s with analog to digital converter/ interface cards from the Nucleus and Oxford Instruments, which were controlled by PCA software. In most of the studies, 4096 channels of the data acquisition were used, even though 8196 were available.

Further information about these systems, showing a schematic block diagram of the instrumental set-up and a typical gamma-ray spectrum of a Venezuelan soil sample can be seen in reference [11].

^{137}Cs Measurements and Activity Determinations

All the ^{137}Cs (Bq kg^{-1}) activity determinations were performed employing the comparator method employing the appropriate reference materials. The 661 keV gamma ray from the ^{137m}Ba radionuclide was used to determine the ^{137}Cs (Bq kg^{-1}) activities. The ^{137}Cs measurements for the samples, the reference material(s) and blanks for the corresponding material were performed in the same manner. The same volume of material was employed in the same geometry, in general about 500 cubic centimeter of sample finely ground in a similar

1l-plastic container filled to the same height; in the case of the measurements in the USB laboratory at the Litoral campus, one-liter Marinelli beakers were employed. The counting times and real times were similar in most cases and did not excess 5% dead time. The counting times ranged from 10 000 to 60 000 seconds of real time, depending the ^{137}Cs activities and sample weights; in general the soils were counted for 20 000 seconds during the working day and the vegetation materials for 50 000 seconds during the after hours of work (nights). More specific information on the various types of the ^{137}Cs measurements can be found in the corresponding original publication references.

The accuracy and precision of all of the methods for the different matrices were determined using the corresponding reference material(s) for n=6 independent sample portions. The reference materials were also employed to estimate the limit of detection of each method. The detection limits were calculated as the value three times the square root of the background of the area used to integrate the 661 keV gamma ray peak.

Finally, it should be noted that the reference materials, the blanks and the background of the spectrometer with its lead housing were routinely measured for quality control purposes.

Results and Discussion

In this section, we shall present the results and their corresponding discussion in chronological order rather than by geographical location on the map of Venezuela in Figure 3, as was the case for the Experimental section. The letters in the (parenthesis) indicate the general locations of the study areas in Figure 3 and their corresponding descriptions in the Experimental section. All the ^{137}Cs activities are expressed as Becquerels per kilogram (Bq kg^{-1}) on a dry weigh basis.

In the first investigation of ^{137}Cs activities in undisturbed surface soils in Venezuela including the islands of Margarita and Los Roques, more than 500 samples were collected and analyzed between 1990 and 1993 [11]. As a result of this country-wide study, about 85% of the surface soil samples were found to contain less than 1.8 Bq kg^{-1} (dry weigh) of ^{137}Cs, the detection limit for these measurements; about 14% were determined to have between 1.8 and 7.0 Bq kg^{-1} of ^{137}Cs and only two had higher values at about 25 Bq kg^{-1} of ^{137}Cs. Thus, ^{137}Cs activity values below 2 Bq kg^{-1} were considered as baseline values and the ^{137}Cs values greater than 10 Bq kg^{-1} as anomalous. In this study, most of the samples that were collected within a few kilometers from the coast and at elevations below 20 m.a.s.l. were determined to contain less than 1.8 Bq kg^{-1} of ^{137}Cs (the detection limit), while the higher ^{137}Cs activity values were found in samples from mountainous areas. The only two anomalous values of ^{137}Cs were determined in soils from a mountainous area on the island of Margarita and at that time was unexplainable since its annual rainfall was very little with respect to the other sampling sites on the mainland. Thus, it was decided to further investigate the surface soils on the island of Margarita in the future.

^{137}Cs Activities in Surface Soils from the Island of Margarita Study Areas (G)

In 1994, more than fifty surface soils were collected from 28 sampling areas which were thought to be undisturbed on the island of Margarita [12]. In general, the ^{137}Cs activity values were below the detection limit of 1.6 Bq kg^{-1} for the analytical method employed. But, the surface soils at three of the sampling areas were found to have relatively high ^{137}Cs activity values, some as much as a factor of ten or more than the soils at the other twenty–five sampling sites. The ^{137}Cs activities for the surface soils collected at the La Sierra Mountain study area (G1) ranged from 2.6 to 48.9 Bq kg^{-1} (seven samples); the sampling elevations were between 350-850 m.a.s.l.; the surface was very inclined, rocky and contained mostly organic litter and; was covered with large trees and bushes, as well as grasses. The ^{137}Cs activities for the surface soils collected at El Maco hill (G2) ranged from 5.1 to 50.6 Bq kg^{-1} (seven samples); the sampling elevations were between 40-120 m.a.s.l.; the surface was slightly inclined, rocky and contained mostly lateritic materials and ; was covered with a few small trees and bushes, some weeds and cactuses. While, at the El Mirador site (G3) the ^{137}Cs activities ranged from 2.2 to 21.7 Bq kg^{-1} (five samples); the elevations were between 120-195 m.a.s.l.; the surface was very inclined with mostly lateritic materials and; was covered with a few small trees and bushes, some grasses, weeds and cactuses.

It was previously reported [49] that there have been conflicting studies with respect to a correlation between elevation and the activity of ^{137}Cs in surface soils, thus another explanation(s) needed to be considered. The summit of La Sierra Mountain was frequently immersed by the clouds, while, the El Mirador site and El Maco hill were known to be foggy daily. Therefore, it was proposed at this time (1996) that the high ^{137}Cs activities in surface soils were due to the direct precipitation from the condensation of the fog and/or clouds during the dry season rather than by rainfall, which was sparse on the Island of Margarita year around [41].

^{137}Cs Activities in Surface Soils from the Altos de Pipe Tropical Cloud Forest (D)

The distribution of the activities of ^{137}Cs in surface soils collected in 1996 at a remote area in a tropical cloud forest at the Altos de Pipe (D) study area ranged from 2 to 14 Bq kg^{-1} [16]. The results showed a correlation between the ^{137}Cs activities and the sampling elevations of the soil samples within this tropical cloud forest. It can be seen in Figure 8, that the activity of ^{137}Cs increased from the lower elevations to the peak of the mountain, where the cloud immersion was denser and more frequent. Similar results were found for both the remote area and the IVIC (commercialized) area. Later on, it was estimated that the altitude of the base of the cloud cover was about 1450 m.a.s.l. from the determined ^{137}Cs activities in the surface soils.

This tropical cloud forest area, specifically the remote area at Altos de Pipe was studied during 1989 and 1990 for its frequency and duration of cloud cover parameters [9], as well as the chemistry of the fog [10]. The summit of this mountain was determined to be in the center of the clouds about one third of the time. It was also reported that higher amounts water input were due to the condensation of the clouds (fog) on the vegetation rather than by rainfall, especially in the dry seasons when there was little water input from rain.

The surface soil sampling sites at the Altos de Pipe (D) study area were grouped according to their elevation and vegetation coverage in Table 3. The ^{137}Cs activities in the surface soils showed an increasing trend at higher elevations, but the soils from sampling sites located in rainforest areas were determined to contain about 20% more ^{137}Cs than those in grassland areas at similar elevations. All of the sampling sites located near the base of the cloud cover between 1400-1500 m.a.s.l. had grassland characteristics; the ^{137}Cs activities in their surface soils were about 35% less than those at the highest elevations near the summit of the mountain. Finally, the sampling sites located below the estimated base of the cloud cover, between 1200-1400 m.a.s.l. resulted with ^{137}Cs activities less than 2.5 Bq kg^{-1}, with most of the values being below the detection limit of 1.8 Bq kg^{-1}. This investigation was carried out in 1995 and was one of first to relate the effect of a tropical cloud forest on the spatial distribution of ^{137}Cs activities in surface soils.

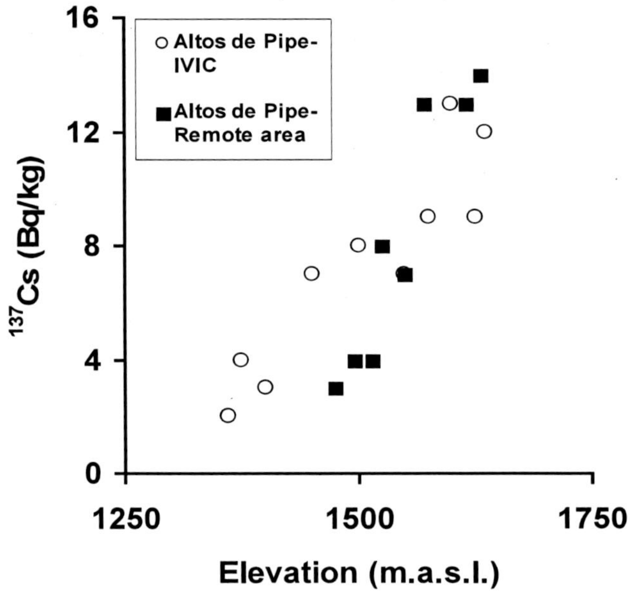

Figure 8. A graph of the ^{137}Cs activity in surface soils versus the elevation of the sampling sites for both a remote area and a commercialized area of the Altos de Pipe (D) tropical cloud forest.

Table 3. The average activity and standard deviation (1σ) of ^{137}Cs (Bq kg^{-1}) in surface soils from different types of areas and elevations at the Altos de Pipe site (D). The base of the cloud cover was estimated to be about 1450 meters above sea level (m.a.s.l.).

Type of Area	Elevation (m.a.s.l.)	^{137}Cs (Bq kg-1)* Mean ± SD
Rainforest	≥1500	10.4 ± 3
Grasslands	≥1500	8.6 ± 2
Grasslands	1400-1500	5.4 ± 2
Grasslands	1200-1400	<2.5**

* Calculated from data in reference [16].

** 2.5 Bq kg^{-1} of ^{137}Cs was the largest activity determined in ten soil samples, most of the values were below the detection limit of 1.8 Bq kg^{-1}

The Migration of ^{137}Cs in Soils on the Island of Margarita (G) and Altos de Pipe (D)

Measurements of the vertical (depth) migration of ^{137}Cs activities in soils that were located in cloud forests and areas affected by fog are presented in Table 4 [50]. Only 65-74% of the ^{137}Cs content was retained in the upper 20 cm layer of the soil profiles with respect to the general principle that 80-90% should be retained [51]. In some cases, the ^{137}Cs even migrated to the 40-50 and 50-60 cm soil layers.

The first vertical profile sampling was performed at the Maco hill site (G2), it showed that only 65% of the ^{137}Cs was retained in the upper 20 cm soil layer, so two other profile samplings were later performed nearby the first one. The results of these migration studies showed that about 75% of the ^{137}Cs was retained in the upper 20 cm soil levels. The major difference between the first profile studied was that the pit contained many large stones that could have helped the ^{137}Cs migrate downward with the water input, since it would not be in direct contact with the absorbing soil components. It should be noted that this study area was not in a cloud forest but rather an arid tropical forest (see the photographs in Figure 5). But, this area was known for its daily fog events, which resulted in condensation on the surface of the lateritic materials and the soils. The another two vertical profile studies were performed near the base of the tropical cloud forest about 400 m.a.s.l at the La Sierra Mountain site (G1) in an area very rocky with about 50% bare spaces with large trees, some weeds and grasses. The retention of the ^{137}Cs in the upper 20 cm soil layers was only 67 and 74%, again much lower than expected; some was even detected in the 50-60 cm soil layers. The profile site at Altos de Pipe (D) was located at the top of the mountain at about 1650 m.a.s.l. elevation in a pristine area of the tropical cloud forest.

Table 4. The ^{137}Cs activity (Bq kg^{-1}) in the surface soils (0-10cm) and the cumulative percentages in the lower soil layers: La Sierra Mountain site (G1), El Maco Hill site (G2) and Altos de Pipe site (D).

	Sampling sites			
	G1a	**G1b**	**G2**	**D**
^{137}Cs*	14.7	34.2	20.8	12.9
	Cumulative percentages (%)			
Depth(cm)				
0-10	44	54	40	55
10-20	67	74	65	70
20-30	80	84	78	78
30-40	91	87	90	85
40-50	100	93	100	92
50-60		100		100

* The ^{137}Cs activity (Bq kg^{-1}) in the surface soils (0-10cm)

Similar to the other sites, about 30% of the ^{137}Cs content migrated below the 20 cm soil level and even some was detected in the 50-60 cm soil layers. The major differences at this site besides being on the mainland instead of a small island were that it was situated in a dense tropical cloud forest which was well developed with large trees and completely covered with vegetation which resulted with large amounts of organic material on the surface, in the mat and litter which covered the soil. Thus, one would expect that the downward migration of the ^{137}Cs would be hindered by the strong retention of the rich organic material/soils.

In general, ^{137}Cs does not migrate downward to below the 40 cm soil level [51], but in all the cases presented in Table 4, more than 10% of the ^{137}Cs in the upper 20 cm soil level migrated below the 40 cm soil level and even some was detected in the 50-60 cm soil layer. Finally, it should be noted that similar results were also reported for ^{137}Cs migration in soils in Brazil [52] and they suggested that it was due to the large amount rainfall in their study areas.

^{137}Cs Activities in Mosses from Tropical Cloud Forests and Foggy Areas

The mean ^{137}Cs activities of mosses and their corresponding nearby surface soils collected in the Altos de Pipe (D) and La Sierra Mountain (G1) cloud forests and along the foggy Caruay River (K) are presented in Table 5. The ^{137}Cs activity range in surface soils for the highest elevations at the Altos de Pipe site (D) was between 4.9 and 11.9 Bq kg^{-1}; near the top of the La Sierra Mountain (G1) the range was between 38.1 and 140.5 Bq kg^{-1} (about an order of magnitude higher than that at the Altos de Pipe site) and; between 2.8 and 7.2 Bq kg^{-1} along the Caruay River (about a factor of two less than that at the Altos de Pipe site). Similarly, the ^{137}Cs activity ranges in the moss specimens in general were higher at the La Sierra Mountain (G1) site than the Altos de Pipe site (D), but only by a factor of two and the difference of the ^{137}Cs activity range at the Altos de Pipe site (D) was about at factor two higher than that along the Caruay River. It was suggested the that the large differences in the ^{137}Cs activities in surface soils at the different sampling locations were due to location, rather than elevation, soil type and/or vegetation. While, the ^{137}Cs activities in the moss samples at the different sampling locations were probably related to the resuspension of ^{137}Cs from the surface material/soils by different processes, such as, wind currents, temperature, surface dryness, etc.

Even though there were large differences in the ^{137}Cs activity ranges in both the surface soils and the moss samples at the three different sampling sites, the ratio of the ^{137}Cs activity of the moss to its respective surface soil at each site were very similar. The ratio of the ^{137}Cs activity of the moss to its respective surface soil at the Altos de Pipe site (D) and its standard deviation (1σ) for n=4 was 2.80 ± 0.2; at the La Sierra site (G1) for n=3 was 0.68 ± 0.1 and; along the Caruay River for n=3 was 2.92 ± 0.6. The ratio of the ^{137}Cs activity of the moss to its respective surface soil at the Altos de Pipe site (D) and along the Caruay River was not significantly different considering their mean values and corresponding standard deviations, even at the 2σ level. But, there was a very large difference between the mean ratio of the ^{137}Cs activity of the mosses to its respective surface soils at the La Sierra site (G1) and the other two sites, this seem to be related to the much higher ^{137}Cs activities in the surface soils.

Finally, a few moss samples and their nearby soils were also collected near Pico Aguila (A) between 3950 and 4050 m.a.s.l., not only were all the soil samples determined to be below the detection limit of 1.5 Bq kg^{-1} for ^{137}Cs, but so were most of the mosses also below the detection limit of 0.9 Bq kg^{-1} for ^{137}Cs. These very low values were not surprising since a previous study of ^{137}Cs in surface soils [11] also resulted with all the values below the detection limit of 1.1 Bq kg^{-1} for ^{137}Cs; the measurement times were increased from 20 000 to 50 000 seconds to decrease the detection limit for the soil samples collected in the Los Andes region. Thus, it was originally concluded that little or no ^{137}Cs fallout was deposited here, but later on it could also be explained that it was wash away or migrate with the large amount of water that runs over this area, especially after an intense rainfall and when snow melts. Another reason could be the large distance in land from the coast line which was affected by the Northeasterly winds.

Table 5. The mean ^{137}Cs activities and their respective standard deviations (1σ) from three independent aliquots measured three times each of the mosses and their corresponding surface soils collected at the Altos de Pipe (D) cloud forest, La Sierra (G) cloud forest and along the foggy Caruay River (K).

Sampling site (elevation)	^{137}Cs activity (Bq kg^{-1}) Moss	Soil	Ratio of ^{137}Cs Moss/soil
Altos de Pipe-D1 (1625 m.a.s.l.)	31.0 ± 1	11.9 ± 3	2.6
Altos de Pipe-D2 (1625 m.a.s.l.)	14.3 ± 1	4.9 ± 1	2.9
Altos de Pipe-D3 (1660 m.a.s.l.)	22.6 ± 3	8.0 ± 3	2.8
Altos de Pipe-D4 (1635 m.a.s.l.)	15.7 ± 1	5.4 ± 1	2.9
La Sierra –G1 (860 m.a.s.l.)	107.9 ± 2	140.5 ± 16	0.8
La Sierra –G2 (850 m.a.s.l.)	28.6 ± 1	42.8 ± 8	0.7
La Sierra –G3 (840 m.a.s.l.)	22.4 ± 2	38.1 ± 6	0.6
Caruay River-K1 (935 m.a.s.l.)	17.9 ± 2	7.2 ± 1	2.5
Caruay River-K2 (888 m.a.s.l.)	12.8 ± 1	4.4 ± 1	2.7
Caruay River-K3 (986 m.a.s.l.)	9.9 ± 1	2.8 ± 1	3.5

^{137}Cs in Surface Soils from the Northern Central Coastal Study Area (E)

The activities of ^{137}Cs, as well as ^{40}K, ^{226}Ra and ^{232}Th were determined in surface soils collected in 1996 from the Northern central coastal region of Venezuela to assess their radiological impact on the environment and man [17]. A cluster analysis, employing the ^{137}Cs activity values above the 0.3 Bq kg^{-1} detection limit for all the surface soils resulted in three groups with the mean ^{137}Cs activities being 1.8, 2.6 and 6.3 Bq kg^{-1}. The differences between the first two groups with the lower activities could not be explained by geographical or environmental parameters, but the highest activity group could clearly be related with the soil samples collected at higher elevations where the climate was very humid and tropical; the vegetation was well developed and frequently immersed by clouds. The relatively high amounts of both clays and organic matter in these surface soils with respect to those in the other two groups helped to retain the ^{137}Cs and the well developed vegetation limited the ^{137}Cs lost from erosion too.

The mean activities of ^{137}Cs in surface soils and their respective standard deviations (1σ) calculated for soils at different elevation ranges at the Osma sampling area near the Northern central coast of Venezuela are presented below in Table 6. It can be clearly seen that the mean activities of ^{137}Cs increased in surface soils that were collected at higher elevations (m.a.s.l.). The Osma sampling area consisted of metamorphic rocks that rose abruptly from the coast to the summit of the mountain range. The soils were predominately sandy at the lower elevations near the coast and their origins were sedimentary and alluvial. While, at higher elevations the soils contained large amount of clay minerals and organic matter. The climate was classified as a tropical semi arid zone near the coast and at the lower elevations (<50 m.a.s.l.); at the higher elevations the climate was classified as tropical humid mountainous with frequent fog cover including a cloud forest at the highest elevations (>500 m.a.s.l.). Thus, again the effect of the direct condensation of water from fog and clouds on the deposition and retention of ^{137}Cs in surface soils had been shown.

Table 6. The mean activities of ^{137}Cs and their respective standard deviation (1σ) in surface soils collected at different elevation ranges at the Osma sampling area near the Northern central coast of Venezuela.

Elevation range (m.a.s.l.)	^{137}Cs (Bq kg^{-1})* Mean ± Standard deviation
<50	2.4 ± 1**
50-99	3.5 ± 2
100-500	5.3 ± 3
>500	6.5 ± 3

*The number of samples in each group was between 6 and 10

** The values below the detection limit of 0.3 Bq kg^{-1} were not included.

^{137}Cs Activities in Surface Soils Collected at Cerro Santa Ana Study Area (B)

The activities of ^{137}Cs in surface soils collected in 1997 at locations within 0.5 km of the coast of the Paraguana Peninsula (B) with elevations less than 25 m.a.s.l. were determined to be between <2 and 10 Bq/kg [19]. In general, these activity values were higher by about a factor of two with respect to those previously investigated near the Northern central coast of Venezuela [17]. It was not possible to explain these high activity values by geographical or climatological factors, since there was little rainfall year around and clouds and fog are rarely if ever present along the coast of the peninsula. It was suggested that possibly some mechanism of the mist that was blown ashore could explain these higher values.

Table 7. The determined ^{137}Cs (Bq kg^{-1}) activities in the surface soils collected at different elevations (m.a.s.l.) at Cerro Santa Ana on the Paraguana Peninsula (B).

Elevation (m.a.s.l.)	^{137}Cs activity (Bq kg^{-1})
95	4
160	10
250	8
300	5
365	6
400	8
450	7
Estimated base of cloud cover	
500	23
520	41
535	49

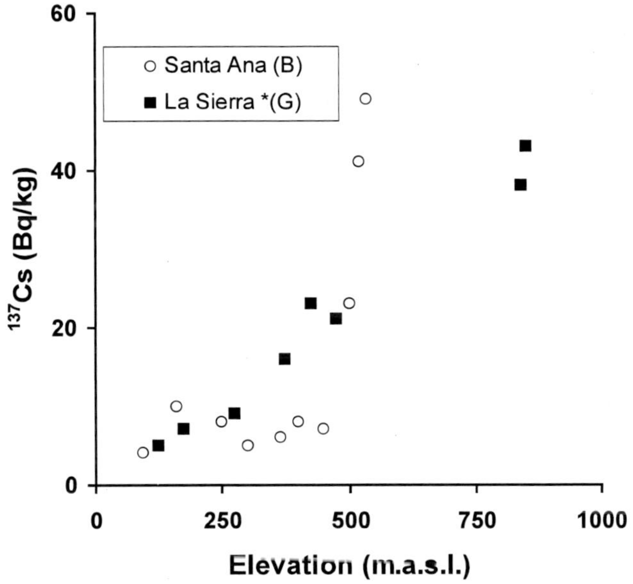

Figure 9. A comparison of the spatial distribution of the ^{137}Cs (Bq kg^{-1}) activities versus elevation at the Santa Ana (B) and La Sierra Mountain (G1) study areas.

Surface soil samples were also collected at nine locations with elevations between 95 and 535 m.a.s.l on Cerro Santa Ana, a small mountain with its peak at about 800 m.a.s.l., which had cloud cover most of the time at the higher elevations. The determined ^{137}Cs activities in these surface soils and the elevations of the sampling sites are presented in Table 7. The first seven sites presented activity values similar to the general background levels of the peninsula, but the site at 500 m.a.s.l was a factor of more than three times higher than the average background levels, thus we had concluded that this was the base of the cloud cover.

At the next sites at 520 and 535 m.a.s.l., the ^{137}Cs activities were 41 and 49 Bq kg^{-1} respectively, about an order higher than the average background levels. Thus, the effect of this low altitude cloud forest on distribution of ^{137}Cs in surface soils was shown to be very significant.

The summit of both of these study areas were less than 1000 m.a.s.l., thus they can be classified as low altitude cloud forests and the sampling sites for the Cerro Santa Ana (B) could only reached up to about 550 m.a.s.l., as the result of the steep cliffs that were encountered at higher elevations. As seen in Figure 9, the graph of the Cerro Santa Ana data (○) exhibited a flat background level (4 ± 2 Bq kg^{-1} of ^{137}Cs) for the first seven sampling sites up to 450 m.a.s.l. and then abruptly increased up to 50 Bq kg^{-1} of ^{137}Cs within the next 100 m.a.s.l. above the estimated base of the cloud cover. While, the graph of the data at the La Sierra Mountain (■) linearly increased from 5 to 43 Bq kg^{-1} of ^{137}Cs from the base of the mountain at about 120 m.a.s.l. to about 880 m.a.s.l. at its peak.

^{137}Cs in Soils Collected in the Henri Pittier National Park Study Area (C)

Surface soils were collected between 1995 and 1997 at different altitudes, about 15 to 25 m from the two roadways that pass through the Henri Pittier National Park (C). Duplicate samples were taken to confirm that the samples were not only representative of the area but also to eliminate samples that may have been disturbed by man or nature. The elevation of the summit of the mountain range where the road from El Limon to Ocumare crossed it was only about 1 100 m.a.s.l.; while, the altitude of the summit of the mountain range where the roadway between Maracay and Choroni passed it was about 1625 m.a.s.l.

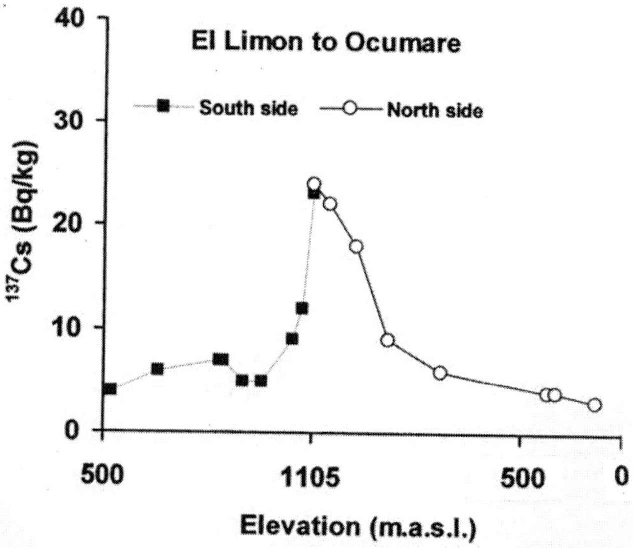

Figure 10. A plot of the ^{137}Cs activities in surface soils collected at different elevations across the mountain range in the Henri Pittier National Park near the roadway between El Limon and Ocumare, showing the local baseline activity and the increased activities of ^{137}Cs at the higher altitudes in the tropical cloud forest.

A graph of the mean activities of ^{137}Cs in the soil samples collected near the roadway from El Limon to Ocumare versus elevation (m.a.s.l.) is presented in Figure 10. The average ^{137}Cs activity and its standard deviation (1σ) was about 5 ± 2 Bq kg^{-1} for the surface soils collected below 1 000 m.a.s.l. on both sides of the mountain range, thus this value was considered the background for this area. While, the average value of ^{137}Cs and its standard deviation (1σ) for soil samples collected above 1000 m.a.s.l. was 17 ± 6 Bq kg^{-1} on the South (land) side and 18 ± 6 Bq kg^{-1} on the North (ocean) side. It can be seen that the ^{137}Cs activities in the surface soil samples from El Limon to Ocumare increased abruptly from 9 to 24 Bq kg^{-1} between the sampling sites above 1 000 m.a.s.l. to the summit, thus the baseline of the cloud cover was estimated to between the 960 and 1050 m.a.s.l. sampling sites. This was in good agreement with visual observations taken many years earlier [43]. While on the North (ocean) side, the abrupt increase of the ^{137}Cs activities in the surface soils were determined to be between the 730 and 885 m.a.s.l. sampling sites. This too was also in agreement with present day visual observations of the cloud cover baseline.

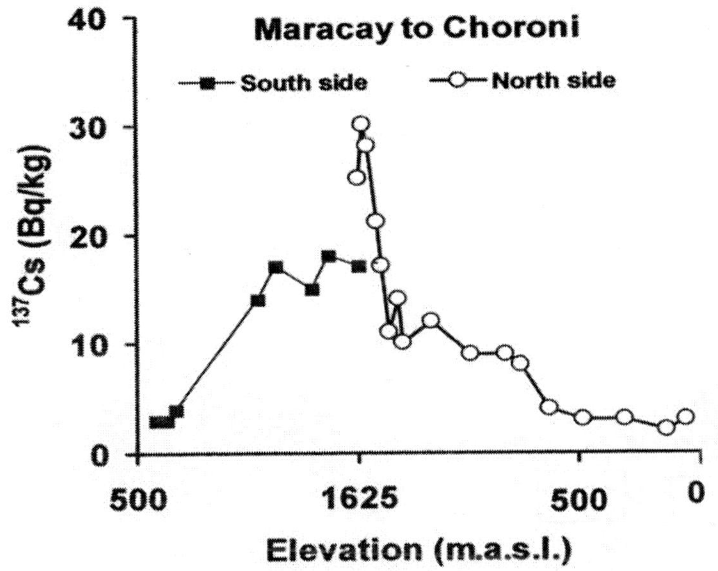

Figure 11. A plot of the ^{137}Cs activities in surface soils collected at different elevations across the mountain range in the Henri Pittier National Park near the roadway between Maracay and Choroni, showing the local baseline activity and the increased activities of ^{137}Cs at the higher altitudes in the tropical cloud forest.

A plot of the mean activities of ^{137}Cs in the surface soil samples collected near the roadway from Maracay to Choroni versus elevation (m.a.s.l.) is presented in Figure 11. The mean value of the ^{137}Cs activity and its standard deviation (1σ) for surface soils collected below 1000 m.a.s.l. was 5 ± 3 Bq kg^{-1} on both sides of the mountain, similar to the values for the local background level of ^{137}Cs near the El Limon to Ocumare roadway, thus confirming that this value was representative for the area below the cloud cover. While the average value of ^{137}Cs (Bq kg^{-1}) and its standard deviation (1σ) for samples collected above 1 000 m.a.s.l. was 15 ± 2 Bq kg^{-1} on the South (land) side and 16 ± 6 Bq kg^{-1} on the North (ocean) side. In Figure 11, it can be seen that the values of ^{137}Cs (Bq kg^{-1}) increased for samples collected above 1000 m.a.s.l from 12 to 30 Bq kg^{-1}. The baseline of the cloud (fog) cover could have been as low as 855 m.a.s.l., but probably was more close to 1000 m.a.s.l., since the activity of ^{137}Cs was about 40% higher at the 1110 m.a.s.l. sampling site. This also was in good agreement with the estimate from visual observations taken many years before [43]. While on the North (ocean) side, the increase of the ^{137}Cs (Bq/kg) activities in the soils were determined to be between the 715 and 820 m.a.s.l. sampling sites. This too was also in agreement with present day visual observations of the cloud cover baseline. Finally, there was no significant difference between the ^{137}Cs activities determined in the surface soils on the South (land) side and the North (ocean) side of the mountain range in general, but the activities increased with elevation (m.a.s.l.), due to the density of the cloud and/or fog.

^{137}Cs in Surface Soils Collected on the Araya (F) and Paria (H) Peninsulas

The activity of ^{137}Cs in surface soils was determined for more than ninety sample sites on the Araya Peninsula (F), the Paria Peninsula (H) and the isthmus in between them in the state of Sucre, Venezuela [38]. In general, the ^{137}Cs activities were higher on the Paria Peninsula (F) than the Araya Peninsula (H) and slightly lower on the isthmus. Even though the peninsulas are only separated by a small isthmus, their environmental characteristics were very different. The sampling sites on the Araya Peninsula (F) resulted with about 80% of the surface soils containing less than 2 Bq kg^{-1} of ^{137}Cs and only one soil sample with the ^{137}Cs activity greater than 4 Bq kg^{-1}. While, the distribution of the ^{137}Cs activities on the Paria peninsula (H) had more than 60% of the values greater than 2 Bq kg^{-1} and about 15% in the 7-10 Bq kg^{-1} activity range. Finally, the distribution of the ^{137}Cs activities in the surface soils on the isthmus resulted with about 35% of the samples having less than 2 Bq kg^{-1} of ^{137}Cs; about 65% were between 2-4 Bq kg^{-1} of ^{137}Cs; and one anomalous value of 7 Bq kg^{-1}. The large difference in the annual rainfall of the peninsulas could have resulted in a greater amount of deposition of the ^{137}Cs on the Paria Peninsula (H), as well as other environmental characteristics, such as the vegetation. Even though these ^{137}Cs activities in the surface soils were relatively low, they were slightly higher with respect to other surface soils investigated along the Northern Venezuelan coast [17]. Finally, these slightly higher ^{137}Cs activities in the surface soils were similar to those determined along the coast line at the Paraguana Peninsula (B) site and could possibly be attributed to fog or mist from the Northeast trade winds. These ^{137}Cs activities in these surface soils can be employed as baseline values to compare the effect of cloud cover in tropical cloud forests along the coast.

^{137}Cs in Soils from the Gran Sabana (L) and the Sierra de Pacarima (M)

Surface soil samples were collected in the Gran Sabana (L) and the Sierra de Pacarima (M) in the state of Bolivar (Venezuela) between 2001 and 2003 for the determination of their ^{137}Cs activity [40]. Both, the Gran Sabana (L) and the Sierra de Pacarima (M) study areas did not include cloud forests. In general, about 85% of the surface soil samples were determined to contain less than 2 Bq kg^{-1} of ^{137}Cs, but many of the samples were probably disturbed by the large amount of seasonal rainfall specially since the soils in general were very poor (infertile).

The ^{137}Cs activities in the surface soils ranged from less than 1.2 Bq kg^{-1} to 14.8 Bq kg^{-1}, but only about 15% of the samples were determined to have values greater than 5 Bq/kg of ^{137}Cs.

In the sampling zone between the Monumento al Soldado Pionero and San Francisco de Yuruani, only eight of the nineteen soil samples were determined to have ^{137}Cs activities higher than 1.2 Bq kg^{-1} and in all cases were located in areas with small bushes and scrubs (which frequently exhibit fog) rather than coarse grasses. While eight of eleven (73%) of the ^{137}Cs values were higher than 1.2 Bq kg^{-1} in the area between Tupea and Kavanayen, where the land surface was generally flat. The four highest values, between 3.7-6.7 Bq kg^{-1}of ^{137}Cs were all located areas, in which fog was produced when there were abrupt changes in the temperature, as well as contained rich organic soils rather than in areas with poor soils covered by coarse grasses. The climate near Kavannayen was classified as very humid pre-mountainous with a higher annual rainfall than the eastern part of this study area which was associated with a humid tropical climate. Only three of the seven soil samples from the area between San Francisco de Yuruani and Parai-tepuey were determined to contain ^{137}Cs activity values higher than the detection limit of 1.2 Bq kg^{-1}, but they included the two highest values (10.2 and 14.8 Bq kg^{-1}) measured in all of the sampling areas of the Gran Sabana (L) and the Sierra de Pacarima (M). The area between San Francisco de Yuruani and Parai-tepuey consisted of rolling hills with coarse grasses and small patches of trees near rivers (which also produced fog) and a few areas with small bushes. The area between San Francisco de Yuriana and Santa Elena de Uairen, had decreasing elevations from 1094 to 863 m.a.s.l with dark reddish brown lateritic soils with a lot of small pebbles on the surface which resulted in the coarse grasses being less populated (less dense). There were only three ^{137}Cs activity values above the detection limit of 1.2 Bq kg^{-1} and they contained between 6.4-8.5 Bq kg^{-1} of ^{137}Cs nearby a river, close by Yuriana. The climate in this area was considered humid low pre-mountainous with lower annual rainfalls between 1500 mm and 2000 mm. Finally, the area between Santa Elena de Uairen and El Pauji in the Sierra de Pacarima (M) in contrast was mostly dense forest. Seven of the twelve soil samples studied here were above the 1.2 Bq kg^{-1} detection limit but the activity range was only between 1.9-3.4 Bq kg^{-1} of ^{137}Cs, even though there was very frequent fog events in this study area.

In general, the higher ^{137}Cs activity values in the surface soils in the Gran Sabana (L) and the Sierra de Pacarima (M) were associated with sample sites that had higher annual rainfall, organic soils in dense forests and in humid and very humid pre-mountainous climates rather than in humid tropical climate. Most of the sampling sites were located at relatively high altitudes, between 1000 and 1500 m.a.s.l. elevations, thus these ^{137}Cs activity values can be used as baseline values for these inland areas, which were not affected by water inputs from cloud forests and/or foggy areas.

^{137}Cs in Surface Soils Collected in the Guri Reservoir Study Area (I)

The ^{137}Cs activities were determined in surface soils collected during 2002 and 2003 surrounding and near by the Guri Reservoir in the state of Bolivar, Venezuela [39]. In general, only about 28% of the soil samples were determined to have activity values of ^{137}Cs greater than 2 Bq kg^{-1} but many of the sample sites were probably disturbed by the seasonal rainfall, as well as flooding in the reservoir areas during the rainy season.

In the sampling area, North of the Guri reservoir, only 5 of the 31 soil samples had ^{137}Cs activities higher than 2 Bq kg^{-1} and their mean value and standard deviation (1σ) was 4.4 ± 1 Bq kg^{-1}. Many of the soil samples that were below 2 Bq kg^{-1} were probably disturbed by man, since the population is relatively high in the Northern area.

In the sampling area on the Eastern side of the reservoir, the elevations were between 188 and 373 m.a.s.l. Similarly, only 5 of the 29 soil samples had ^{137}Cs activities higher than 2 Bq kg^{-1} but their mean value and standard deviation (1σ) was only 2.7 ± 1 Bq kg^{-1}. Many of these surface soils were collected in places that probably would be located underwater when the reservoir was at its highest water level.

With respect to the sample sites on the Western side of the reservoir with similar elevations between 153-374 m.a.s.l., a much higher percentage of the samples were above 2 Bq kg^{-1} of ^{137}Cs, actually nine of the 13 soil samples had ^{137}Cs activities higher than 2 Bq kg^{-1} and their mean value and standard deviation (1σ) was almost double at 4.8 ± 2 Bq kg^{-1}. These higher values were explained by the difference in the soil types; the soils on both sides of the reservoir were high mineralized, but many of the samples on the Western side were above rich mineral deposit of iron ores. The soils on the Western side of the reservoir were very different than those on the Eastern side; they were classified as Untisols on the West side and Entisols on the East side [44]; the organic content and ratios of clay to sand fractions were also different. Finally, there had been much more leaching of the soils on the Eastern side by the water that covered the area when the reservoir was full. Similar results and explanations have been reported by other authors before [52-54].

Most of the surface soil samples Southeast of the reservoir were below 2 Bq kg^{-1} of ^{137}Cs, but the mean value and standard deviation (1σ) was 9.3 ± 11 Bq kg^{-1}, much higher than the other areas because of anomalously high values between 8.3 and 33.4 Bq kg^{-1} at the base of Sierra de Lema, which was frequently foggy and considered in a cloud forest at higher elevations.

It was previously reported [55] that the annual ion deposition by cloud droplets capture in a cloud forest in the temperate zone was between 61% and 81%. Thus, it could be much more during the dry seasons in the tropical zone. It had also been shown that the ionic concentrations were higher in cloud water than in rain [10, 56, and 57]. Thus, rainfall combined with cloud capture droplets can lead to exceptionally high rates of deposition of harmful substances. In areas, where trees were on the edges of cloud forests it was also reported that the direct deposition was on an average of three times higher and up to fifteen times higher than those in the interior [58].

^{137}Cs in Mosses, Lichens and Other Vegetation along the Caruay River (K)

The activity of ^{137}Cs was determined in surface soils, mosses, lichens and other vegetation samples collected in 2004 along the banks of a 20 km section of the Caruay River, which was characterized by daily fog in the mornings [26]. Samples were also collected and analyzed for ^{137}Cs from the river towards the town of Kavanayen for comparison purposes. The range of activity values for the surface soils collected near by the Caruay River was from <1.2 Bq kg^{-1} of ^{137}Cs to 14.1 Bq kg^{-1} of ^{137}Cs. While, the highest activity value in the surface soils near Kavanayen was only 4.3 Bq kg^{-1} of ^{137}Cs with respect to 14.1 Bq kg^{-1} of ^{137}Cs from a soil collected near the river bank. The mean ^{137}Cs values for the surface soils along the river were more than double of those near Kavanayen, 5.4 ± 4 Bq kg^{-1} of ^{137}Cs versus 2.3 ± 1 Bq kg^{-1} of ^{137}Cs. The difference of the vegetation of the areas, dense forest along the river with respect to the areas with some bushes and grasses near Kavanayen could be an explanation. But, another factor that could explain the higher values along the river was the differences in the temperatures of the water and air, which created fog and moisture daily, as well as the mist created by the water falls (see Figure 6).

The ^{137}Cs activities in the mosses ranged from 9.9 to 17.9 Bq kg^{-1} with a mean value of 13.4 ± 4 Bq kg^{-1}; all the moss samples were found along the river. The ratios of ^{137}Cs activity in the mosses versus their corresponding surface soils were between 2.4 and 3.5 Bq kg^{-1}; these values were compared to those from other study areas in Venezuelan cloud forests [27]. The ^{137}Cs activities in the lichens ranged from 9.1 to 29.8 Bq kg^{-1}; the two values from samples collected along the river were about three factors higher than the one sampled near Kavanayen. The ^{137}Cs activity ratios between the lichen and their corresponding soil were between 2.4 and 5.4 Bq kg^{-1}; the highest ratio being near Kavanayen. This suggested that the more organic rich soil along the river was a better accumulator and /or retainers of ^{137}Cs than the sandy soils near Kavanayen.

^{137}Cs in Soils and Vegetation Collected at the Sierra de Lema Study Area (J)

Surface soils, black mat, palm tree leaves and trunks, fruit bushes leaves and its` fruit and fern leaves were collected in 2003 at the El Mirador (Lookout) study area at the base of the Sierra de Lema mountain range [15]. The determined ^{137}Cs activity values ranged from 16 to 31 Bq kg^{-1} in the surface soil samples; 21-32 Bq kg^{-1} for the black mat specimens; 26-38 Bq kg^{-1} for the palm leaves; 17-31 Bq kg^{-1} for the palm trunks and 18-27 Bq kg^{-1} for the fruit bush leaves. The ^{137}Cs activity values for the whole fruit were between 23-31 Bq kg^{-1}, but for the center of the fruit (the edible parts) the mean value was 52 Bq kg^{-1}; the mean value of the ^{137}Cs activity for the fern leaves was 72 Bq kg^{-1}. Thus, most of the ^{137}Cs activity values determined in the surface soils, black mat and vegetation samples from the El Mirador (Lookout) were considered anomalously high for ^{137}Cs with respect to those of other values found in samples collected near the equator and in other areas of Venezuela. Only the center of the fruit from the *Clusia grandiflora* bushes and the fern leaves had relatively high activity ratios, about a factor of three and could be considered as bio monitors that concentrated and retained the ^{137}Cs. These anomalously high ^{137}Cs activities had been principally attributed to the direct water input from the cloud forest effect and fog, as well as the rich organic soils that acted as sinks. The abrupt daily changes in the temperature of the diabase rock surface possibly also contributed to these anomalous ^{137}Cs activities, similarly as the rocky lateritic surface at the El Maco hill (G2) study area on the island of Margarita.

Conclusion

The most significant environmental pathways of man-made (artificial) radionuclides had been well known since the publication of SCOPE report 50, *Radioecology after Chernobyl: Biogeochemical Pathways of Artificial Radionuclides* [59] in 1993. This report was the consensus among leading members of the international scientific community and stated that

"The involvement of precipitation processes may act as a concentration mechanism leading to areas of enhanced deposition."[60].

In this chapter, it was demonstrated that the deposition of ^{137}Cs in tropical cloud forests was greatly enhanced in surface soils and vegetation by a precipitation mechanism of condensing water directly from the clouds in addition to the conventional rainfall process, especially during the dry seasons. Similarly, it was shown that direct condensation from fog also enhanced the deposition of ^{137}Cs. It was suggested that the slightly higher ^{137}Cs activities in surface soils along the coast could have been due to mist that was blown from the ocean, but this needs to be further investigated.

It was concluded that the distribution of ^{137}Cs activity in surface soils in tropical cloud forests could be related to the density of the cloud immersion and that it also could be employed to estimate the base of the cloud cover over long time periods.

In conclusion, it is very important to determine baseline reference values for radioactive fallout in ecosystems that include different precipitation processes that enhance the deposition of fallout and not just in the areas that man wants to inhabit! If not, future monitoring of these ecosystems may result in reporting very high activity values, which may be interpreted as anomalous and suggest that a large amount of ^{137}Cs has recently entered the local environment.

Finally, it should be noted that enhancement of the deposition of fallout by different precipitation processes is similar for other atmospheric pollutants which could be transported globally when dispersed into the upper atmosphere. Thus, atmospheric pollution in general is a global problem and knows no boundaries!

Acknowledgments

The author is proud to acknowledge the following persons for their participation in the different parts of research presented in this chapter: Pedro Rosales, Owaldo O`Carias, Pedro Cordoves, Jhonny Gonzales, Gustabo Cordovez, and Carlos Cordoves for their assistance in the field work; Pedro Rosales, Pedro Cordoves and Carlos Bastidas for preparing the samples in the laboratories and performing the high resolution gamma ray measurements; Eder Pena and Guillermina Aguilar, who identified the species of the vegetation specimens. I also like to thank my collaborators at the University of Simon Bolivar, who also participated in investigating ^{137}Cs in surface soils: Lazlo Sajo-Bohus, Daniel Palacios and David Castro, as well as Franco Urbani at the University Central of Venezuela.

Jesus Montenegro, who assisted with the preparation of this chapter, Dr. Daniel Palacios, who reviewed the technical aspects of this chapter and the editorial staff of Novascience Publishers should also be acknowledged and thanked, as well as, the publishers of the *Journal of Radioanalytical and Nuclear Chemistry* for their copyright permissions.

Many thanks to the Instituto Venezolano de Investigaciones Cientificas-IVIC (The Venezuelan National Scientific Research Institute) for their financial and infrastructural support for more than 30 years. As well as, The Venezuelan National Science Foundation, CONICIT as it was called then for their many research contracts, especially an oriented research grant PC-075 in 1989 and RP VII 7600026 that were related to this chapter.

Finally, special thanks to Pedro Rosales, who was not only my principal co-investigator with respect to the environmental radioactivity studies presented in this chapter, but also a friend for more than the twenty-five years, he .spend under my supervision in various laboratories at IVIC before I retired as an Investigator Emeritus in 2004.

References

[1] Schmitt, G. In : *Mechanisms and Effects of Pollutant Transfer into Forests*, Georgii, H.W. (editor) ; Kluwer, Dordrecht, 1989, pp. 213-220.

[2] Weathers, K.C., Likens, G.E., Bormann, F.H., Eaton, J.S., Bowden, W.B., Anderson, J.L., Cass, D.A., Gallow, J.N., Keene, W.C., Kimball, K.D., Huth, P., Smiley, D., *Nature* 1986, 319, 657-658.

[3] Steinhardt, U., Fassbender, H.W., *Turrialba* 1979, 29, 175-182.

[4] Hamilton, L.S., *Mountain Research and Development* 1995, 15, 259-266.

[5] Cloud forest (n.d.). The American Heritage Dictionary of the English Language, Fourth Edition. Retrieved March 31, 2008, from Answers.com Web site: http://www.answer.com/topic/cloud-forest

[6] Cloud forest (n.d.). Wikipedia. Retrieved March 31, 2008, from Answers.com Web site: http://www.answer.com/topic/cloud-forest

[7] Cloud forest, Cs-137, soils and vegetation. Retrieved March 2, 2008 from Google Quicksearch Web site: http://www.goolge. com/webhe?sourceid =firefox

[8] Cloud forest and Cs-137. Retrieved March 2, 2008 from ISI Web of Knowledge site: http://apps.isiknowledge.com

[9] Gordon, C.A., Herrera, R., Hutchinson, T.C., *Atmospheric Environment* 1994, 28, 317-322.

[10] Gordon, C.A., Herrera, R., Hutchinson, T.C., *Atmospheric Environment* 1994, 28, 323-337.

[11] LaBrecque, J.J., *Journal of Radioanalytical and Nuclear Chemistry* 1994, 178, 327-336.

[12] LaBrecque, J.J., Rosales, P.A., *Journal of Trace and Microprobe Techniques* 1996, 14, 203-211.

[13] The Use of Gamma Ray Data to Define the Natural Radiation Environment, *IAEA-TECDOC-566*, International Atomic Energy Agency : Vienna Austria, 1990, 48pp.

[14] Dowdall, M., Gwynn, J.P., Moran, C., Davids, C., O`Dea, J., Lind, B., *Journal of Radioanalytical and Nuclear Chemistry* 2005, 266, 217- 223.

[15] LaBrecque, J.J., Alfonso, J.A., Cordoves, P.R., *Journal of Radioanalytical and Nuclear Chemistry* 2008, accepted, presently on-line only.

[16] LaBrecque, J.J., Rosales, P.A., *Journal of Radioanalytical and Nuclear Chemistry* 1997, 220, 59-64.

[17] Palacios, D., Castro, D., Perez, M., Urbani, F., Sajo-Bohus, L., LaBrecque, J.J., *Journal of Radioanalytical and Nuclear Chemistry* 1998, 241, 69-73.

[18] LaBrecque, J.J., Rosales, P.A., Cordoves, P.R., *Journal of Radioanalytical and Nuclear Chemistry* 2002, 253, 87-92.

[19] LaBrecque, J.J., Rosales, P.A., Cordoves, P.R., *Journal of Radioanalytical and Nuclear Chemistry* 2001, 263, 563-566.

[20] Fredriksson, L., *Forsvarets Forskningsanstalt, FOA 4 Rapport* 1963, A4319-4623.

[21] Lopatkina, A.P., *Geochemical International* 1964, No. 4, 788- 795.

[22] Idiz, E.F., Carlisle, D., Kaplan, I.R., *Applied Geochemistry* 1986, 1, 573- 582.

[23] Dowdall, M., O`Dea, J., *Radiochimica Acta* 1999, 87, 109-114.

[24] LaBrecque, J.J., Rosales, P.A., In : *Practical Methods fo Environmental Analysis*, Burk, R., Clement, R.E: (editors), Polyscience Publications, Ottawa, Canada, 1966, p 49-52.

[25] Rosales, P.A., LaBrecque, J.J., Cordoves, P.R, In : *Jornadas Venezolanas de Tecnicas Analiticas No Convencionales para un Desarrollo Sustenable*, Livinalli, P.L. (editor), Universidad de Simon Bolivar, Sede El Litoral, Venezuela, 1996, p 260-265.

[26] LaBrecque, J.J., Rosales, P.A., Cordoves, P.R., *Journal of Radioanalytical and Nuclear Chemistry* 2007, 273, 401-404.

[27] LaBrecque, J.J., Rosales, *Journal of Radioanalytical and Nuclear Chemistry* 1998, 231, 139-142.

[28] Conti, M.E., Cecchetti, G., *Environmental Pollution* 2001, 114, 471-492.

[29] Berg, T., Royset O., Steinnes, E., Vadset, M., *Environmental Pollution* 1995, 88, 67-77.

[30] Ruhling, A., Tyler, G., *Bot. Notiser* 1968, 121, 321- 342.

[31] Ruhling, A., Tyler, G., *Applied Ecology* 1971, 8, 497- 507.

[32] Nifonnnntova, M., *Science of the Total Environment* 1995, 160, 749-752.

[33] Rucklidge, J., Kilius, L., Fuge, R., *Nuclear Instruments and Methods in Physics Research* 1994, B92, 417-420.

[34] Steinnes, E., *Science of the Total Environment* 1995, 160, 243-249.

[35] Golbev, A.V., Golubea, V.N., Krylov, N.G., Kuznetsova, V.F., Mavrin, S.V., Aleinikov, A. Yu., Hoppes, W.G., Surano, K.A., *Journal of Environmental Radioactivity* 2005, 84, 333- 342.

[36] Rodigro, A., *Science of the Total Environment* 1999, 233, 359-367.

[37] Knops, J.M.H., Nash III, T.H., Boucher, V.I., Schlesinger, W.I., *Lichen* 1991, 23, 309-314.

[38] LaBrecque, J.J., Cordoves, P.R., *Journal of Radioanalytical and Nuclear Chemistry* 2003, 258, 227-231.

[39] LaBrecque, J.J., Cordoves, P.R., *Journal of Radioanalytical and Nuclear Chemistry* 2005, 265, 91-94.

[40] LaBrecque, J.J., Cordoves, P.R., *Journal of Radioanalytical and Nuclear Chemistry* 2004, 262, 375-381.

[41] *Imagen Atlas de Venezuela ; Una Vision Espacial*, edited by Petroleos de Venezuela, S.A. Caracas, Venezuela, 1992, 271pp.

[42] Ewel, J.J., Madriz, A., Tosi, J.A., In : *Zonas de Vida de Venezuela*, Ministerio de Agricultura y Cria, Fondo Nacional de Investigaciones Agropecuarias, Editorial Sucre, Caracas, Venezuela, 1976.

[43] *La Selva Nublada de Rancho Grande, Parque Nacional Henri Pittier ; El Ambiente Fisco, Ecologia Vegetal y Anatomia Vegetal*, Huber, O. (editor), Fondo Editorial Acta Cientifica Venezolana, Caracas, Venezuela, 1986, 288pp.

[44] Rabinovich, J.E., In : *Modelo Guri, Analisis de un conflicto de intereses en el uso de recursos naturales en una cuenca tropical,* Centro de Ecologia, IVIC, Caracas, Venezuela, 1977.

[45] *Guia Ecologica de la Gran Sabana*, Huber, O., Febres, G.,(editors), The Nature Conservancy, Editorial Arte, Caracas, Venezuela, 2000, 189pp.

[46] *Parques Nacionales y Monumentos Naturales de Venezuela*, edited by Instituto Nacional de Parques, Omnia Industrias Graficas, Madrid, Spain, 1978, 192pp.

[47] Schubert, C., Huber, O., *La Gran Sabana*, Cuadernos Lagoven, Refolit C.A., Caracas, Venezuela, 1989,108pp
[48] Measurement of Radionuclides in Food and the Environment-A Guidebook, *Technical Report Series No. 295,* International Atomic Energy Agency, Vienna, Austria, 1986, 169pp.
[49] McGee, E.J., Colgeen, P.A., Dawson, D.E., Rafferty, B., *Analyst* 1992, 117, 461-464.
[50] LaBrecque, J.J., Rosales, P.A., *Journal of Trace and Microprobe Techniques* 1996, 14, 213-221.
[51] Arnalds, O., Cutshall, N.H., Nielsen, G.A., *Health Physics* 1989, 57, 955-958.
[52] Schuch, L.A., Nordemann, D.J.R., Zago, A., Dallpai, D.L., Godoy, J.M., Recequilo, B., *Journal of Radioanalytical and Nuclear Chemistry* 1994, 177, 101-106.
[53] Thiry, Y., Myttenaere, C., *Journal of Environmental Radioactivity* 1993, 18, 247-257.
[54] He, Q., Walling, D.E., *Applied Radiation and Isotopes* 1997, 48, 677-690.
[55] Lovett, G.M., Rienners, W.A., Olson, R.K., *Science* 1982, 218, 1302-1308.
[56] Fowler, D., Cape, N.D., Leith, I.D., Choularton, T.W., Gay, M.J., Jones, A., *Atmospheric Environment* 1988, 22, 1355-1362.
[57] Saxena, V.K., Strogner, R.E., Hendler, A.H., DeFelice,T.P., Yeh, R.J-Y., Lin, N-H., *Tellus* 1989, 41B, 92-109.
[58] Weathers, K.C., Lovett, G.M., Likens, G.E., *Atmospheric Environment* 1995, 29, 665-672.
[59] SCOPE Report 50 : *Radioecology after Chernobyl : Biogeochemical Pathways of Artificial Radionuclides*, Warner,F., Harrison, R.M. (editors), John Wiley & Sons : Chichester, UK,1993, 367pp.
[60] Warner, F., *Chemistry International* 1994, 16, 25-31.

Reviewed by Dr. Daniel Palacios, Universidad Simon Bolivar, Sartenejas, Baruta, Venezuela

In: Nuclear Chemistry: New Research
Editor: Axel N. Koškinen, pp. 51-81
ISBN: 978-1-60456-957-5

Chapter 2

CERAMIC IN ANCIENT SOCIETIES: A ROLE FOR NUCLEAR METHODS OF ANALYSIS

M. Isabel Prudêncio
Instituto Tecnológico e Nuclear, EN 10, 2686-953 Sacavém, Portugal

Abstract

The application of the instrumental neutron activation analysis (INAA) in support of provenance research of archeological ceramics has been largely used over the past few decades. INAA is a sensitive technique useful for quantitative multi-element analysis of major, minor, and trace elements. Among these, trace elements can be particularly useful to determine provenance of the mineral components in the paste of archaeological ceramics or to distinguish different ceramic groups, better differentiating diverse raw materials. The chemical characterization of the regional potential raw materials is particularly important in the case of the lack of kilns structures in archaeological excavations, and if the ceramic artifacts study points to local productions. This is the case of Bracara Augusta (nowadays Braga), a Roman town located in the north of Portugal, recognized as a center of production and distribution of pottery. In this chapter chemical data (major and specially trace elements) of the different types of clay materials of Braga region obtained by instrumental neutron activation analysis are used to establish the geochemical profiles of raw materials, which can be used as indicators of provenance of several ceramic types found in Bracara Augusta. Among the chemical elements studied rare earth elements play an important role differentiating resources and their correlation with ceramics. The results obtained suggest that during Roman times two main resources were used to produce different types of ceramics: (1) the sedimentary clays of the Prado basin (located near Braga) to produce common wares, amphorae, as well as imitations of "Pompejanish-roten slip", and "Terra Sigillata Hispanica" and "African Red Slip Ware" forms; and (2) sedimentary kaolin deposits located 30 km west of the town, near the coast, for the production of the "Bracarense" ceramics, a fine yellow-pale paste covered with a brownish yellow slip. Until the finding of archaeological evidence of pottery workshops in the area, one may assume that potters, already settled in Bracara Augusta, may have went further west to look for especial clays to answer specific needs, functional and aesthetic. The chemical characterization of the ceramics pastes and their comparison with ceramics of well established origin also point to imports of several types of ceramics from other locations of the Roman Empire, such as Italic "Pompejanish-roten" production (Campania) as well as imitations of "Pompejanish-roten slip" produced in Lucus Augusti (NE Braga).

1. Introduction

Ceramics enclose messages about the society that manufactures and uses that must be clarify within a broader context of the ancient society, its cosmological structures, ritual behavior, overall productive systems, resources exploitation strategies and the interactions with neighboring areas. The objectives of ceramic research in archaeological studies are at inferring how the ceramic might have been used and determining the chronology, and the location and techniques involved in its manufacture - that is what for, when, where and how was a ceramic piece done. In this way pottery can be seen as a source of insights into people and cultures.

Identification of archaeological ceramics based on visual inspection may not allow distinguish different production sites or different productions of the same site. Technological properties of pottery – the physical, mineralogical, and chemical characteristics of ceramic materials – constitute an irreducible minimum of detail on ceramic objects supplementing morphological and stylistic criteria. Pottery was the first synthetic material humans created, and the discussion of its history must begin with the recognition of clay itself as a raw material. For potters the term clay means the basic ingredient in ceramic manufacture, composed of plastic particles (clay) and natural non-plastic grains. In archaeology the term ceramics usually refers to cooking and serving utensils and art objects manufactured of clay. The chemical analysis of archaeological ceramics and raw materials in support of provenance research has grown rapidly over the past few decades. Technological information together with classical archaeological approaches has been largely used in solving a broad variety of questions and an overview of these types of studies is well described and summarized in the literature (Rice, 1987, Chappell, 1991, Hector, 1992, and Velde & Druc, 1999) and a review of ceramic production, provenance and use was recently published (Tite, 2008). Compositional profiles of ceramics and raw materials (clays and temper) are used to trace individual artifacts from their find spot to their origin. Chemical analysis, together with statistical data treatment, has been used extensively to supplement archaeological investigations when provenance studies or socioeconomic aspects are concerned. The information obtained is used by archaeologists to investigate questions regarding the location of production areas, the identification of routes of trade and exchange of raw materials and artifacts, and the mobility patterns in ancient times.

In chemical-based provenance studies of archaeological ceramics important issues must be considered. Among the materials and processes involved in making a pot, the most important is clay and its manipulation to make the vessel body. Hence a discussion of pottery must include clay and its origin, composition and properties. Therefore it is important to know how the different elements make clays, what the leftovers are in the clay-making process, and the geochemical and mineralogical fingerprints which can give one a clue to the geographic origin of the clay and temper materials found in the ceramics. Thus, the use of resources is very important to deciphering the provenance of a ceramic object, i.e., the geographic area where the object was produced, especially when no archaeological evidence of pottery workshops are found. The resources should be compared with care to the possibilities of a given production site. Today clay and other ingredients used for ceramic production can come from sites thousands of kilometers from the production site. However, in the past, the potter was more tied to the local/regional resources for his production, since in general clays did not travel more than several kilometers. In this way the potter would have adapted local resources to answer specific needs. In ancient times, sources of clays tended to be those easily available. Soils or surface sediments are likely candidates, which can be adapted to form a ceramic without further treatment. By their natural mixture of plastic and non-plastic components, they can be used to form pottery. If the proportions of clay, silt and sand are not adequate, tempering by mixtures of materials of the same source or from different sources can be done. All these procedures during manufacture, together with eventual modifications of the ceramic composition during use and burial, can lead to compositional alterations than can difficult the establishment of provenance.

Technological data are independent of other common categorizations of pottery used by archaeologists, such as styles, type classes, or shape categories, yet they can be used for comparing such groupings or creating new ones (compositional groups). In combination with ceramic ecology, ceramic technology can, for example, characterize properties of both resources and pottery and permit comparisons between them. Furthermore, these objectives can be achieved on broken fragments as well as intact vessels. Compositional data provide a basis for understanding many questions about production techniques and organization, functional relationships between specific resource manufacturing combinations, and patterns of local, regional, or extra regional distributions of pottery. Thus, better understand a culture, time period or human interaction. These are the main research issues of ceramic technological (and ecological) analysis.

The methodological approaches to study archaeological ceramics vary depending on the nature of the raw materials used and the manufacturing techniques. Research began by a careful examination of the broken pottery fragments that archaeologists recover in their excavations. Ceramic characterization is done with three main objectives – classification, production technology and provenance. The more commonly used techniques to characterize archaeological vessels focus on the chemical and mineralogical constituents. There are several methods that can be employed. However it should be noted that special considerations in applying these methods to archaeological pottery must be taken into account. The objectives of the analysis must be defined so they can be translated into compositional (geochemical and mineralogical) terms.

Pottery was important in trade, and the composition of pottery is strongly related to the source of clay and the recipe of the fabrication. This is highly site specific and, although similar in style and appearance, in critical cases it is possible to distinguish among the products of different producers by determining the chemical composition.

Within their composition, ceramics preserve a record of their raw materials, so chemical analysis of the ceramics pastes, determining the larger number of chemical elements as possible, is particularly useful in provenance studies. In this way, large data sets are generated, which need to be processed by computer with statistical programs. One important thing to not forget is that statistical analysis is only a tool to help us with very large amount of data matrixes (variables and samples) and remember that the results obtained are an approximation of the reality. The attribution of one pot to one particular site or compositional group is based on statistical probabilities. The results obtained by statistical analysis must be checked taking into account geochemical (and mineralogical) considerations.

In some cases, an important step prior to any statistical analysis is normalization of the chemical elements contents. To compensate for grain-size and mineralogy effects on trace element concentrations, thus diminishing erroneous interpretation of ceramic provenance, a common approach used in determining regional geochemical baselines is to normalize geochemical data using one element as grain-size and mineralogical proxy, that is to express the ratio of the concentration of a given element to that of the normalizing factor (Sanei et al. 2001; Prudêncio et al., 2007b). Normalization in archaeological ceramics studies is defined as a procedure to compensate for the influence of natural (geological and ceramic burial time) and anthropogenic (technology of production) processes on the measured variability of the concentration of elements, emphasizing the importance of taking into account geochemical behavior of the element chosen for normalization. So, the selection of the normalizing element should be not purely based in statistical considerations (Prudêncio et al., 2006; Dias & Prudêncio, 2008). In this way, normalization of elemental concentration to an immobile, conservative element provides deeper insights about other elements distribution than the consideration of the absolute concentration itself alone. The criteria used to choose the normalization factor must be supported by geochemistry. Considering that it is well established that more relevant information can be derived from the concentration of any element when it is compared with some reference element, the question arises is which element should be chosen to normalize concentration. When dealing with ancient pottery production, whenever a possible anthropogenic input of a grain-size proxy is suspected, preliminary investigation of the nature of the added temper is recommended before choosing the reference element. The variable of time burial is important, as it may induce element mobilization / lixiviation. Therefore, the best option appears to be the use of a conservative element, and among these, Sc appears to be the more appropriate to normalize chemical data then other elements; even their variability is higher, since the raw materials are related with a specific geological context which has to be considered. Sc^{3+} easily hydrolizes and is structurally combined in clay minerals and micas, being a good tracer of phyllosilicates, particularly in sediments containing Al-silicates in all size fractions. Normalization of chemical elements contents to Sc was successfully employed to distinguish production centers using raw materials within the same geological context of estuarine sediments of the Tagus River, Portugal (Dias & Prudêncio, 2008).

Several multivariate statistical methods can be employed by using programs like Statistica (Statsoft, 2003). Each value in the series is standardized: x = (x-M)/SD, where M and SD are the overall mean of each variable and the standard deviation for the untransformed series, respectively. Clustering methods, using the absolute concentration of the chemical elements as variables, or the ratios of elements, like element/Sc, are commonly employed, including hierarchical methods such as joining tree clustering and non-hierarchical, particularly the k-means clustering. Principal components analysis (PCA), involving a transformation of the dataset on the basis of eigenvector methods to determine the magnitude and direction of maximum variance in the dataset distribution in hyperspace, is also commonly employed. The PCA transformation provides a new basis for viewing the entire data distribution to reveal structure not readily observed when plotting the original variables (Baxter, 1994; Sneath and Sokal, 1973). The results obtained with different statistical methods must be integrated and interpreted taking into account the significance of correlations between the chemical elements (variables) found in ceramics and their meaning in terms of raw materials composition, and possible alterations due to the production technologies (sieving, sedimentation, add of temper, mixing clays, etc.) and to eventual modifications during the use of pots and post-depositional processes.

A number of different analytical techniques have been applied with varying degrees of success, but all of them need to have multielement capability and sufficient sensitivity to detect traces of elements in the various matrices. Although a number of techniques have been employed to characterize archaeological materials, the analytical method with one of the longest and most successful histories of application for provenance research has been neutron activation analysis (NAA). The real success story of this technique, however, comes from the investigation of ceramics dating from throughout archaeological times. NAA is a sensitive technique useful for quantitative multi-element analysis of major, minor, and trace elements, which concentration can have different meanings in minerals or in rock descriptions, as well as in ceramics. Major elements may give an idea of the type of mineral material which makes up the ceramic. Minor elements indicate the variations between different sources of material and major mineral abundance, which varies according to its geologic source. Trace elements can be used to determine provenance of the mineral components in the paste or to distinguish different ceramic groups, better differentiating diverse raw materials even within the same geological context.

Among trace elements, rare earth elements (La-Lu, Y and Sc) distributions can be particularly useful in distinguishing clayey materials resources. Among rare earth elements (REE), lanthanide elements, La (Z=57) through Lu (Z=71), are a particularly coherent group of elements that is usually used in geochemistry. Ytrium (Z=39) is commonly included because its similar chemical behavior; scandium (Z=21) is small enough to have a distinct chemical behavior. In this chapter lanthanides will be referred as REE. The subtle variations in the properties of REE make them sensitive to mineral/melt equilibria, as well as to weathering conditions after the breakdown of primary minerals and the formation of new mineral phases, sedimentary sorting, and diagenesis (McLennan, 1989). These elements have very similar chemical and physical properties, which is the result of the nature of their electronic configurations. The dominant oxidation state is the +3 state, and there is a small but steady decrease in ionic radius with increasing atomic number (for a given co-ordination number). The REE, therefore, tend to occur in nature as a group. They are lithophile, in that they concentrate predominantly in the silicate rather than the metal or sulphide phases when they coexist. The differences existing among the REE lead to differences in their relative behavior in response to the chemical environment, making this group particularly useful in geochemistry since they can be a pointer of the genesis processes of the rocks and minerals and subsequent alterations.

The REE are usually divided in two or three groups: (a) light REE (LREE) - those from La to Sm (i.e. lower atomic numbers and masses) and heavy REE (HREE) - those from Gd to Lu (higher atomic number and masses); (b) LREE - La to Nd, middle REE (MREE) - Pm to Ho, and HREE - Er to Lu. The fractionation between LREE and HREE is measured by the $(La/Yb)_{ch}$ ratio, i.e. the ratio between the La and Yb concentrations normalized relative to chondrites (Haskin, 1971). As referred above, REE occur in a trivalent state in most of the temperature-pressure conditions of the earth. There are two exceptions in nature to this: Ce and Eu. Under oxidizing conditions, Ce^{3+} may be oxidized to Ce^{4+} leading to a decrease in the ionic radius of about 15%. This reaction occurs on a large scale in the marine environment, associated with the formation of manganese nodules, with a consequent depletion of Ce in ocean waters. On a smaller scale, this reaction also occurs in superficial environments during weathering (Prudêncio et al. 1993; 1995; Gouveia et al. 1993). The Ce anomaly degree is measured by the Ce/Ce* (Ce is the concentration measured and Ce* is the value obtained at the cerium position by straight-line interpolation between the plotted points for La and Nd). Under reducing conditions, europium may exist in the divalent state, resulting in an increase of the ionic radius of about 17%. Europium is the only element for which a significant proportion of the ions in igneous systems are likely to be present in valences other than 3+, leading to an anomalous behavior. Anomalous behavior of Eu can be expected for many minerals, but is most pronounced for plagioclases, because Eu^{2+} is similar to Sr^{2+} and substitutes much more easily for Ca and Na in the large feldspar site than does Eu^{3+}. The amount of feldspars (particularly plagioclase) and the fate of Eu after the breakdown of these primary minerals during weathering and sedimentary processes have an important control of the Eu anomaly of residual or sedimentary clayey materials. Thus Ce and Eu anomalies may occur depending on several factors, and may be used as indicators of geological formations. The sum of the elements concentrations ($\sum$ TR) can also be used for comparison purposes.

A more controversial control on REE distributions in geological samples is the so-called "tetrad" or "double-double" effect in which REE representing quarter, half, three-quarter, and completely filled 4f shells have increased stability with resulting anomalous behavior. In geological samples it has been suggested that the tetrad effect is most apparent in natural waters and other aqueous liquids, and in rocks it results from interaction with aqueous solutions (McLennan, 1994). Sea water and precipitates from sea water would process a "W-type" to the REE pattern; the interaction water-rocks would result in "M-type" effects to the REE pattern of the rock (Masuda et al, 1987).

A very large proportion of REE contents may be found within heavy minerals. There are minerals in which the REE are essential constituents (e.g., bastnaesite, monazite, xenotime, aeschynite, allanite), although the chemical mechanisms and limits of REE substitution in some rock-forming minerals (e.g., zircon, apatite, titanite, garnet) are also derived (Burt, 1989). These minerals are resistant to weathering and may be concentrated during sedimentary sorting processes. Many heavy minerals have high concentrations of elements, other than the REE, that are found in trace levels in most residual and sedimentary clayey materials. Accordingly, in some cases it is possible to evaluate the influence of heavy mineral fractionations on REE patterns by examining other geochemical data like, Hf, Zr and Th contents (McLennan, 1989). Also a correlation between REE contents and the clay minerals present in sediments has been found, pointing to the incorporation of REE especially in kaolin minerals (Prudêncio et al., 1993; Marques, 2007).

With the exception of inductively coupled plasma- mass spectrometry (ICP-MS) on liquid samples, NAA offers sensitivities that are superior to those possible by all other analytical methods. Even compared with ICP-MS, INAA is more sensitive and accurate method for certain trace elements such as middle and heavy REE, particularly when these elements are present in very low amounts in the samples as shown by Dai Kin et al. (1999) in a comparative study using instrumental NAA (INAA) and ICP-MS. As mentioned above, REE are very good indicators of geological contexts, particularly the differentiation between light REE (LREE) and heavy REE (HREE). Thus, the precise and accurate determination of both LREE and HREE is fundamental for a good characterization, INAA being the most appropriate method. Besides the difficulty of achieving complete digestion of geological samples for analysis by ICP-MS is known the same happens with artifacts like archaeological ceramics.

The INAA technique involves the irradiation of a sample by neutrons to make the sample radioactive followed by gamma spectrometry to determine the amounts of different elements present in the sample. INAA has a number of advantages over most other analytical methods when investigating archaeological specimens. In fact the preparation of archeological materials, particularly ceramics, for analysis by INAA is extremely easy in most instances a representative sample may be only a portion of the sample weighed and place in an appropriate container, which is very important when dealing with cultural materials that must be preserved. INAA is a sensitive, precise and accurate technique for quantitative multi-element analysis and the potential of this method as an archaeological tool to establish the provenance of archaeological ceramics was first recognized by Robert Oppenheimer in 1954 (Sayre and Dodson 1957; Harbottle 1976). Since then numerous works have been done based on this method. Although nuclear reactors are becoming less available, INAA applications on archaeological materials proceed, still being considered one of the most appropriate methods to analyze artifacts. A brief history of NAA and its application to archaeology, and a description of the physics behind the analytical method can be found in Glascock & Neff, 2003. After fifty years of successful NAA applications to archaeology, a special issue of Archaeometry (vol. 49, 2007) was published with contributions describing the history of several NAA laboratories and case studies of archaeological artifacts. Among these, several case studies concerning ceramics provenance were presented (Dias & Prudêncio, 2007; Hancock et al., 2007; Kilikoglou et al., 2007). A brief review of the 50th anniversary of NAA applications was done by Pollard (2008).

The application of the INAA method in support of provenance research has been largely used over the past few decades in the Instituto Tecnológico e Nuclear (ITN). The first analyses of archaeological ceramics were conducted in the late 1970's by using the Portuguese research reactor (RPI), and proceeded generating a large database for ceramics and raw materials from different chronologies and archaeological sites (Cabral & Gouveia, 1979; Cabral et al., 1983, 1988; Dias et al., 2001; 2002; 2003a, 2003b; 2005a; 2005b; Prudêncio et al., 1989; Prudêncio et al., 2003; 2006). In the late 1990's, a research group especially devoted to archaeometry (measurements techniques in archaeology), particularly compositional and dating studies (TL-OSL) of ceramics and raw materials, was formed in ITN – "Applied Geochemistry & Luminescence on Cultural Heritage" (GeoLuC). INAA has been used not only to help answer questions related to ceramics provenance but also to study the degradation of stones in archaeological monuments and historical buildings, and to determine the provenance of the stones in order to perform conservation products tests and restoration of buildings with a similar geological material (Prudêncio et al. 1995; 1998; 2000; Romão et al. 2000; Sequeira Braga et al., 2000; Nasraoui, 2002). INAA has also been used for precise and accurate determination of the contents of natural radioactive elements such as potassium, rubidium, thorium and uranium aiming to evaluate the dose rate, which is fundamental for TL-OSL dating of cultural materials and archaeological contexts, as well as geological contexts (Richter et al. 2003; Prudêncio et al, 2007a).

In this chapter the INAA technique (comparative method) used in the ITN laboratory of the Chemistry Unit to characterize archaeological ceramics and clayey materials is described. The chemical data obtained for shards and potential raw materials used for their production of the NW Iberian Peninsula are discussed, particularly for ceramic production during the Roman occupation of Bracara Augusta and vicinities, where no archaeological evidence like kilns or wastes has been found. The importance on the application of techniques of nuclear chemistry to ceramics and raw materials to solve archaeological problems mainly related to the location of production sites in this area during Roman times is enhanced, which is fundamental to pursue studies of distribution and exchange. Among the chemical elements studied, special attention is paid to the importance of the precise and accurate determination of the REE contents by nuclear methods, allowing geochemical patterns in ceramics and raw materials that can be used to identify provenance.

Chemical Characterization of Archaeological Ceramics and Clays by INAA

The theory of neutron activation analysis can be found in Parry, 1991. In this chapter a brief description of procedures for archaeological ceramics and clays in the INAA laboratory of the Chemistry Unit of ITN is given.

To obtain powder samples of ceramics, the procedure is as follows: the inner and outer surfaces of the shard are scraped using a drill burr made of tungsten carbide (Fig. 1). The scraped specimens are subsequently brushed clean, washed, boiled for 30 min in deionized water, and then dried for several hours at 30°C.

After drying, a planetary agate mortar for small samples is used to crush and homogenize ~2 g of pottery sample into a fine powder (<53 μm). Potential source clays (500-1000 g depending on the grain size) are dried at 30°C and ground in agate mortars into a fine powder (<53 μm).

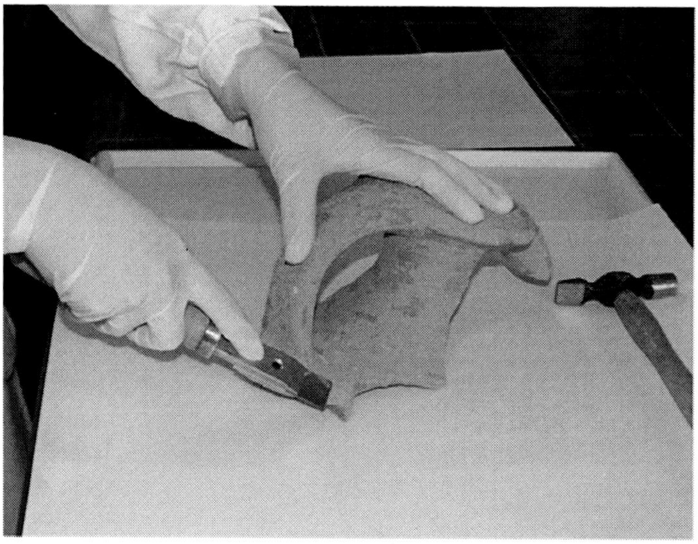

Figure 1. Photograph showing the scraping of the surfaces of the shard.

For INAA, aliquots of ~1 g of powder of ceramics are then dried in an oven at 110°C for 24 h and stored in a desiccator until the samples can be weighed for irradiation. The same procedure is performed for clays and reference materials. Two multi-element reference materials are normally used. All powdered samples (ceramics, clays and standards) are prepared by weighing 200-300 mg of powder into cleaned high-density polyethylene vials (Fig. 2).

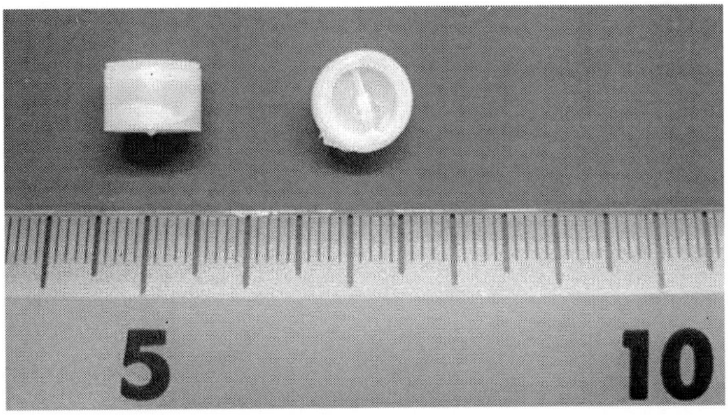

Figure 2. Photograph of the polyethylene vials used for irradiation (scale in cm).

Long irradiations (7 h) on pottery and clay specimens are performed in batches of 20 unknowns along with four standards (Fig. 3) in the core grid of the Portuguese Research Reactor at a thermal flux of 3.346 x 10^{12} $cm^{-2}s^{-1}$; ϕ_{epi}/ϕ_{th} = 1.4%; ϕ_{th}/ϕ_{fast} = 12.1. The bundles are rotated continuously during irradiation to ensure that all samples received the same exposure to neutrons. Even so, Fe flux monitors are placed in appropriate plastic containers for irradiation together with the samples for neutron flux variation corrections.

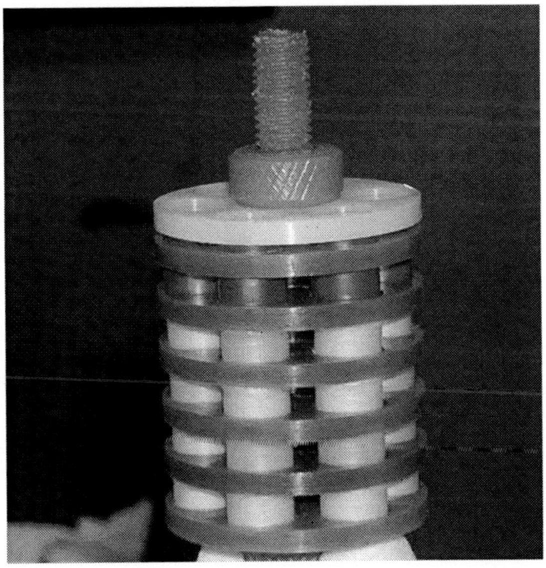

Figure 3. Container used for long irradiations in the RPI.

Two γ-ray spectrometers are used: (1) one consisting of a 150 cm^3 coaxial Ge detector connected through a Canberra 2020 amplifier to an Accuspec B (Canberra) multichannel analyzer. This system has a FWHM of 1.9 keV at 1.33 MeV; and (2) the other consisting of a low energy photon detector (LEPD) connected through a Canberra 2020 amplifier to an Accuspec B (Canberra) multichannel analyzer. This system has a FWHM of 300 eV at 5.9 keV and of 550 eV at 122 keV (Fig.4).

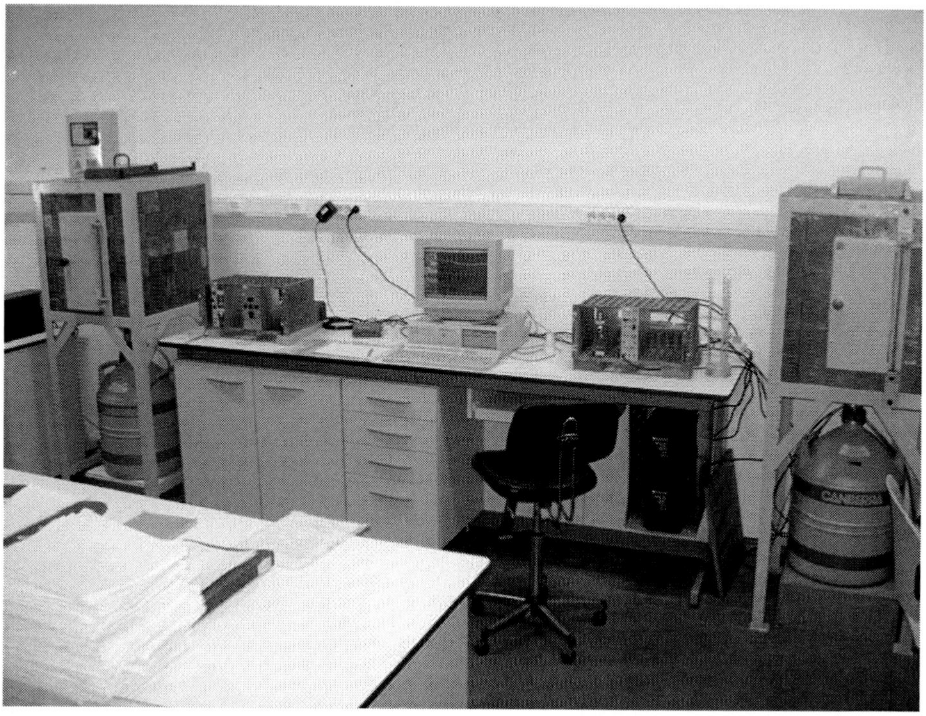

Figure 4. Counting room of the INAA laboratory of the Chemistry Unit at ITN.

Different cooling times were selected to determine radionuclides with different half-lives. These cooling times and energies of γ rays, as well as the type of detector used are given in Table 1. Na, K, Fe, Sc, Cr, Co, Zn, As, Br, Rb, Cs, Ba, La, Ce, Nd, Sm, Eu, Tb, Yb, Lu, Hf, Ta, Th and U abundances are obtained. Relative precision and accuracy are, in general, to within 5%, and occasionally within 10%.

Table 1. Radionuclides, cooling times, energies of γ rays, and type of detector used.

Element	Radionuclide	Decay time (d)	Energy of γ-rays used (keV)	Detector Coaxial Ge(Li)	LEPD
Na	^{24}Na	3	1368	x	
K	^{42}K	3	1524	x	
Sc	^{46}Sc	27	889	x	
Cr	^{51}Cr	27	319	x	x
Fe	^{59}Fe	27	192;1099	x	x
Co	^{60}Co	27	1332	x	
Zn	^{65}Zn	27	1115	x	
As	^{76}As	3	559	x	
Br	^{82}Br	3	776	x	
Rb	^{86}Rb	27	1077	x	
Zr	^{95}Zr	27	756	x	
Sb	^{124}Sb	27	1690	x	
Cs	^{134}Cs	27	796	x	

Table 1. Continued

Element	Radionuclide	Decay time (d)	Energy of γ-rays used (keV)	Detector Coaxial Ge(Li)	LEPD
Ba	^{131}Ba	27	496	x	x
La	^{140}La	3	1596	x	
Ce	^{141}Ce	27	145	x	x
Nd	^{147}Nd	27	91; 531	x	x
Sm	^{153}Sm	3	103	x	x
Eu	^{152}Eu	27	1408	x	x
Tb	^{160}Tb	27	881; 1178	x	x
Yb	^{169}Yb	27	177; 198	x	x
Lu	^{177}Lu	27	208	x	x
Hf	^{181}Hf	27	482	x	
Ta	^{182}Ta	27	1222	x	x
Th	^{233}Th	27	300; 312	x	x
U	^{239}Np	3	106; 228	x	x

The spectra are processed by using the programs GELI and DAISY, adapted respectively from GELIAN and OLIVE programs (Op de Beek 1972/74). In general, two pairs of standards are used in order to obtain an internal calibration.

Corrections are made for the interference from uranium fission. It is well known that the presence of U in samples interferes in the determination of several elements concentration by instrumental neutron activation analysis. This is partly due to the identity of some ^{235}U fission products with radionuclides formed by radioactive capture of neutrons on those elements (nuclear interference), and also to the fact that the γ-rays emitted by some fission products cannot be resolved from γ-rays emitted by radionuclides of interest produced by (n,γ) reaction (spectral interference). The extent to which the interference may affect the accuracy of the determination depends on the ratio of the concentrations of the element to be determined and of U in the sample, as well as on the irradiation conditions and cooling time. In this chapter corrections for the determination of rare earth elements (REE), zircon and barium were done. The determination of the concentration of the light rare earth elements (LREE), Zr and Ba in geological materials containing U must be done with great care in this respect. A numerical procedure to correct for U interference in this determination is necessary, based on the Runge-Kutta algorithm, for solving the differential equations governing the growth and decay of nuclides during and after the irradiation. The interference factors for element X is defined as the ratio $F = N^*(^{\alpha}X)/N(^{\alpha}X)$, where $N(^{\alpha}X)$ = number of nuclei of the radionuclide $^{\alpha}X$ produced by (n,γ) reaction in 1 gram of the natural element X; $N^*(^{\alpha}X)$ = the number of nuclei of the same radionuclide produced by fission of ^{235}U in 1 gram of natural U. The values of $N(^{\alpha}X)$ and $N^*(^{\alpha}X)$ may be obtained by solving the systems of differential equations that govern the number of nuclei $^{\alpha}X$ present at time t, during and after the irradiation. Experimental values were obtained by irradiating an uranyl nitrate solution with U_3O_8 previously verified by alpha-spectrometry as being of natural isotopic composition and

REE, Zr and Ru standards also in the form of solution prepared with pure REE oxides and Zr and to account Ru chlorides. The determination of Ba based on the reaction $^{130}Ba(n,\gamma)^{131}Ba$ must take into account the spectral interference from fission due to ^{103}Ru. The results obtained in the experiments performed in the RPI showed a good agreement between experimental and calculated values. Especial attention must be paid to the irradiation conditions, since the correction factors depend particularly on the ϕ_{epi}/ϕ_{th} ratio (Martinho et al., 1991; Gouveia et al., 1987). More details of the analytical method can be found elsewhere (Gouveia et al. 1988; Prudêncio et al, 1986; 1988).

NW Iberian Peninsula - Production and Trade of Pottery in Bracara Augusta

Bracara Augusta, Braga (NW Portugal), was a Roman town recognized as a centre of production and distribution of pottery to a wide regional area (Martins & Delgado, 1995). The morphological analysis of Roman ceramics established the existence and chronology of different production types used during the five centuries of Roman occupation (1st – 5th century). The location of the town is shown in Fig. 5.

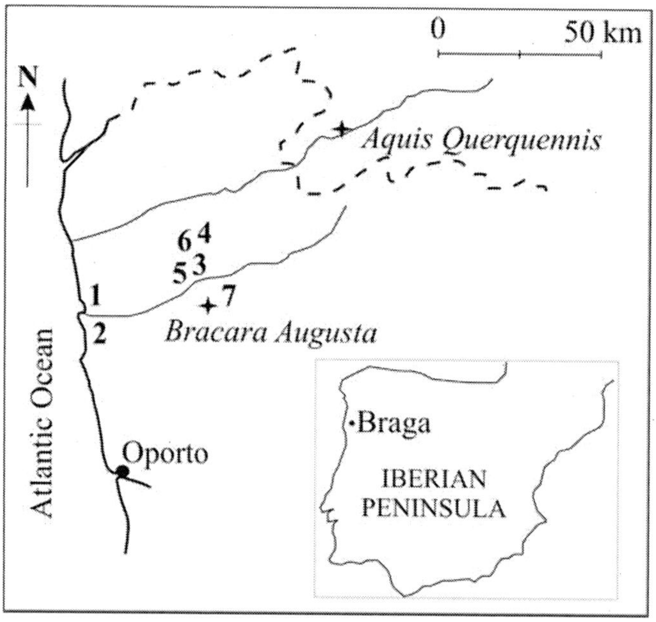

Figure 5. Map of the studied area, with the location of the archaeological sites - *Bracara Augusta* and *Aquis Querquennis*, and raw materials sampling. 1 – Campados; 2 – Barqueiros; 3 – Prado; 4 – Bustelo; 5 – Espinheira, Ucha; 6 – Quebrosas; 7 – Braga.

The study of thousands of shards found in Braga allowed the identification of different types of Roman pottery, such as Bracarense, polished fine gray painted ceramics, late gray ceramics, local amphorae production, non-vitrifiable red slip ware and also several types of common ware (Martins & Delgado, 1989-90a; 1989-90b). The location of production sites was first made by the so-called criterion of abundance or gravity model (Rice, 1987), since there is little evidence of pottery workshops in the Braga region. In this way a compositional study of the different types of ceramics and the inventory and the compositional features of regional clay deposits is crucial in order to identify local/regional productions and also imports. The chemical characterization by using INAA, and complemented with mineralogical studies, of the pastes of different types of ceramics found in Bracara Augusta has been done in an attempt to confirm the morphological classification by macroscopic observation of the pastes. Chemical-based provenance studies can be performed aiming to identify the raw materials used to produce the several types of ceramics, and thus delimiting the geographic source region by tracing the clay source used to infer the most probable location of production. While provenance studies may be able to identify the kinds of clay material used in a particular kind of pottery, they do not provide information on the cultural or socioeconomic context, such as the locations of workshops, which depends on the finding of kilns, deposits of raw materials, wasters, etc. at a site. Thus, the approach adopted for the establishment of the existence of ceramic production in Bracara Augusta (town and vicinities areas) has been a combination of the abundance of the types of potteries (forms and paste types) and comparisons of the ceramic composition with clay resources available in the area to establish local/regional production, and thus creating "ceramics reference groups" to be used for provenance studies. Besides, an important achievement is to distinguish imports from imitations of imports locally produced (Prudêncio et al, 2006, Oliveira, 1997; Oliveira et al, 2005; Gomes, 2000; Gaspar, 2000).

Ceramics from Bracara Augusta

Among the ceramics found in Bracara Augusta, special attention is paid in this chapter to several types of ceramics, namely the "Bracarense" (B), common wares (C), the "non vitrifiable red slip" wares including imitations of "Terra Sigillata Hispanica" and "African Red Slip Ware" forms and stamp-types (IS), and imitations of "pompejanish-roten" slip (IP). Also chemical composition of ceramics of known origin, namely Lugo (Galicia, Spain), Cadiz (Andaluzia, Spain), and Italy (Oliveira, 1997; Oliveira et al., 2005; Gaspar, 2000; and Dias et al, 2003c) are compared with ceramics found in Braga in order to identify imports and imitations of imports.

The "Bracarense" ceramics paste is characterized by a very fine- to medium-grained yellow-pale paste covered with a brownish yellow slip that is clearly different from other ceramic types found in the town (Alarcão and Martins, 1976). A similar fine yellowish paste is also found in some common wares and special ceramic pieces, suggesting the use of the same clay source type used to produce ceramics of different typologies (Oliveira, 1997; Prudêncio et al., 2006). The designation – Bracarense - was due to the fact that it was located and identified first in Bracara Augusta and vicinity. However, shards with a similar fine paste were found later on in other archaeological sites of the NW Iberian Peninsula region, such as Aquis Querquennis (Galiza, Spain) (Fig. 5). In fact, the textural and mineralogical features of the Bracarense ceramics from both archeological sites, Bracara Augusta and Aquis Querquennis, are very similar, pointing to the possible use of the same type of raw material. This finding together with the absence of kilns near Braga, question the production centre's location. Aquis Querquennis is a large archaeological complex, including several structures, namely the military camp, a possible inn facing the XVIII vía, one thermal waters bath complex, local pre-Roman settlements and Roman villae.

As referred above, some pieces of common wares and other typologies present a fine yellowish paste like the one used to produce the so-called "Bracarense" ceramics type. However, the majority of the "Common ware" found in Bracara Augusta present a reddish-brown to yellowish-brown and pale brown paste. The same paste type is also found in ceramics of the so-called "non vitrifiable red slip" ware, which is also very frequent in Braga and vicinity. The same technique of manufacture and the typical non vitrifiable red slip occurred in different types that lasted from the 1st to the middle of the 5th century. This large span of time can explain the significant particularities of the fabric and shapes which determined their inclusion in different groups, according to well defined criteria (Delgado, 1993/94).

A multivariate statistical approach, using the chemical contents obtained by INAA, of the several ceramics found in large amounts in the Braga region point to two main compositional groups of pastes, each of them including different typologies: (i) Group 1 - the fine yellowish paste mostly found in Bracarense ceramics – the "Bracarense paste" – of Bracara Augusta and Aquis Querquennis , and also in common wares and other typologies of Bracara Augusta; and (ii) Group 2 - the "common ware paste" found in the majority of the common wares of Bracara Augusta and also in imitations of "Terra Sigillata Hispanica", "African Red Slip Ware", and stamp-types (all these three types made with the non vitifiable red slip technique), and ceramics with imitations of the "pompejanish-roten" slip found in this Roman town.

The chemical composition obtained by INAA showed that the "Bracarense paste" differs from the "common-ware paste" mainly due to a lower REE fractionation and a greater intermediate and heavy REE relative to the light REE, as can be seen in the chondrite-normalized REE patterns (Haskin et al., 1971) of both pastes groups presented in Fig. 6 and 7. Greater U and Ta, and smaller Fe, Na and Ba contents were also found. The Aquis Querquennis pottery shards analyzed present the same geochemical pattern as those from Bracara Augusta.

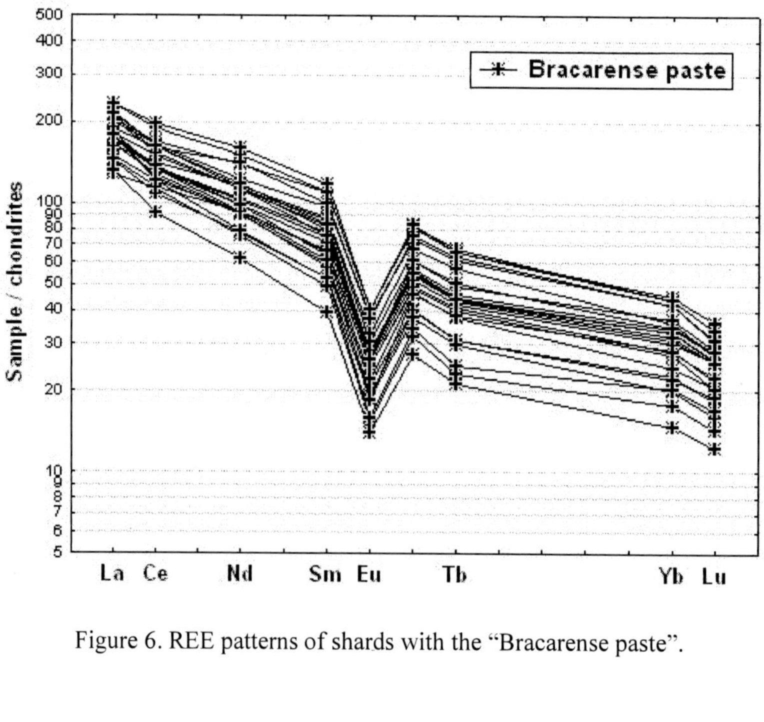

Figure 6. REE patterns of shards with the "Bracarense paste".

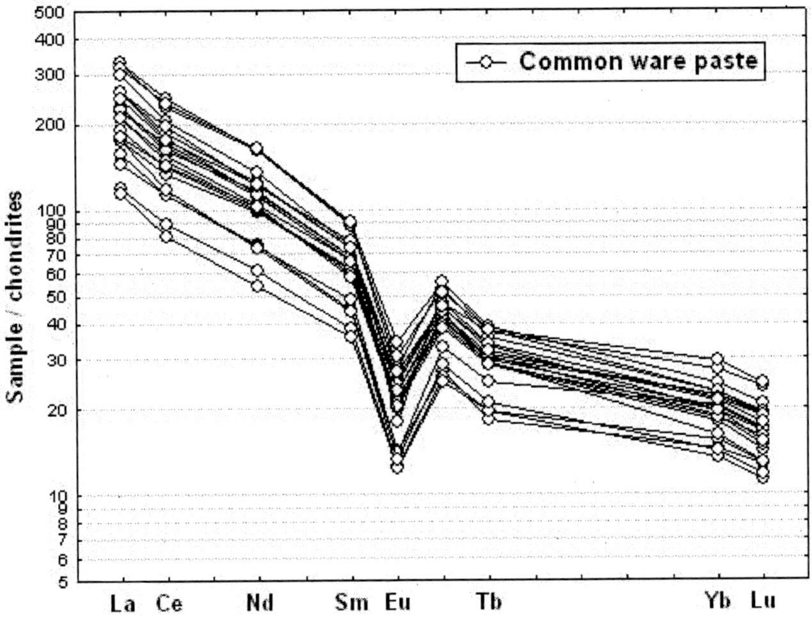

Figure 7. REE patterns of shards with the "common ware paste".

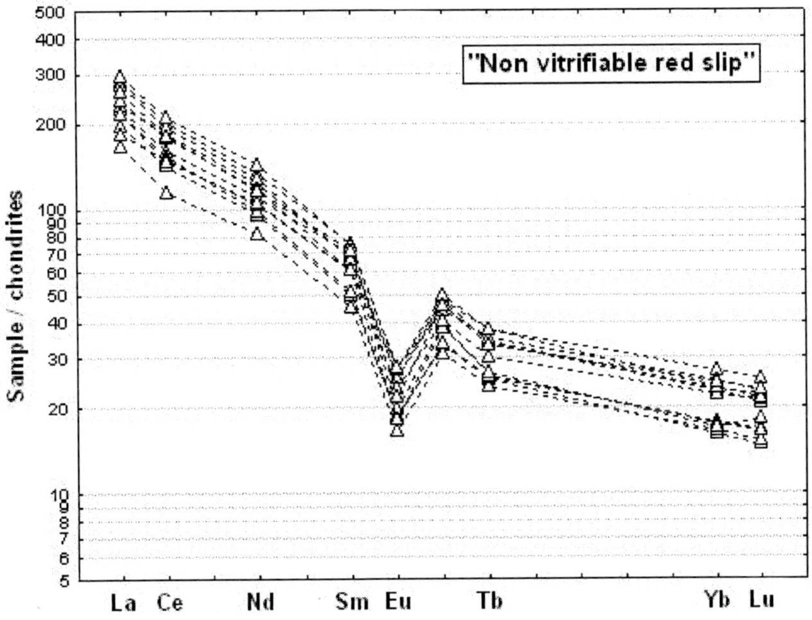

Figure 8. REE patterns of shards of the "non vitrifiable red slip" wares – imitations of "*Terra Sigillata Hispanica*", "African Red Slip Ware" and stamp-types.

The composition of the pastes of "non vitrifiable red slip wares" (IS) analyzed is similar to the common ware paste, as shown by the REE patterns (see Fig. 7 and 8). Assuming that the production of the majority of the common wares is local, the similarity presented by the IS ceramics suggests that the same clay resource was used in the Braga region to produce imitations of imports. It should be noted that the paste of ceramics with "non vitrifiable red slip" present in general higher contents of Fe and other transition metals of the 1st transition series (Fig. 9) which can be due to their incorporation into the paste (for instance as iron oxides) during the technological procedures to obtain the red slip.

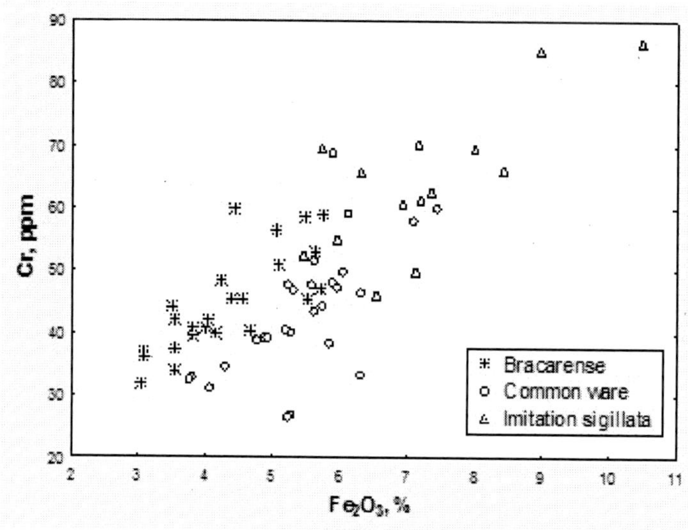

Figure 9. Fe vs. Cr contents plot of the "Bracarense paste", the "common ware paste" and the "non vitrifiable red slip" wares (IS).

In this way, two main types of pastes appear to exist in shards of different typologies found in large amounts in Bracara Augusta and vicinity (the "Bracarense paste" and the "common-ware paste"), pointing to local/regional productions with the use of two main types of local/regional raw materials. The absence of archaeological evidence of ceramic production justifies the clay tracing to confirm local/regional productions.

Local and Regional Raw Material Sources

The regional clay materials between the city of Braga and the shoreline were previously studied by Sequeira Braga (1988), Sequeira Braga et al. (2002) and Ferreira et al. (2000). These previous geological studies, including the inventories of regional clay materials and their mineralogical characterization have been used in tracing the clays. Clay materials are more concentrated in the Cávado river basin (NW Portugal), between the city of Braga and the shoreline, corresponding to sedimentary deposits, as well as residual clays derived by weathering from granites and schist.

The strategy procurement for clays to study was directed to the more adequate raw materials. In this way, fine grain size and the yellow-pale color characteristics of the "Bracarense paste" was taken into account, and kaolin deposits located near the coast were collected. In the case of the brownish "common ware paste", clay materials deposits located near the city of Braga were selected for compositional analysis. The sampling locations of the clays are shown in Fig. 5: (a) residual kaolin (#1); (b) sedimentary kaolin (#2); (c) paleoweathered schist (#4,5); (d) sedimentary deposits (small temporary basins) (#6); (e) Prado sedimentary formation (#3); and (f) clays derived by weathering of granite (#7).

The chemical results obtained for the clay materials clearly differentiate the samples according to the geological environments. A first multivariate statistical approach of the clay materials using the chemical contents obtained by INAA as variables (except Br due to its mobile character) show that samples gather according to the nature of the deposit (Fig. 10): (a) near Braga - (1) clays derived by weathering of schist (Espinheira and Bustelo) and sedimentary deposits of small temporary basins (Ucha and Quebrosas); (2) clays derived by weathering of the Braga granite and the upper level of the Prado formation; (3) Prado sediments; and (b) near the coast - (4) residual kaolin; and (5) sedimentary kaolin.

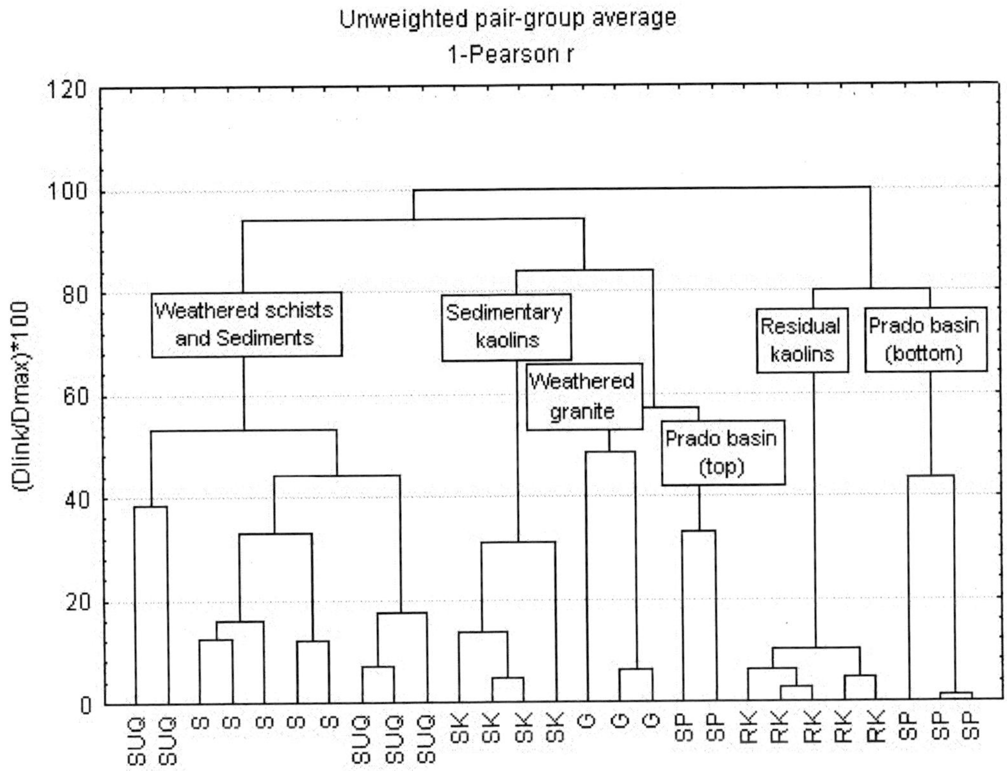

Figure 10. Phenogram resulting from the joining tree clustering applied to clay materials, using the Pearson correlation coefficient and the UPGMA amalgamation method.

Among the chemical elements studied, REE patterns well characterize and differentiate the two main compositional groups of ceramics pastes found in large amounts in Braga - the Bracarense and the common ware paste (see Fig. 6 and 7). The REE patterns of the clay materials are shown in Fig. 11-14. The REE patterns of residual kaolin (Fig. 11) are clearly different of the remaining clay materials studied, including the sedimentary kaolin of Barqueiros, as well as the clayey materials collected in different geological environments near the city of Braga. Residual kaolin samples present low REE contents and relative enrichment of the MREE relative to the LREE and the HREE. Also, the results obtained point to the existence of Yb-Lu anomalies, with inflexion of the REE distribution curves in Yb, which can be explained by the "tetrad" effect (McLennan, S.M., 1994; Masuda et al., 1987). Concerning the sedimentary kaolin, REE are more concentrated and the MREE and HREE contents vary more than the LREE, particularly La and Ce (see Fig. 11).

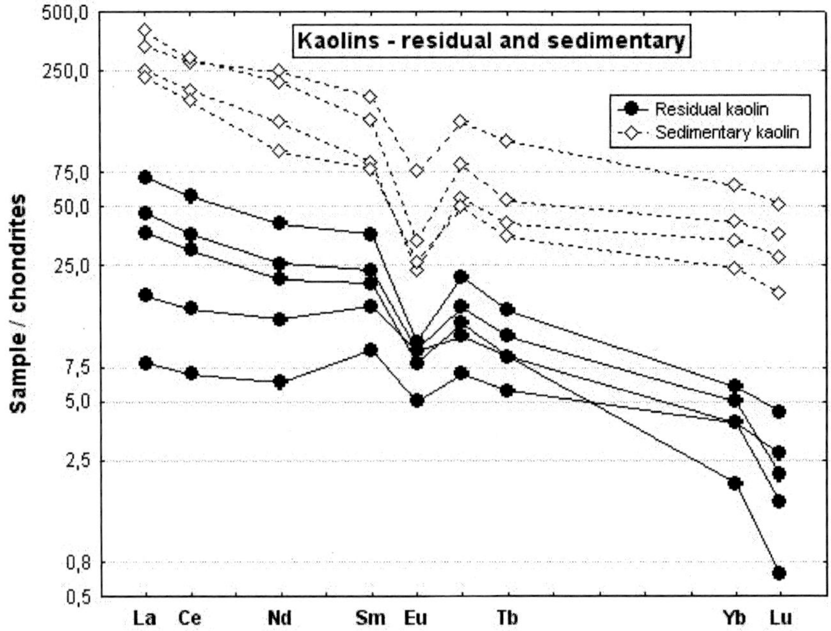

Figure 11. REE patterns of residual and sedimentary kaolin deposits located near the coast.

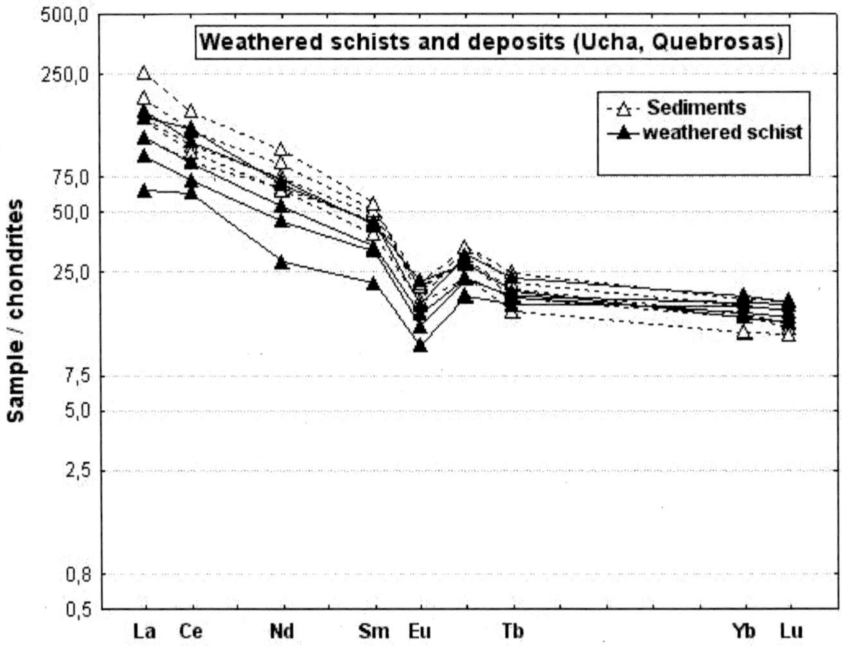

Figure 12. REE patterns of weathered schist and sediments of small sedimentary basins near Braga.

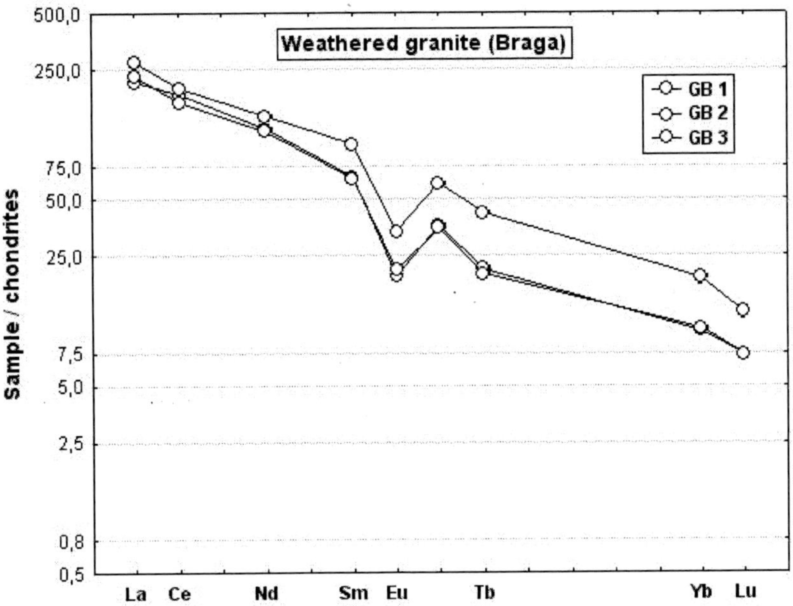

Figure 13. REE patterns of weathered granite located near Braga.

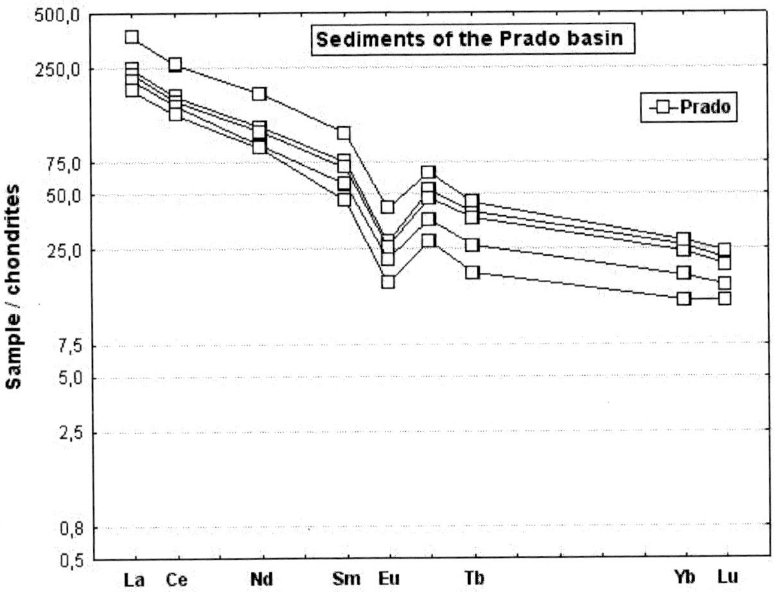

Figure 14. REE patterns of sediments of the Prado basin located near Braga.

The REE patterns of the clay samples from the vicinities of Braga are shown in Fig. 12-14. The sedimentary deposits of Ucha and Quebrosas show an enrichment of the LREE relative to the HREE when compared to the clays derived by weathering of schist (Fig. 12). The weathered granite materials present a higher LREE/HREE ratio due to lower HREE contents (Fig. 13). The sediments of the Prado sedimentary basin (Fig. 14) present in general higher REE contents, when compared with the small basins of Ucha and Prado. However,

La/Yb ratios are not as high as observed for the clay materials derived by weathering of the granite of Braga. In this way, the REE patterns of the Prado sediments point to a more important contribution of granites as source materials than of schist. On the contrary, the geochemical pattern of the sediments from the small basins of Ucha and Quebrosas point to a composition more similar to the weathered schist.

Thus, the REE patterns of the kaolin samples, both residual and sedimentary, collected near the shoreline are clearly different of the patterns found for the clay materials derived by weathering of granites and schist, as well as sedimentary deposits located near Braga. Also a clear distinction can be made of the several deposits studied in both regions. Among the other elements studied, thorium and chromium contents are also good indicators, and can be used to differentiate the geological formations.

The residual kaolin can be clearly differentiated due to lower contents of Hf, HREE, Sc, Co, and particularly Cr. Low Fe concentrations were also found, but of the same magnitude of those found in the sedimentary kaolin. High contents of K, Cs, and particularly Rb also differentiate the residual kaolin.

Ceramics and Clays

Multivariate statistical analysis of the clays and the ceramics by using the chemical elements contents as variables (except Br due to its possible modifications after the ceramic manufacture) showed that the ceramics with the "Bracarense paste" gather with the sedimentary kaolin. The composition of "common ware paste" presents more similarities with the clay materials collected near Braga. However, using a multivariate statistical approach it is not possible to relate the ceramics compositional group with one particular type of raw materials (see Fig. 15). For this purpose geochemical considerations must be considered. As referred above, some trace elements are good indicators to distinguish the clay deposits, namely the REE, Th and Cr contents. In this way these elements were used in an attempt to identify the most probable raw materials used to obtain the "common ware paste", and among the clay materials occurring near Braga, the results obtained so far point to the clays of the Prado sedimentary basin (see Fig. 16).

Thus, the results obtained so far suggest that in Roman times, clays from the Prado sedimentary basin located near Braga appear to have been used to produce ceramics of different typologies with the "common ware paste". On the other hand, the similarity between the "Bracarense paste" and the sedimentary kaolin deposits occurring along the NW Iberian Peninsula point to the procurement of this type of special clays to obtain a fine paste to produce different types of pots, including the Bracarense ceramics and also other special pots like thin walls from the Augustus Tiberius period, as well as common wares. Although no ceramic kilns were found in Braga, given the large number of occurrences of ceramics with these two types of pastes, as well as the importance of the site, production centers probably did exist in the area. To produce pottery with finer pastes, the procurement of appropriate raw materials may have led to the exploitation of the kaolin deposits which occur near the coast of the north of Portugal and Galiza. White fine clays are rare and valuable, not only nowadays, but also in antiquity. This might explain its use in the area, even though the source is not very close (~40 km away) to the Bracara Augusta archeological site, especially considering that the manufacture could have taken place in Bracara Augusta. However the transport of raw materials by boat up the river should be considered.

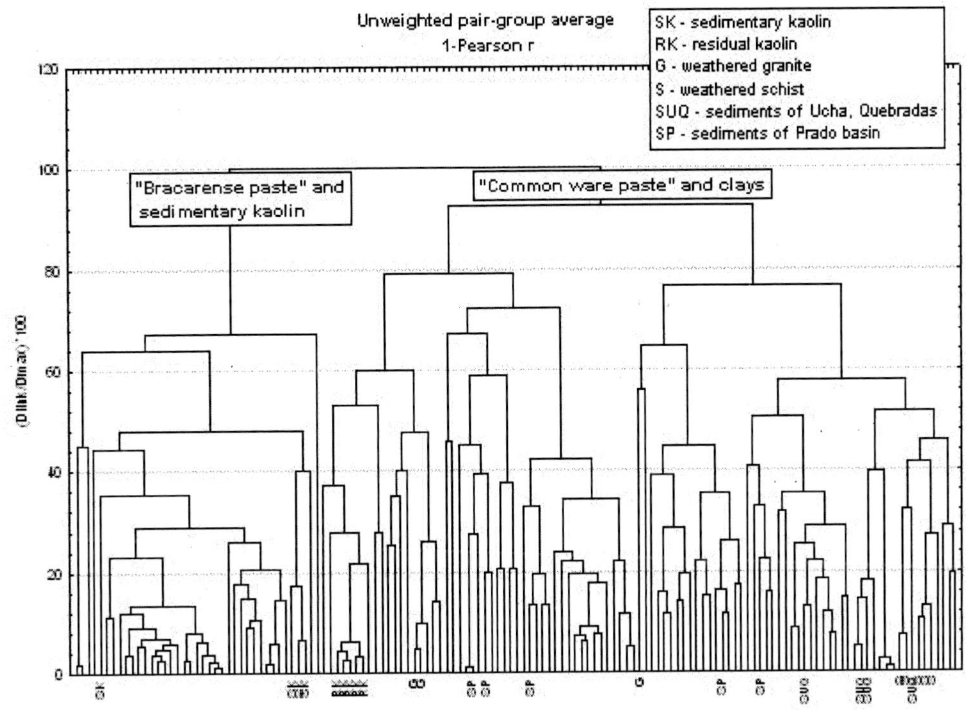

Figure 15. Phenogram resulting from the joining tree clustering applied to clay materials and ceramics from Braga, using the Pearson correlation coefficient and the UPGMA amalgamation method.

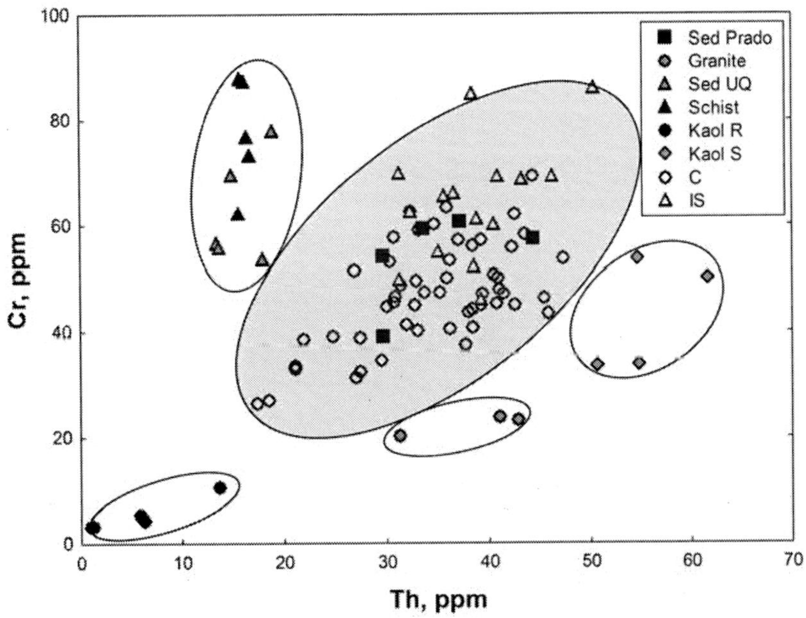

Figure 16. Th vs. Cr plot of the "common ware paste" compositional group and the clay deposits. The grey ellipse includes "the common ware paste", the "non vitrifiable slip" wares and the clays of the Prado basin.

The Bracarense ceramics from Aquis Querquennis and Bracara Augusta archeological sites present the same geochemical and mineralogical features, pointing to the use of the same type of raw material. The differences found between the ceramics from both sites can be explained by compositional variations along the same sedimentary kaolin deposit which is widespread along the NW Iberian Peninsula coast, as well as differences in clay processing and/or differences in the pottery use and burial environment. A comparison of Bracarense ceramics from the two sites and regional clay materials reinforces the suggestion that sedimentary kaolin were exploited in their manufacture. Mullite, spinel and hematite were the newly crystallized minerals in the shards, indicating firing temperatures near 900°C, inferred by the reaction series of the kaolin minerals and associated minerals (illite and goethite) during firing (Prudêncio et al., 2006).

Concerning the imports, the chemical analysis of the pastes of samples of Italic production (Pompejanish-roten Platen) found in Bracara Augusta and of imitations of the "pompejanish slip" also found in this town confirmed that the imitations have a paste with distinct chemical composition. Two different types of composition were identified – one similar to the "common ware paste" pointing to a local production of these imitations, and the other with a completely different composition, suggesting a different provenance. In this way a comparative study with ceramics from other Roman sites, namely Lugo (L) and Cadiz (CD) in Spain, and also Italic ceramics (I) including the one found in Bracara Augusta, was performed. The result of the joining tree clustering using the chemical contents as variables and the UPGMA amalgamation method is shown in Fig. 17. The sample of Pompejanish-roten Platen (P) found in Bracara Augusta is similar to other ceramics from Campania, Italy. The Imitations of "pompejanish-roten" slip (IP) found in Bracara Augusta gather with ceramics from Lugo (located NE of Braga) indicating that these imitations were produced in the Roman town of Lucus Augusti and imported to Bracara Augusta, except one shard analyzed which appears to have been produced in Bracara Augusta since the paste is similar to the "common ware paste". The REE patterns of the Italic shard and the imitations of the "pompejanish slip" found in Braga are shown in Fig 18, together with the REE patterns of ceramics from Lugo. The REE pattern presenting the lower REE contents and the higher negative Eu anomaly corresponds to the imitation with a composition similar to the "common ware paste" of Braga and the one with the smaller negative Eu anomaly corresponds to the Italic ceramic found in Braga. Despite some similarities found between composition of the Italic and Cadiz ceramics (see Fig. 17), the REE patterns differentiate well these two groups of ceramics due to lower REE contents and a slightly lower $(La/Yb)_N$ ratio as shown in Fig. 19 and 20.

Significant differences among the several provenances were also found for other chemical elements determined by INAA, as shown in Fig. 21-23.

The study confirmed the importation of "Pompejanish-roten Platen" from Italy (Campãnia) as well as imitations of this production made in Lucus Augusti. Also evidence of production in Bracara Augusta of imitations of "Pompejanish-roten Platen" and forms of "Terra Sigillata Hispanica", "African Red Slip Ware" and stamp types was found.

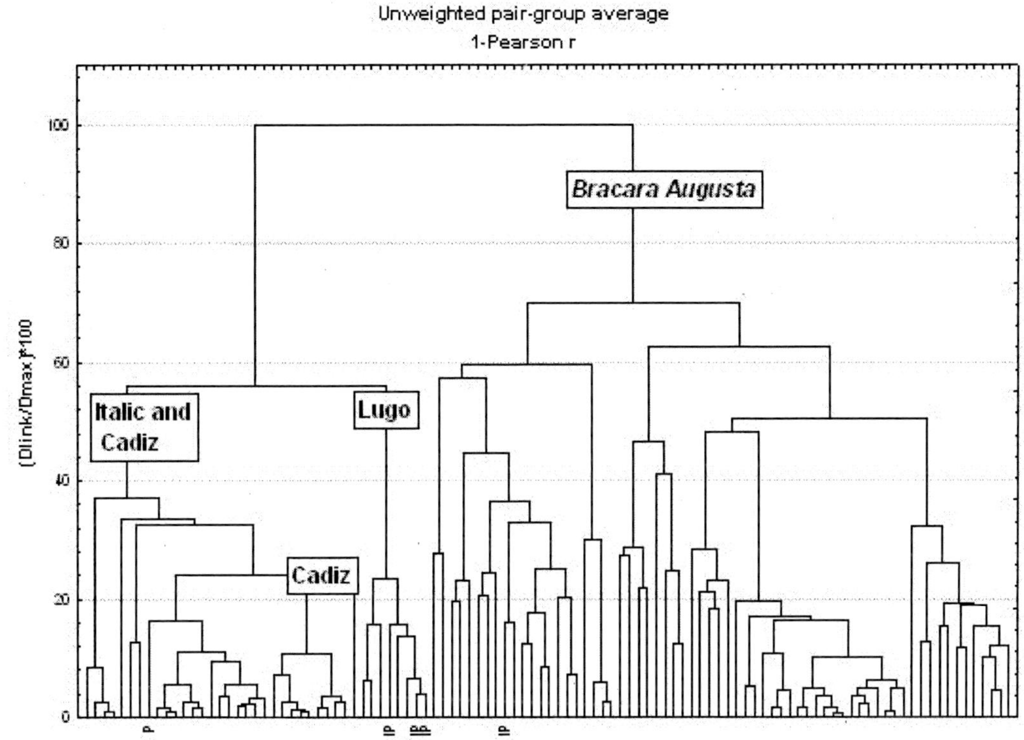

Figure 17. Phenogram resulting from the joining tree clustering applied to ceramics from Braga and imports, using the Pearson correlation coefficient and the UPGMA amalgamation method.

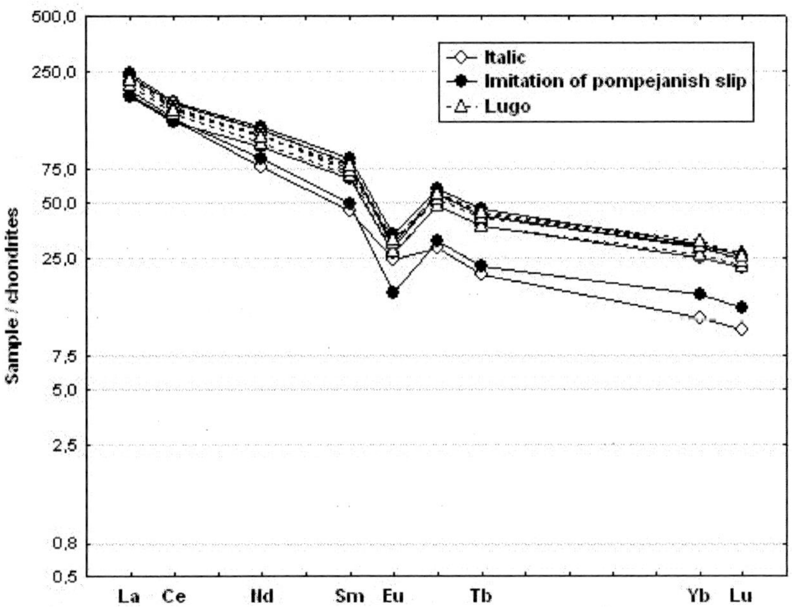

Figure 18. REE patterns of shards with imitation of the pompejanish slip and ceramics from Lugo (Spain).

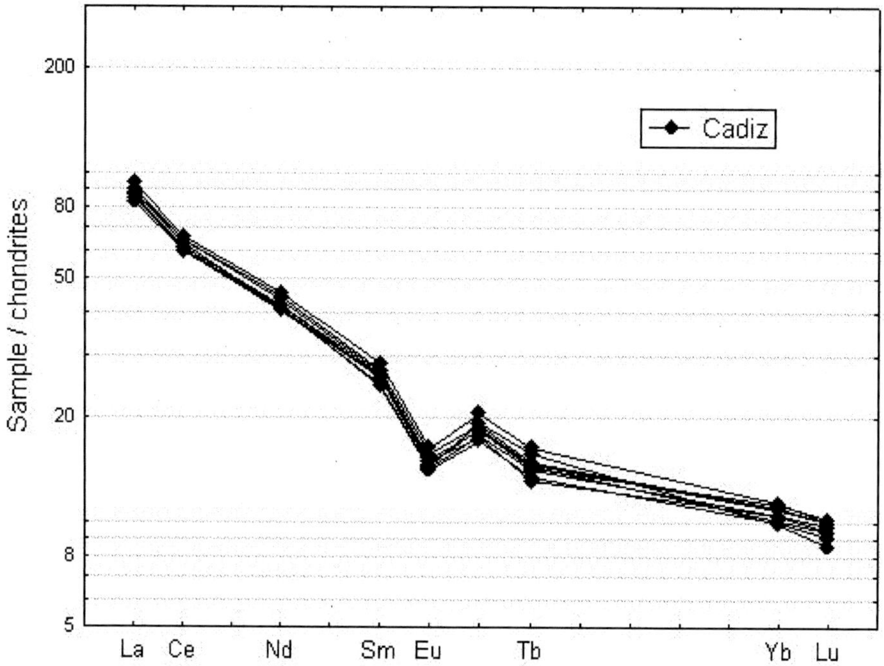

Figure 19. REE patterns of ceramics from Cadiz (Spain).

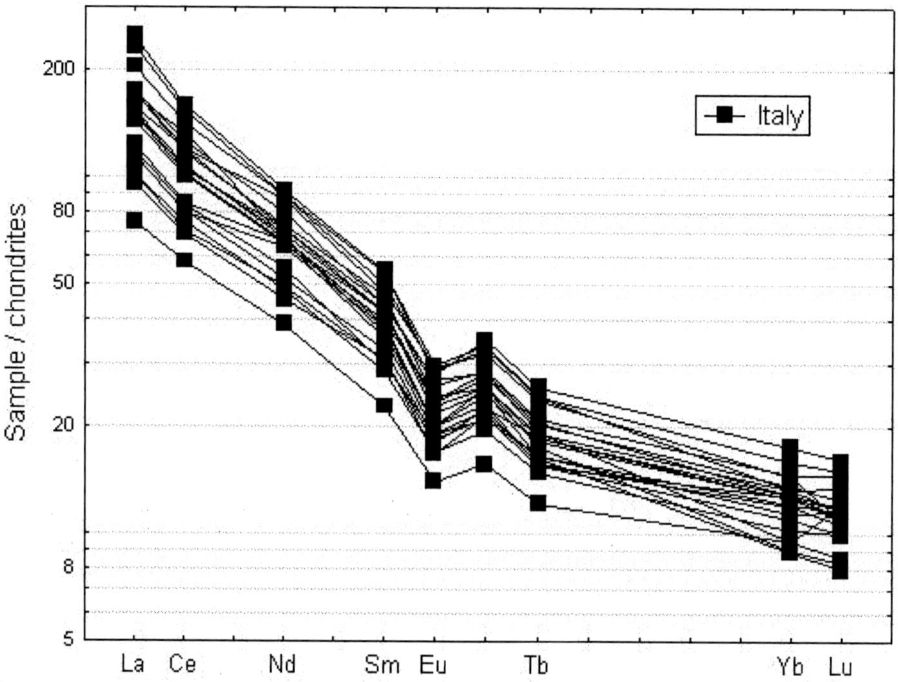

Figure 20. REE patterns of ceramics from Campania (Italy).

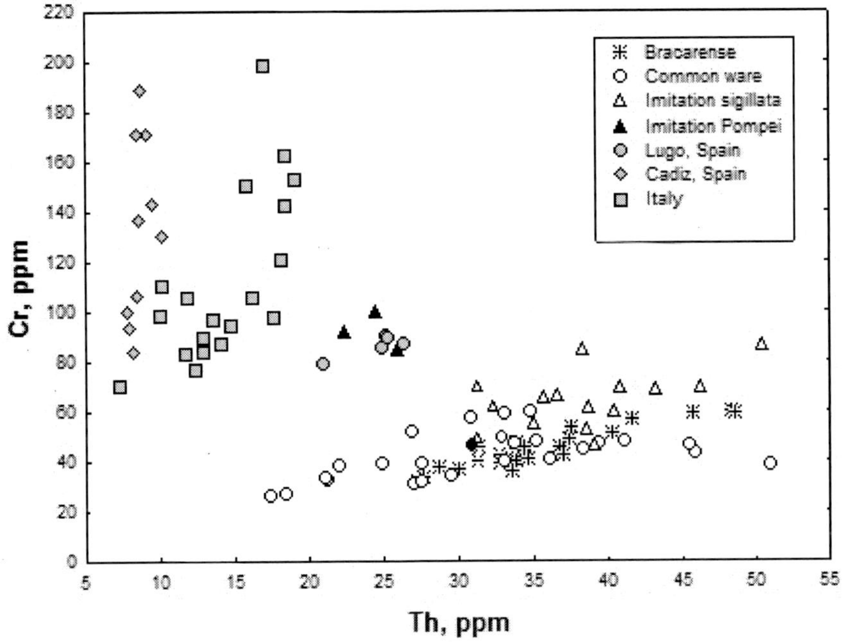

Figure 21. Th vs. Cr plot of the ceramics from Cadiz, Italy, Lugo, and ceramics found in Braga.

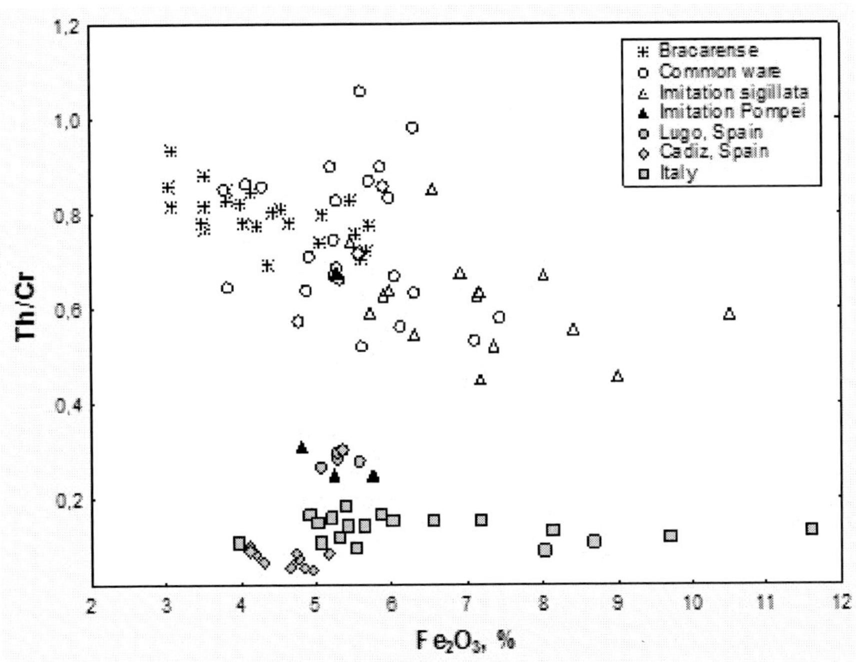

Figure 22. Fe vs. Th/Cr plot of the ceramics from Cadiz, Italy, Lugo, and ceramics found in Braga.

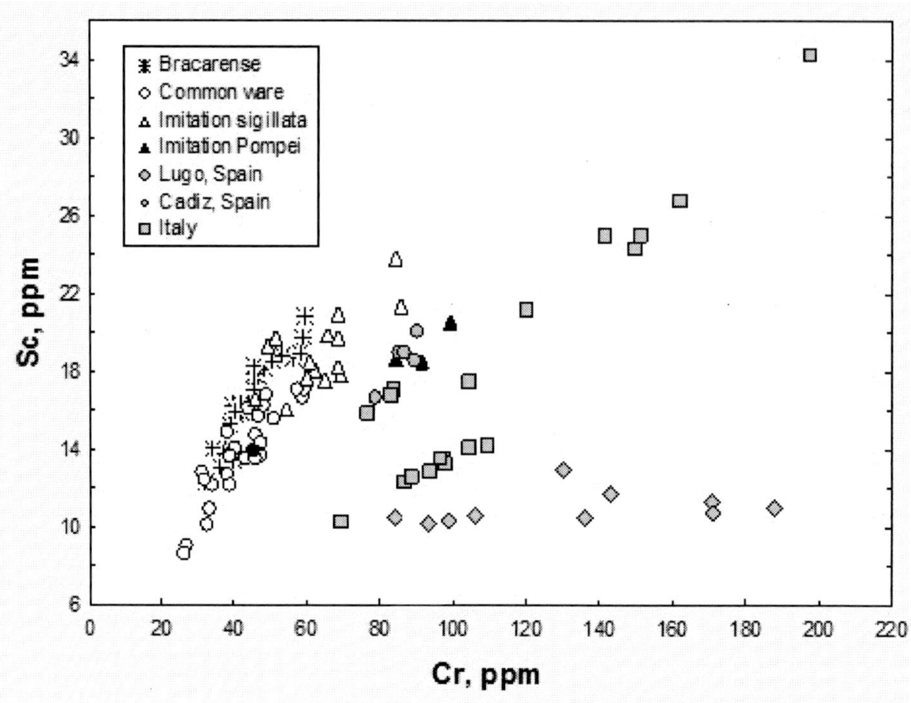

Figure 23. Sc vs. Cr plot of the ceramics from Cadiz, Italy, Lugo, and ceramics found in Braga.

Conclusion

Chemical analysis by instrumental neutron activation analysis of the paste of archaeological ceramics found in the Braga region (NW Portugal) showed that there are two main compositional groups of ceramics – the "common ware paste" and the "Bracarense paste". A comparative study with the chemical composition of regional clays allowed identifying local/regional productions, and distinguishing regional productions and imports:

- The clays of the Prado sedimentary basin, located near Braga, were used to obtain the "common ware paste" to produce most of the common wares typologies and also imitations of different types of imports of other regions of the Roman Empire ("Terra Sigillata Hispanica", "African Red slip ware", stamp types, and "Pompejanish-roten slip").
- The sedimentary kaolin deposits occurring along the north coast of Portugal and Galiza (Spain) appear to have been used to obtain the "Bracarense paste" to produce the Bracarense ceramics and other types of ceramics. The shards collected in Aquis Querquennis have the same geochemical pattern than those found in Bracara Augusta, i.e. they all appear to have been manufactured with the same clay type, reinforcing the suggestion that sedimentary kaolin deposits were exploited in their manufacture.
- The compositional analysis confirmed the importation of "Pompejanish-roten Platen" from Campania (Italy) as well as of ceramics with imitations of the "pompejanish-roten slip" made in Lucus Augusti.

It is important to enhance that different geochemical patterns found in clays from different geological contexts obtained by the precise and accurate results of INAA, consists on an important basis for the comparison with ceramics since the identification of the clays used for certain productions can contribute to delimit a potential area of clay resources and thus establishing provenance. This is particularly useful when no archaeological evidence of production exists. Among the chemical elements obtained by INAA, the precise and accurate determination of REE contents and other trace elements with diverse geochemical behavior is very important for the characterization of ceramic productions and their relation with raw materials, which is crucial to pursue studies of provenance establishment and trade routes reconstruction in ancient times.

Technological styles incorporate materials and behavior. Compositional analysis of archaeological ceramics and raw materials, particularly the chemical characterization by using INAA, together with classical archaeological methods is a powerful approach to solve questions dealing with provenance, and also some indications of the technology and use in ancient times. Clay analysis plays a crucial role in studies of ancient ceramics, contributing with answers to better understanding resource exploitation strategies, production technologies, and regional and interregional interaction patterns. The vessels are made essentially by clays, the composition of the ceramics reflecting the origin of the clay materials used and conclusions may be inferred concerning the type of raw materials and the establishment of the geographic area of production. However, a special care has to be paid to the interpretation of the results obtained by the compositional analysis of the archaeological ceramics, since the composition of the broken fragments that archaeologists recover in excavations is the combination of several parameters, starting with the initial composition of the raw materials, its manipulation to make the vessel, the use and the burial environment.

The application of nuclear methods of chemical analysis has been contributing to a large extent to resolve questions related with the production of ceramics in Bracara Augusta. The chemical results obtained allowed to establish two compositional groups of local/regional productions that can be used as "ceramic reference groups" in further studies involving imports and exports of pottery of this Roman town.

Acknowledgments

Grateful acknowledgements are made to Doctor José Marques, Head of the Portuguese Research Reactor (RPI) Sacavém, and also to all the staff for their assistance with the neutron irradiations. The financial support of the Fundação para a Ciência e a Tecnologia, Portugal (Project PRAXIS/HAR/13047: Production and technology of Roman ceramics in Bracara Augusta) is gratefully acknowledged.

References

Alarcão, A.M.; Martins, A.N. *Conímbriga*, 1976, XV, 91-109.

Baxter, M. *Exploratory Multivariate Analysis in Archaeology*; Edinburgh University Press: Edinburgh, 1994.

Burt, D.M. In Geochemistry and Mineralogy of Rare Earth Elements, *Reviews in Mineralogy*, 1989; Vol. 21, pp. 259-307.

Cabral, J.M.P.; Gouveia, M.A. *Portugal Physics*, 1979, 10, 101-116.

Cabral, J.M.P.; Gouveia, M.A.; Alarcão, A.M.; Alarcão, J. *Journal of Archaeological Science,* 1983, 10, 61-71.

Cabral, J.M.P.; Prudêncio, M.I.; Gouveia, M.A.; Morais Arnaud, J.E. In: *Proceedings of the 26th International Archaeometry Symposium*; Farquhar, R.M.; Hancock, R.G.V.; Pavlish, L.A.; Ed.; Archaeometry Laboratory, University of Toronto, 1988, pp 172-178.

Chappell, J. *The potter's complete book of clay and glazes.* Watson-Guphill Publications: New York, 1991, 416p.

Dai Kin, F.; Prudêncio, M. I.; Gouveia, M. A.; Magnusson, E. *Geostandards Newsletter,* 1999, 23, 47-58.

Delgado, M. *Cadernos de Arqueologia*. Braga. 1993-94, 10/11, 113-149.

Dias, M.I. Prudêncio, M.I., Feliu, M.J. & Morais, R. (2003c). Importações gaditanas em Bracara Augusta (Portugal). V Congreso Ibérico de Arqueometria, Puerto de Santa Maria, Spain. Extended Abstracts, p. 133-144.

Dias, M.I.; Prudêncio, M.I. *Archaeometry*, 2007, 49, 2, 383-393.

Dias, M.I.; Prudêncio, M.I. *Microchemical Journal*, 2008, 88, 136-141.

Dias, M.I.; Prudêncio, M.I.; Gouveia, M.A. In: A Clay Odissey, *Proceedings of the 12th International Clay Conference,* Universidad Nacional del Sur, Departamento de Geologia, Bahia Blanca, Argentina; Dominguez, E.; Mas, G.; Cravero, F. (eds), Elsevier: Amsterdam, 2003a; pp 65-70.

Dias, M.I.; Prudêncio, M.I.; Gouveia, M.A. In: GARB, Sítios Islâmicos do Sul Peninsular; IPPAR/MC, Portugal; *Junta da Extremadura – Consejaría de Cultura*, Spain; Ed.; 2001, 257-81.

Dias, M.I.; Prudêncio, M.I.; Rocha F. In: *Applied Study of Cultural Heritage and Clays*; Pérez-Rodríguez, J.L.; Ed.; Sociedade Española de Arcillas (SEA), CSIC: Madrid, 2003b, pp 187–200.

Dias, M.I.; Prudêncio, M.I.; Valera, A.C.; Lago M.; GOUVEIA, M.A. *Geoarchaeological and Bioarchaeological Studies*, 2005a, 3, 161-164.

Dias, M.I.; Prudêncio, M.I.; Valera, A.C.; Sequeira Braga, M.A.; Gouveia, M.A. *British Archaeological Reports, International Series*, 2002, 1011, 253-264.

Dias, M.I.; Valera, A.C.; Prudêncio, M.I. In: Understanding People through their Pottery; Prudêncio, M.I., Dias, M.I. & Waerenborgh, J.C.; Ed.; *Trabalhos de Arqueologia, Série Monográfica Trabalhos de Arqueologia*; IPA: Lisboa, Portugal, ISBN 972-8662-25-4, 2005b; Vol. 42, pp. 41-50.

Ferreira, N.; Dias, G.; Meireles, C.; Sequeira Braga, M.A. Carta Geológica de Portugal, escala 1/50 000. *Notícias Explicativas da Folha* 5-D, Braga, 2ª edição, IGM: Lisboa, 2000.

Gaspar, A. Contribuição para o Estudo de Cerâmicas dos Séc. V-VI DC de Braga. MsC dissertation, University of Minho, Braga, Portugal, 2000.

Glascock, M.D.; Neff, H. Measurement. *Science and Technology*, 2003, 14, 1516-1526.

Gomes, A. *Cerâmicas Pintadas da Época Romana*: Tecnologia, Tipologia e Cronologia. MsC dissertation, University do Minho, Braga, Portugal, 2000.

Gouveia, M.A., Prudêncio, M.I., Figueiredo, M.O., Pereira, L.C.J., Waerenborgh, J.C.; Morgado, I.; Pena, T.; Lopes, A. (1993). *Chemical Geology*, 1993, 107, 293-298.

Gouveia, M.A.; Prudêncio, M.I.; Cabral, J.M.P. *Journal of Radioanalytical and Nuclear Chemistry, Articles*, 1988, 124, 1, 227-233.

Gouveia, M.A.; Prudêncio, M.I.; Freitas, M.C.; Martinho, E.; Cabral, J.M.P. *Journal of Radioanalytical and Nuclear Chemistry, Articles*, 1987, 114, 2, p. 309-318.

Hancock, R.G.V.; Pavlish, L.A.; Aufreiter, S. *Archaeometry*, 2007, 49, 2, 229-243.

Harbottle, G. Activation Analysis in Archaeology; *Radiochemistry*, **3**; Newton, G.A.; Ed.; 1976, The Chemical Society: London.

Haskin, L.A., Helmke, P.A., Paster, T.P. & Allen, R.O. (1971). *Activation Analysis in Geochemistry and Cosmochemistry*: 201-218, A.O. Brunfelt and E. Steinnes, eds., Universitetsforlaget, Oslo.

Hector, N. Chemical characterization of ceramic pastes in archaeology; Monographs in World Archaeology 7; Prehistory Press: Madison, Wisconsin, 1992.

Kilikoglou, V.; Grimanis, A.P.; Tsolakidou, A.; Hein, A.; Malamidou, D.; Tsirtsoni, Z. *Archaeometry*, 2007, 49, 2, 301-319.

Marques, R. Geoquímica e Mineralogia de Argilas do Cretácico de Taveiro e Aveiro, Portugal. MsC dissertation, University of Aveiro, Portugal, 2007.

Martinho, E.; Gouveia, M.A.; Prudêncio, M.I.; Reis, M.F.; Cabral, J.M.P. Applied Radiations and Isotopes, 1991, 12, 11, 1067-1071.

Martins, M.; Delgado, M. *Cadernos de Arqueologia*, Braga, 1989-90a, II, 6/7, 41-186.

Martins, M.; Delgado, M. *Cadernos de Arqueologia*, Braga, 1989-90b, II, 6/7, 11-38.

Martins, M.; Delgado, M. In: *Actas do Colóquio Internacional de Arqueologia - Los Finisterres Atlânticos em la Antiguedad* (Época Preromana y Romana); Gijón, 1995, pp 121-128.

Masuda, A.; Kawakami, O.; Dohmoto, Y.; Takenaka, T. *Geochem. Journal*, 1987, 21, 119-124.

McLennan, S.M. *Geochimica and Cosmochimica Acta*, 1994, 58, 2025-2033.

McLennan, S.M. In Geochemistry and Mineralogy of Rare Earth Elements, *Reviews in Mineralogy*, 1989; Vol. 21, pp. 169-200.

Nasraoui, M., Waerenborgh, J.C., Prudêncio, M.I.; Bilal, E., *Journal of Cultural Heritage*, 2002, 3, 127-132.

Oliveira, F.; Sequeira Braga, M.A.; Prudêncio, M.I.; Delgado, M.; Gouveia, M.A. In: Understanding People through their Pottery; Prudêncio, M.I., Dias, M.I. & Waerenborgh, J.C.; Ed.; *Trabalhos de Arqueologia*, Série Monográfica Trabalhos de Arqueologia; IPA: Lisboa, Portugal, ISBN 972-8662-25-4, 2005; Vol. 42, pp. 80-89.

Oliveira, F.M.S. Contribuição para o estudo da cerâmica fina de Braga. A cerâmica "dita bracarense". MsC dissertation, University of Minho, Braga, Portugal, 1997.

Op De Beek, J. GELIAN program, Institute for Nuclear Sciences, University of Ghent, 1972.

Op De Beek, J. OLIVE program, Institute for Nuclear Sciences, University of Ghent, 1974.

Parry, S.J. Activation Spectrometry in Chemical Analysis; *Chemical Analysis Series* 119; Wiley: New York, 1991.

Pollard, A.M. *Archaeometry* 2008, 50, 2, 191-193.

Prudêncio, M.I., Figueiredo, M.O., Waerenborgh, J.C., Pereira, L.C.J., Gouveia, M.A., Silva, T.P., Morgado, I., Costa, I.R. & Rodrigues, J.D. (1995). In: The Study of Marble and Other Stones Used in Antiquity, *proceedings of the 3rd International Symposium of the Association for the Study of Marble and Other Stones Used in Antiquity, ASMOSIA III,*

Athens, Greece; Maniatis, Y.; Herz, N.; Basiakos, Y.; Ed.; Archetype: London, 1995, pp 285-289.

Prudêncio, M.I., Gouveia, M.A.; Cabral, J.M.P. *Journal of Trace and Microprobe Techniques*, 1988, 6, 1, 103-111.

Prudêncio, M.I., Marques, R., Rebelo, L., Cook, G.T., Cardoso, G.O., Naysmith, P., Freeman, S.P.H.T., Franco, D., Brito, P., Dias, M.I. *Radiocarbon*, 2007a, 49, nº2, 1145-1151.

Prudêncio, M.I.; Braga, M.A.S.; Gouveia, M.A. *Chemical Geology*, 1993, 107, 251-254.

Prudêncio, M.I.; Cabral, J.M.P. *Journal of Radioanalytical and Nuclear Chemistry, Articles*, 1988, 123, nº1, 309-320.

Prudêncio, M.I.; Cabral, J.M.P.; Tavares, A. In: *Archaeometry: Proceedings of the 25th International Symposium;* Maniatis, Y.; Ed., Elsevier: Amsterdam; 1989, pp 503-514.

Prudêncio, M.I.; Dias, M.I.; Raposo, J.; Gouveia, M.A.; Fabião, C.; Guerra, A.; Bugalhão, J.; Duarte, A.L.; Sabrosa, A. In: *Ceramic in the Society*; Di Pierro, S., Serneels, V. & Maggetti, M.; Ed.; 2003, pp 245-253.

Prudêncio, M.I.; Gonzales, M.I.; Dias, M.I.; Galan, E.; Ruiz, F. *Journal of Arid Environments*, 2007b, 69, 285-298.

Prudêncio, M.I.; Gouveia, M.A.; Cabral, J.M.P. *Geostandards Newsletter*, 1986, X, 1, 29-31.

Prudêncio, M.I.; Gouveia, M.A.; Sequeira Braga, M.A. *Clay Minerals*, 1995, 30, 239-248.

Prudêncio, M.I.; Nasraoui, M.; Trindade, M.J.; Sequeira Braga, M.A.; Figueiredo, M.O. In: *Proceedings of the 9th International Congress on Deterioration and Conservation of Stone*; Fassina, V.; Ed.; Elsevier: Amsterdam, 2000, Vol. 1, pp 165-170.

Prudêncio, M.I.; Oliveira, F.; Dias, M.I.; Sequeira Braga, M.A.; Delgado, M.; Martins, M. *Clays and Clay Minerals*, 2006, 54, 5, 638-649.

Prudêncio, M.I.; Waerenborgh, J.C.; Gouveia, M.A.; Trindade, M.J.; Alves, C.A.S., Sequeira Braga, M.A. In: Water - Rock Interaction; Arehart, G.; Hulston, J.; Ed.; A.A. Balkema: Rotterdam, 1998, pp 391-394.

Rice, P.M. *Pottery Analysis*, A source book; The University of Chicago Press: Chicago, 1987.

Richter, D.; Zink, A.; Przegietka, K.; Cardoso, G.O.; Gouveia, M.A.; Prudêncio M.I. Ancient TL, 2003, 21, 1-7.

Romão, P.; Prudêncio, M.I.; Figueiredo, M.O.; Trindade, M.J.; Nasraoui, M.; Gouveia, M.A.; Silva, T. In: *Proceedings of the 9th International Congress on Deterioration and Conservation of Stone*; Fassina, V.; Ed.; Elsevier: Amsterdam, 2000, Vol. 1, pp 493-497.

Sanei, H.; Goodarzi, F.; Flier-Keller, E. J. Environ. Monit. 2001, 3, 27-36.

Sayre, E.V.; Dodson, R.W. *Am. J. Arcaheol.*, 1957, 61, 35.

Sequeira Braga, M.A. Arenas e depósitos associados da bacia de drenagem do rio Cávado (Portugal) e contribuição para o estudo da arenização. PhD thesis, Universidade do Minho, Portugal, 1988, 325 pp.

Sequeira Braga, M.A.; Paquet, H.; Begonha, A. *Catena*, 2002, 49, 41-56.

Sequeira Braga, M.A.; Prudêncio, M.I.; Delgado Rodrigues, J.; Alves, C.A.S.; Costa, D.; Silva, T.; Trindade, M.J.; Waerenborgh, J.C.; Nasraoui, M.; Gouveia, M.A. In: *Proceedings of the 9th International Congress on Deterioration and Conservation of Stone*; Fassina, V.; Ed.; Elsevier: Amsterdam, 2000, Vol. 1, pp 235-245.

Sneath, P.H.A., Sokal, R.R. *Numerical Taxonomy*. The Principles and Practice of Numerical Classification; W.H. Freeman and Company: San Francisco, 1973.

StatSoft, Inc. (2003). STATISTICA (data analysis software system), version 6. www.statsoft.com

Tite, M.S. *Archaeometry*, 2008, 50, 2, 216-231.

Velde, B.; Druc, I.C. *Archaeological Ceramic. Materials. Origin and utilization*; Springer-Verlag: Berlin, Heidelberg, 1999.

In: Nuclear Chemistry: New Research
Editor: Axel N. Koskinen, pp. 83-98
ISBN: 978-1-60456-957-5

Chapter 3

EFFECT OF BIOSUBSTANCES ON THE ROS FORMATION BY DEPLETED URANIUM

***Akira Nakajima*[1,*], *Emiko Matsuda*[1] *and Yuto Ueda*[2]**
[1]Division of Chemistry, Department of Medical Science,
[2]Department of Psychiatry, Faculty of Medicine, University of Miyazaki,
5200 Kihara, Kiyotake, Miyazaki 889-1692, JAPAN

Abstract

The characteristics of the ROS, such as hydroxyl radical, formation in depleted uranium (DU)-hydrogen peroxide (H_2O_2) systems and the effect of biosubstances on the radical formation were studied in detail. The hydroxyl radical formation in the UO_2^{2+}-H_2O_2 system can be described as a stepwise reaction process including the reduction of UO_2^{2+} to UO_2^{+} by H_2O_2 and the Fenton-like reaction of UO_2^{+} with H_2O_2. A part of UO_2^{+} will be reduced to UO^{2+} through disproportionation under appropriate pH conditions. The UO^{2+} ion is oxidized back to UO_2^{2+} through a couple of Fenton-like reactions under the coexistence of H_2O_2, which suggests that twice the amount of the hydroxyl radical will be formed in the UO^{2+}-H_2O_2 system than in the UO_2^{2+}-H_2O_2 one. The behaviors of biosubstances, such as saccharides, amino acids, and microbial cells, in the UO_2^{2+}-H_2O_2 system could be explained by the direct elimination of the hydroxyl radical and the inhibition of the hydroxyl radical formation through the coupling of the UO_2^{2+} ion with these compounds. Catechins, especially epigallocatechin, enhanced the hydroxyl radical formation by the reduction of UO_2^{2+} to UO^{2+}. Mimosa and quebracho tannins enhanced the hydroxyl radical formation by the reduction of UO_2^{2+} to UO^{2+}, while Chinese gallo, mylaboram, and chestnut tannins depressed the hydroxyl radical formation by strongly coupling with the UO_2^{2+} ion. The behaviors of vitamins, such as C and E, eliminated spin adducts in addition to the hydroxyl radical, which caused complicated behaviors in the spin-trapping procedure in the UO_2^{2+}-H_2O_2 systems. In summary, some biosubstances enhance the ROS formation in the DU-H_2O_2 system, and some of them reduce it, which suggests positive and negative effects, respectively, on the chemical toxicity of DU.

---

* E-mail address: akanaka@med.miyazaki-u.ac.jp. Tel. & Fax: +81-(0)985-85-0993. (Corresponding author)

Introduction

Recently, the chemical toxicity of depleted uranium (DU) has become the focus of considerable interest due to its increasing use for military applications [1]. The main route of potentially hazardous exposure is inhalation. The uranyl ion derived from DU enters the systemic circulation, slowly cleared to the blood, and then excreted through the kidneys [2, 3]. Results obtained from long-term studies using small animals concluded that a high dose of DU causes toxicity in the kidney tubules [4]. Uranium-238, the major radioisotope of DU, should decay with 4.51×10^9 years of half-life to produce α particles that will be absorbed around the irradiated surface. These considerations indicated that the radiation dose of DU is smaller than those of natural radon and its decay products. DU should, therefore, reveal a similar chemical toxicity to ordinal heavy metals, such as copper, cadmium, and lead. As the epidemiology of DU has been extensively investigated [5-7], the information on the molecular aspects of its chemical toxicity has become progressively clear. McQueney and Markham reported that uranium inhibits S-adenosylmethionine (AdoMet) synthesis by AdoMet synthase [8]. Hamilton *et al.* studied the hydroxyl radical formation in DU-hydrogen peroxide (H_2O_2) systems and proposed a mechanism including the reduction of UO_2^{2+} to UO^{2+} and a Fenton-type reaction [9]. Miller *et al.* showed DU-catalyzed oxidative DNA [10]. Both results support the idea that reactive oxygen species (ROS), such as hydroxyl radicals, in the system cause DU toxicity. In contrast, Yazzie *et al.* reported a DNA strand cleavage by uranyl acetate in the presence of ascorbate [11] and discussed the possibility of the direct genotoxicity of DU rather than through a Fenton-like reaction. These results indicate the diversity of the chemical toxicity of DU. In order to obtain further information on this chemical toxicity, the authors investigated hydroxyl radical formation in DU-H_2O_2 systems in detail [12-15].

In the present work, to analyze the reaction mechanism between DU and H_2O_2 and obtain information on the effect of DU-H_2O_2 system on biosystems, the characteristics of the hydroxyl radical formation in the DU-H_2O_2 systems and the effect of biosubstances on the hydroxyl radical formation, therefore, were studied in detail.

Experimental

Chemicals

5,5-Dimethyl-1-pyrroline N-oxide (DMPO) was obtained from DOJINDO Ltd. (Kumamoto, Japan). During manufacture, a purity of more than 99% was obtained by repeated vacuum distillation and checked by gas chromatography. Depleted uranium (DU) as uranyl nitrate hexahydrate, $UO_2(NO_3)_2 \cdot 6H_2O$, was obtained from Merck & Co., Ltd. (Darmstadt, Fed. Rep. Germany), and other chemicals (guaranteed reagents), from Nacalai Tesque, Inc. (Tokyo, Japan) and Wako Pure Chemical Industries, Ltd. (Tokyo, Japan).

A uranous (UO^{2+}) solution was prepared as follows [16, 17]: 1 g of granular zinc was added in 10 ml of a uranyl nitrate solution (a mixture of 2.5 ml of 4×10^{-2} M uranyl nitrate and 7.5 ml of 5 M HCl); the mixture was allowed to stand for 1 h and then filled up to 25 ml with water. The formation of UO^{2+} was ascertained by UV-VI absorption maximal peaks at 429, 494, 549, and 648 nm [18]. The final concentration of the UO^{2+} stock solution was 4 ×

10^{-3} M. Carefully distilled, deionized, and Chelex 100-treated water was used for the preparation of all solutions to avoid the formation of DMPO-OH by contaminated iron ion [19, 20].

Electron Paramagnetic Resonance Measurements

Uranyl nitrate (final concentration, 1.0×10^{-3} M), DMPO (final concentration, 0.1 M), and H_2O_2 (final concentration, $1.0 \times 10^{-5} \sim 0.1$ M) were mixed. The solution pH was adjusted by using 0.1 ~ 1 M HNO_3 and NaOH solutions. The pH of the reaction mixture without pH adjustment was 4.0. After aspirating the mixture into a capillary tube, its ESR spectrum was immediately recorded using an X-band ESR spectrometer (JES TE-100, JEOL, Ltd., Tokyo, Japan) controlled by a WIN-RAD ESR data analyzer (Radical Research, Inc., Tokyo, Japan) under the following conditions: microwave power, 5 mW; microwave frequency, 9.42 GHz; magnetic field, 335.3 mT; field sweep width, ± 5 mT; field modulation, 100 kHz; modulation width, 0.079 mT; sweep time, 1 min; response time, 0.1 s. Metal ions, such as UO^{2+}, VO^{2+}, and Fe^{2+} (final concentration, 1.0×10^{-3} M), were also mixed with DMPO (final concentration, 0.1 M) and H_2O_2 (final concentration, 0.1 M), of which the ESR spectrum was measured. Uranyl nitrate (final concentration, 1.0×10^{-3} M), DMPO (final concentration, 0.1 M), H_2O_2 (final concentration, $1.0 \times 10^{-5} \sim 0.1$ M), and desired amounts of biosubstances were mixed and then used for ESR measurements. A preliminary examination confirmed that the zinc ion in the UO^{2+}-H_2O_2-DMPO solution did not affect the formation of DMPO-OH.

Desired amounts (final concentration, 0.1 - 100 mM) of vitamin E (VE) dissolved in diethyl ether were poured into a micro-tube, and the solvent was then evaporated. A sodium dodecyl sulfate (SDS, sodium lauryl sulfate) solution was added into the micro-tube, and VE was dispersed in the SDS solution [21]. Desired amounts of UO_2^{2+}, H_2O_2, and DMPO were added to the tube. The reaction mixture was mixed vigorously and then used for the ESR measurements.

A solution of H_2O_2 (final concentration, 1.0×10^{-2} M) and DMPO (final concentration, 10^{-2} M) in a capillary tube was irradiated with a glass fiber-type UV irradiator (RUVF-203S, Radical Research, Inc., Tokyo, Japan; Hg-Xe lamp, wavelength, 365 nm; power, 200 W) for 2 ~ 6 s. After the irradiation, the ESR spectrum of the mixture solution was recorded. The final concentrations of H_2O_2 and DMPO for the UV irradiation experiments were adjusted to 10^{-2} M to obtain almost the same ESR signal intensities for DMPO-OH as those in the UO_2^{2+}-H_2O_2-DMPO solution.

The concentration of the DMPO-OH radical was calibrated using TEMPOL as a standard after double integration of the observed ESR signal.

Determination of Uranium Peroxide

Uranyl nitrate (final concentration, 1.0×10^{-3} M) and H_2O_2 (final concentration, 1.0×10^{-2} M) were mixed, and the pH was adjusted by using 0.1 ~ 1.0 N HNO_3 and NaOH solutions. After the mixture had stood for 1 h, the resultant precipitate, uranium peroxide (metastudite), $UO_4 \cdot 2H_2O$, or $UO_2 \cdot 2H_2O_2$ [18], was filtered using a membrane filter (pore size, 0.2 μm). The amounts of uranium peroxide were estimated from the residual uranium in the filtrate determined by spectrophotometry using Arsenazo III [22].

Optical Absorption Spectral Measurements

Reaction solutions of uranyl or uranous ion (final concentration, 1.0×10^{-3} M) and H_2O_2 (final concentration, 1.0×10^{-1} M) were mixed. The optical absorption spectra of both mixture solutions were recorded in the range of wavelengths from 200 to 800 nm by using a UV-VIS spectrometer (V-550, JASCO, Tokyo, Japan). As uranyl ion, UO_2^{2+}, formed precipitate (uranium peroxide) above pH 1, its optical absorption spectrum could measure only below pH 1.

Determination of the Dissolved Oxygen Concentration in the Reaction Mixture

The concentration of oxygen in the reaction solutions containing hydrogen peroxide and uranium ion was monitored by an electrochemical method using a water quality meter (WQC-1A, DKK-TOA Corp., Tokyo, Japan). The time dependency of the oxygen concentration was also monitored using this system.

Results

Hydroxyl Radical Formation in the DU-H_2O_2 System

The reaction solution composed of the UO_2^{2+} ion (final concentration, 1.0×10^{-3} M), DMPO (final concentration, 0.1 M), and H_2O_2 (final concentration, 0.1 M) revealed an ESR signal with four lines (intensity ratio, 1:2:2:1) (Fig. 1a). From its g-value (= 2.0066) and hyperfine coupling constants (a_N = 1.49 mT, a_H = 1.49 mT), the signal was identified to be a DMPO-OH radical. The hydroxyl radical formation could be confirmed by the formation of spin adducts, DMPO-CH_3 (Fig. 1b, a_N = 1.60 mT, a_H = 2.29 mT) and DMPO-CH_2OH (Fig. 1c, a_N = 1.55 mT, a_H = 2.21 mT), in the UO_2^{2+}-H_2O_2-dimethylsulfoxide (DMSO) and UO_2^{2+}-H_2O_2-methanol (MeOH) mixture solutions. The formation of the hydroxyl radical in the UO_2^{2+}-H_2O_2 solution was already reported by Hamilton *et al.* [9], and the present results reconfirmed their results. As shown in Fig. 1c (asterisks), a signal of DMPO-OCH_3 (a_N = 1.45 mT, $a_{\beta H}$ = 0.94 mT, $a_{\gamma H}$ = 0.14 mT), being formed by the nucleophilic addition of MeOH, was also observed. The detection of DMPO-OCH_3 suggested the formation of DMPO-OH caused by the nucleophilic addition of water to DMPO, as in the presence of the Fe^{3+} ion [19, 20]. However, in the present system, the concentration of the DMPO-OCH_3 radical was so small that it did not affect the main reactions.

The addition of organic solvents also revealed the formation of DMPO-OH, DMPO-CH_3, and DMPO-CH_2OH (Fig. 1d-f) in the UO^{2+}-H_2O_2-DMPO-organic solvent solution (pH 1), where UO^{2+} is the reduction form of UO_2^{2+}. These results indicated that the UO^{2+} ion produced a hydroxyl radical in the reaction with H_2O_2 in a manner similar to the Fenton reaction [23]. The signal intensity of the DMPO-OH in the UO^{2+}-H_2O_2-DMPO solution was about 4 - 5 times larger than that in the UO_2^{2+}-H_2O_2-DMPO solution.

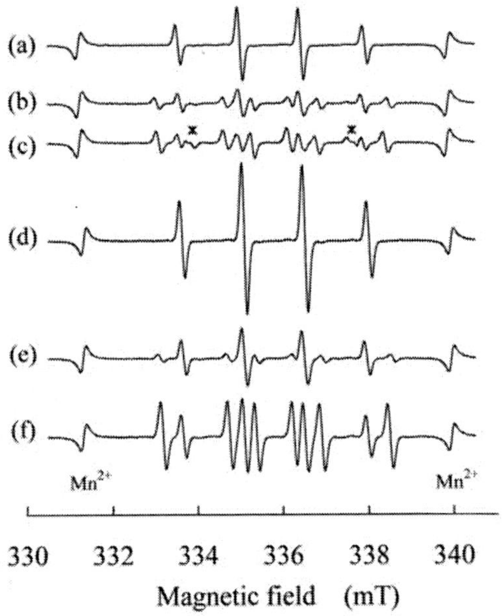

Figure 1. ESR spectra of aminoxyl radicals observed in the UO_2^{2+}-H_2O_2-DMPO (pH 4) and the UO^{2+}-H_2O_2-DMPO solutions, twenty minutes after mixing of 10^{-3} M UO_2^{2+} or UO^{2+}, 0.1 M H_2O_2 and 0.1 M DMPO. (a) UO_2^{2+}-H_2O_2-DMPO solution (pH 4) without organic solvent, (b) with 25 % DMSO, (c) with 25 % MeOH, (d) UO^{2+}-H_2O_2-DMPO solution (pH 1) without organic solvent, (e) with 25 % DMSO, (f) with 25 % MeOH.

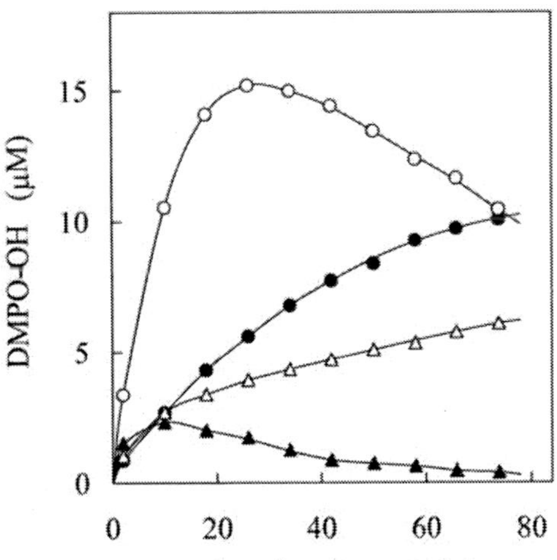

Figure 2. Effect of saline and buffer salts on the DMPO-OH formation in the UO_2^{2+}-H_2O_2-DMPO solution. Uranyl ion (10^{-3} M), H_2O_2 (0.1 M), DMPO (0.1 M), and 0.9 % sodium chloride or 0.1 M of each buffer were mixed. Closed circle indicates control, open circle, saline, closed triangle, HEPES, and open triangle, phosphate buffer.

The ESR signal intensity of DMPO-OH in the UO_2^{2+}-H_2O_2-DMPO solution was affected by saline and buffer solutions (Fig. 2). Phosphate ion forms precipitate with uranyl ion, which causes a decrease in the DMPO-OH formation. The biochemical buffer, N-2-hydroxyethyl-piperazine-N'-2-ethanesulfonic acid (HEPES), forms a yellow complex with the uranyl ion, showing a significant decrease in the DMPO-OH formation. The DMPO-OH formation in the UO_2^{2+}-H_2O_2-DMPO solution containing sodium chloride was enhanced within the first 20 min and then decreased rapidly. These results suggest that the use of saline and buffer solutions should be given significant attention for DMPO spin trapping in a UO_2^{2+}-H_2O_2 solution. In subsequent experiments, therefore, saline and buffer solutions should not be used.

Kinetics of the Hydroxyl Radical Formation in the DU-H_2O_2 System

As shown in Fig. 3, the ESR signal intensity of DMPO-OH in the UO_2^{2+}-H_2O_2-DMPO solution (pH 4) gradually increased with time, the time dependency of which was [1 - exp(-0.033t)] [13]. The amount of DMPO-OH in the solution at 1 h after the reaction was started was about 10^{-5} M. The concentration of oxygen also gradually increased with time. Although the time dependency of the DMPO-OH formation in the UO_2^{2+}-H_2O_2-DMPO solution (pH 1) was the same as that in the UO_2^{2+}-H_2O_2-DMPO solution (pH 4), the amounts of DMPO-OH reached about 2.5×10^{-5} M. In contrast, the amount of DMPO-OH in the UO^{2+}-H_2O_2-DMPO solution (pH 1) increased rapidly within first 20 min and then gradually decreased with time. The time dependency of the DMPO-OH formation in the initial stage in the UO^{2+}-H_2O_2-DMPO solution (pH 1) was formulated as [1 - exp(-0.153t)] [13], the rate constant of which was five times that in the UO_2^{2+}-H_2O_2-DMPO solution (pH 1 and 4). In addition, the maximum signal intensity of DMPO-OH increased to about 4×10^{-5} M, indicating that the UO^{2+} ion forms DMPO-OH far more easily than the UO_2^{2+} ion in the reaction with H_2O_2. The absorption maxima at 429, 494, 549, and 648 nm derived from the UO^{2+} ion [18] disappeared within few minutes after the addition of H_2O_2, and then a small peak at 415 nm derived from UO_2^{2+} ion [18] appeared (Fig. 4). These results confirmed that UO^{2+} ion was oxidized to UO_2^{+} by H_2O_2 through a Fenton-type reaction. As the absorption peaks for the UO_2^{+} ion were not observed either, the ion is, therefore, unstable and oxidizes to UO_2^{2+} immediately.

As a reference, analogous ESR measurements were carried out for a reaction solution containing 10^{-3} M of Fe^{2+} and VO^{2+} ions (pH 4), which exhibits pronounced activity for the production of hydroxyl radical from H_2O_2 [23, 24]. In the case of the Fe^{2+} ion, the signal intensity of the DMPO-OH radical reached its maximum (4.7×10^{-5} M) at 10 min after the reaction started and then quickly decreased (Fig. 3). In the case of the VO^{2+} ion, the formation of the radical was too fast for the initial reaction stage of the DMPO-OH radical formation to be recorded. The maximum amounts of DMPO-OH in the UO^{2+}-H_2O_2-DMPO solution were comparable to those of the Fe^{2+}-H_2O_2-DMPO solution. These results indicated that the UO^{2+} ion has a similar activity to that of Fe^{2+} and VO^{2+} ions for hydroxyl radical formation from H_2O_2, while the UO_2^{2+} ion has characteristically low activity. Hamilton *et al.* reported that the order of the amount of the hydroxyl radical formation in the metal-H_2O_2 systems was VO^{2+} > Fe^{2+} > UO_2^{2+}, UO^{2+} [9]. As shown in Fig. 3, the order represents the DMPO-OH radical formation only in the first 20 min. Excess metal ion reductively decomposed DMPO-OH in the VO^{2+}-H_2O_2 and Fe^{2+}-H_2O_2 systems, causing a time-dependent decrease of DMPO-OH, whereas, in the UO_2^{2+}-H_2O_2 system, the amounts of DMPO-OH increased progressively, and the order of the amount of DMPO-OH formed became $UO^{2+} > VO^{2+} > Fe^{2+} > UO_2^{2+}$ at 1 h after the reactions started.

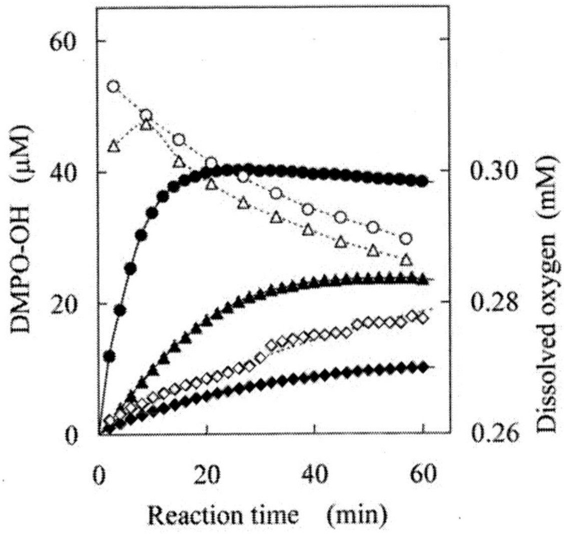

Figure 3. Time course of DMPO-OH formation in metal ion-H_2O_2-DMPO solutions. Metal ion (1 mM), H_2O_2 (100 mM), and DMPO (100 mM) were mixed. Closed diamond indicates the UO_2^{2+}-H_2O_2-DMPO solution (pH 4), closed triangle, UO_2^{2+}-H_2O_2- DMPO solution (pH 1), closed circle, UO^{2+}-H_2O_2-DMPO solution (pH 1), open circle, VO^{2+}-H_2O_2-DMPO solution, and open triangle, Fe^{2+}-H_2O_2-DMPO solution, open diamond, dissolved oxygen.

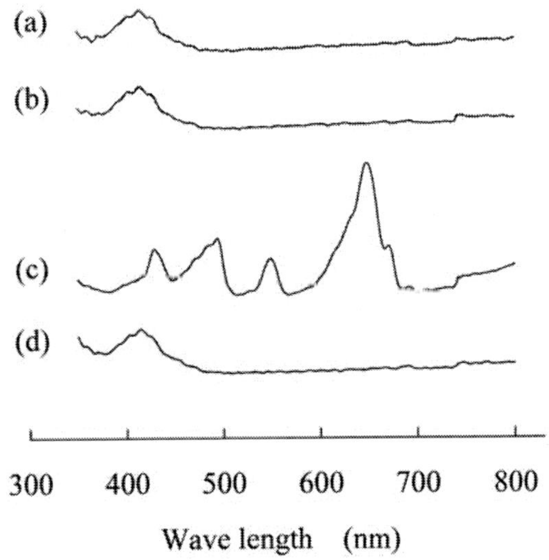

Figure 4. Optical absorption spectra of UO^{2+} and UO_2^{2+} solutions. Spectra were recorded (a) for UO_2^{2+} (1.0 mM, pH 1) solution, (b) for UO_2^{2+} solution (1.0 mM) with H_2O_2 (100 mM), (c) for UO^{2+} solution (1.0mM, pH 1), (d) for UO^{2+} solution (1.0mM, pH 1) with H_2O_2 (100 mM).

Profile of the Reaction Occurring between DU and H_2O_2

Hamilton *et al.* reported that the hydroxyl radical formation was increased in the narrow pH range around 0.6 (the pH of metal solution) [9]. Our previous results also showed that the ESR signal intensity of DMPO-OH in the UO_2^{2+}-H_2O_2-DMPO solution showed a maximum (7.9×10^{-5} M) at pH 1.5 and a rapid decrease above and below that pH [13]. As the final pH of reaction solution was adopted in this experiment, the pH range was different from that of Hamilton *et al.* In the pH range of 4 to 8, only small amounts of DMPO-OH were observed (10^{-5} M, about 1/7 of those at pH 1.5). In contrast, the DMPO-OH signal intensity in the UV-irradiated-H_2O_2 system increased with the pH, reached a plateau above pH 2, and then gradually decreased. These results indicated that the decrease of the DMPO-OH signal intensity in the UO_2^{2+}-H_2O_2-DMPO solution below pH 1.5 is caused by instability of the DMPO-OH due to irreversible protonation of the radical center. The reaction solution containing 10^{-3} M of UO_2^{2+} and 0.1 M of H_2O_2 resulted in the precipitation of uranium peroxide, $UO_4{\cdot}2H_2O$ [18]. At pH 3.6 and 2.6, the precipitation ratios were found to be 98 and 95 %, respectively, close to that at pH 4. The ratio showed a dramatic decrease to 6 % at pH 1.6. This observation relates to the significantly low DMPO-OH signal intensity in the UO_2^{2+}-H_2O_2-DMPO solution at pH 4.

The concentration of the DMPO-OH radical observed 1 h after the reaction was started in the UO_2^{2+}-H_2O_2-DMPO solution (10^{-3} M of UO_2^{2+}) revealed a steep increase and became constant above 2.5×10^{-4} M H_2O_2 [13]. This constant value was almost equal to that observed in the UO_2^{2+}-H_2O_2-DMPO solution (10^{-3} M UO_2^{2+} and 0.1 M H_2O_2). These observations indicate that excess H_2O_2 above 2.5×10^{-4} M depresses hydroxyl radical formation by the precipitation of uranium peroxide, as reported above. The solubility product of $UO_4{\cdot}2H_2O$, $[UO_2^{2+}][H_2O_2]$, can be estimated to be 2.5×10^{-7} M^2 at pH 4.

Effects of Biosubstances on the Hydroxyl Radical Formation in the UO_2^{2+}-H_2O_2 System

The effects of various biosubstances, such as saccharides, amino acids, proteins, microbial cells, catechins, tannins, and vitamins, on the hydroxyl radical formation in the UO_2^{2+}-H_2O_2 solution were examined using the median inhibitory dose (mg/ml), ID_{50} [12-15], which indicates concentrations of substrates in which the amount of DMPO-OH is half of that without a substrate [25].

The order of ID_{50} for proteins was ovalbumin << milk casein << gelatin in the UO_2^{2+}-H_2O_2-DMPO solution and ovalbumin << gelatin < milk casein in the UV-irradiated H_2O_2-DMPO solution. The ID_{50} of amino acids was almost the same value in both the UO_2^{2+}-H_2O_2-DMPO and the UV-irradiated H_2O_2-DMPO solutions. The ID_{50} of saccharides in the UO_2^{2+}-H_2O_2-DMPO solution was larger than that in the UV-irradiated H_2O_2-DMPO solution. The ID_{50} of microbial cells, such as *Arthrobacter nicotianae* (IAM 12342) and *Zooglea nicotianae* (IAM 12136), had a similar value to that of milk casein. As a whole, ovalbumin indicated the lowest value of ID_{50}, which suggests that ovalbumin has the highest ability of hydroxyl radical elimination among these biosubstances.

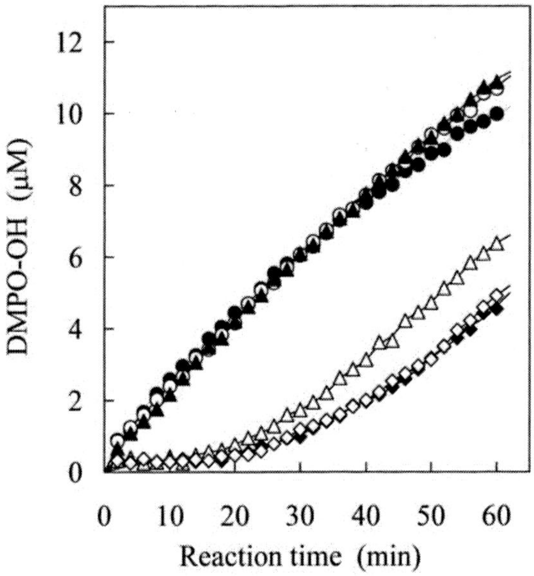

Figure 5. Effect of vitamin C (VC) on the time course of DMPO-OH formation in the UO_2^{2+}-H_2O_2-DMPO (pH 4) solution. Uranyl ion (1 mM), H_2O_2 (100 mM), and DMPO (100 mM) were mixed. Closed circle indicates the control, open circle, VC 0.025 mM, closed triangle, VC 0.125 mM, open triangle, 0.25 mM, closed diamond, 0.375 mM, and open diamond, 0.5 mM.

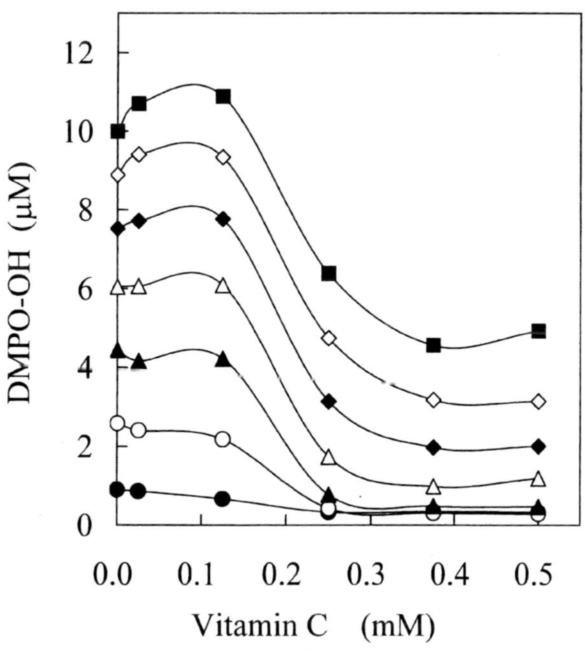

Figure 6. Effect of vitamin C (VC) on the DMPO-OH formation in the UO_2^{2+}-H_2O_2-DMPO (pH 4) solution remade from Fig. 5. Closed circle indicates 2 min, open circle, 10 min, closed triangle, 20 min, open triangle, 30 min, closed diamond, 40 min, open diamond, 50 min, and closed square, 60 min.

Catechins, such as catechin (CA), epicatechin (EC), epigallocatechin (EGC), and epigallocatechin gallate (EGCG), enhanced the formation of DMPO-OH in the UO_2^{2+}-H_2O_2-DMPO solution [12, 14, 15]. In particular, the signal intensity of DMPO-OH with EGC was six times larger than that without EGC. The order of the DMPO-OH formation was ECG >> EGCG > EC > CA > GA. On the other hand, the order of the hydroxyl radical elimination abilities (ID_{50}) in the UV-irradiated H_2O_2-DMPO was GA > EGCG > CA > EGC > EC. EGC has a strong metal-reducing ability, which causes the enhancement of DMPO-OH formation.

The effects of tannins, such as mimosa tannin (MMT), myrabolam tannin (MBT), quebracho tannin (QBT), ganbir tannin (GBT), Chinese gallotannin (CGT), chestnut tannin (CNT), and persimmon tannin (PST), on the hydroxyl radical formation in the UO_2^{2+}-H_2O_2-DMPO solution differed with different kinds of tannins [12, 14, 15]. Condensed tannins, such as MMT and QBT, increased the DMPO-OH formation, and hydrolyzable tannins, such as GBT and PST, decreased it. The order of the amounts of DMPO-OH at 60 min after the reaction started was MMT > QBT ~ Control > GBT > PST > CNT > MBT > CGT. On the other hand, the order of the hydroxyl radical elimination abilities (ID_{50}) in the UV-irradiated H_2O_2-DMPO was MBT > CNT, PST, CGT, GBT > QBT > MMT. MMT and QBT have strong metal-reducing abilities, which cause the enhancement of DMPO-OH formation in a manner similar to EGC. The hydroxyl radical elimination indicates a reverse change from the metal-reducing abilities.

As vitamins, such as C (VC) and E (VE), are powerful anti-oxidants for humans, the effects of these vitamins on the hydroxyl radical formation in the UO_2^{2+}-H_2O_2 and the UO^{2+}-H_2O_2 solutions were examined. Figure 5 shows the time lag of the DMPO-OH formation in the UO_2^{2+}-H_2O_2-DMPO solution (pH 4) with a higher VC concentration above 0.125 mM. Figure 6, which was made from Fig. 5 and shows the VC concentration horizontally, indicates slight increases of the DMPO-OH signal at less than 0.125 mM VC, an abrupt decrease of the DMPO-OH signal at 0.125 - 0.25 mM VC, and a slow decrease above 0.25 mM. Similar results were also observed in the UO_2^{2+}-H_2O_2-DMPO (pH 1) and UO^{2+}-H_2O_2-DMPO (pH 1) solutions. The abrupt decrease indicated the elimination of the hydroxyl radical by VC. These results suggested that VC was used for the elimination of both the hydroxyl radical and DMPO-OH and that VC was changed to dehydroascorbate in the initial stage. After all VC was consumed, DMPO-OH regenerated. The behavior of VC in the UO_2^{2+}-H_2O_2-DMPO solution, therefore, became very complicated.

As shown in Fig. 7, the time courses of the DMPO-OH formation in the UO_2^{2+}-H_2O_2-DMPO-SDS solution (pH 4) indicated peculiar profiles with the concentration of VE. Figure 8, which was made from Fig. 7 and shows the VE concentration horizontally, indicates abrupt decreases of DMPO-OH in the concentration range of 0.1 - 1 mM and an increase of DMPO-OH in the concentration range of 10 - 100 mM. These results suggested that VE should eliminate DMPO-OH in addition to the hydroxyl radical in a manner similar to that of VC. The regeneration of DMPO-OH above 10 mM of VE suggested that DMPO-OH was increased by the reduction of UO_2^{2+} to UO^{2+} by excess amounts of VE.

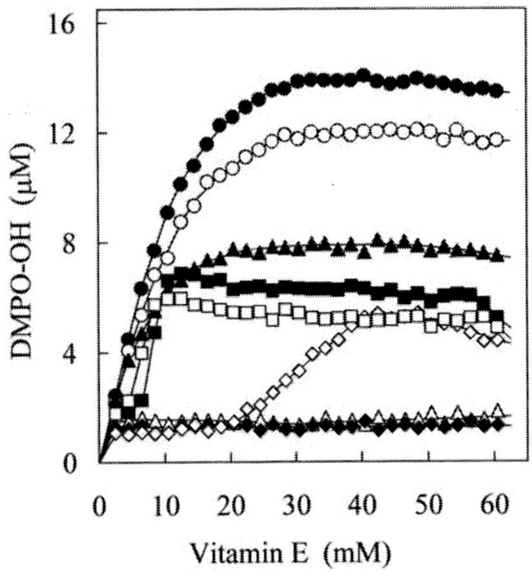

Figure 7. Effect of vitamin E (VE) on the time course of DMPO-OH formation in the UO_2^{2+}-H_2O_2-DMPO-ether vaporized-SDS solution (pH 4). Desired amounts of vitamin E (VE) dissolved in diethyl ether was poured into a micro-tube and then the solvent was evaporated. Sodium dodecyl sulfate (SDS, sodium lauryl sulfate) solution was added into the micro-tube, and VE was dispersed in SDS solution. Uranyl ion (1 mM), H_2O_2 (100 mM), and DMPO (100 mM) were mixed. Closed circle indicates the control, open circle, VE 0.1 mM, closed triangle, VC 0.5 mM, open triangle, 1 mM, closed diamond, 5 mM, open diamond, 10 mM, closed square, 50 mM, and open square, 100 mM.

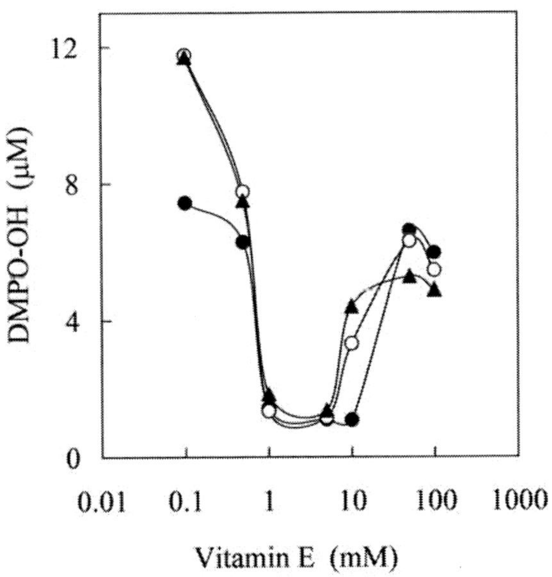

Figure 8. Effect of vitamin E (VE) on the DMPO-OH formation in the UO_2^{2+}-H_2O_2-DMPO-ether vaporized-SDS solution (pH 4) remade from Fig. 7. Closed circle indicates 10.5 min, open circle, 30.5 min, and closed triangle, 60.5 min.

Discussion

A possible mechanism of the hydroxyl radical formation in the DU-H_2O_2 system was evaluated from the previous [13] and present results. Uranyl ion, UO_2^{2+}, being the highest oxidation state of uranium, could not react with H_2O_2 to produce a hydroxyl radical. The reduction of UO_2^{2+} to UO_2^{+} and/or UO^{2+}, therefore, should be indispensable. The production of molecular oxygen in the solution suggested the reduction of UO_2^{2+} to UO_2^{+}. In the UO^{2+}-H_2O_2-DMPO solution, the hydroxyl radical was also formed in amounts about twice that in the UO_2^{2+}-H_2O_2-DMPO solution (pH 1). Possible reactions in the UO_2^{2+}-H_2O_2-DMPO solution are as follows:

(a) reduction of UO_2^{2+} to UO_2^{+}

$$2UO_2^{2+} + H_2O_2 \rightarrow 2UO_2^{+} + 2\,H^{+} + O_2\uparrow \tag{1}$$

(b) disproportionation of UO_2^{+} to UO_2^{2+} and UO^{2+}

$$2UO_2^{+} + H_2O \rightarrow UO^{2+} + UO_2^{2+} + 2OH^{-} \tag{2}$$

(c) Fenton-type reaction of UO_2^{+}

$$UO_2^{+} + H_2O_2 \rightarrow UO_2^{2+} + \bullet OH + OH^{-} \tag{3}$$

(d) Fenton-type reaction of UO^{2+}

$$UO^{2+} + H_2O_2 \rightarrow UO_2^{+} + \bullet OH + H^{+} \tag{4}$$

(e) spin-trapping reaction

$$\text{DMPO} + \bullet OH \rightarrow \text{DMPO-OH} \tag{5}$$

(f) precipitation of uranium peroxide

$$UO_2^{2+} + H_2O_2 + 2H_2O \rightarrow UO_4{\cdot}2H_2O\downarrow + 2H^{+} \tag{6}$$

The disproportionation, shown in Equation 2, was so slow at pH 2 - 4 that Equation 4 should be negligible in the UO_2^{2+}-H_2O_2-DMPO solution [18]. The reaction scheme in the UO_2^{2+}-H_2O_2-DMPO solution, therefore, should be composed of Equations 1, 2, and 3 and can be summarized as follows:

$$UO_2^{2+} + (3/2)H_2O_2 \rightarrow UO_2^{2+} + \bullet OH + H_2O + (1/2)O_2\uparrow \tag{7}$$

The optical absorption spectra of the UO_2^{2+}-H_2O_2 solution indicated the conservation of the peak at 415 nm for UO_2^{2+} ion throughout the reaction, which supports Equation 7. The proposed reaction scheme indicates that the redox reaction of UO_2^{2+} ion should be repeated in the presence of H_2O_2. The uranyl ion, therefore, produces a hydroxyl radical continuously for several hours in the presence of 10^{-6} - 0.1 M of H_2O_2.

In the UO^{2+}-H_2O_2-DMPO solution, two Fenton-type reaction steps should occur. Thus, Equations 3 and 4 were combined into the following equation:

$$UO^{2+} + 2\,H_2O_2 \rightarrow UO_2^{2+} + 2\bullet OH + H_2O \quad (8)$$

Equations 7 and 8 can reasonably explain why the amounts of DMPO-OH in the UO^{2+}-H_2O_2-DMPO solution (pH 1) were about twice those in the UO_2^{2+}-H_2O_2-DMPO solution (pH 1). The hydroxyl radical-trapping reaction of DMPO is very rapid (Equation 5; rate constant $k_0 = 2.1 \times 10^9\ M^{-1}s^{-1}$) [26], and the trapping ratio of the hydroxyl radical is almost 100 % [27], which supports the concentration of DMPO-OH directly corresponding to the hydroxyl radical formation. As the present experiments were conducted under the presence of excess H_2O_2, the initial reaction stage of the DMPO-OH radical formation was sufficiently regarded as a pseudo first-order reaction. The rate constant of the radical formation in the UO^{2+}-H_2O_2 solution, therefore, can be estimated to be 0.153 s^{-1} [13]. The initial reaction stage of the DMPO-OH radical formation in the UO_2^{2+}-H_2O_2-DMPO solution at pH 4 is also regarded as a pseudo first-order reaction, the rate constant of which is estimated to be 0.033 s^{-1} [13]. As the rate constant of the Fenton-type reaction, shown in Equation 4, should almost be 0.153 s^{-1} or faster, the rate constant of the reduction of UO_2^{2+} to UO_2^{+}, shown in Equation 1, becomes the rate-determining step, and its rate constant should be 0.033 s^{-1}.

Strong reductants, such as $Na_2S_2O_4$, reduce UO_2^{2+} to UO^{2+} at a stretch (Equation 9), and the hydroxyl radical was produced through Equation 8.

$$UO_2^{2+} + 2H^+ + 2e^- \rightarrow UO^{2+} + H_2O \quad (9)$$

Larger amounts of DMPO-OH were, therefore, formed.

As described in the previous section, some biosubstances [BSH] inhibited the hydroxyl radical formation, and others enhanced its formation. Two inhibitory actions of biosubstances for the formation of DMPO-OH in the UO_2^{2+}-H_2O_2-DMPO solution should be possible, i.e., the direct elimination of the hydroxyl radical (Equation 10) and the inhibition of the hydroxyl radical formation through coupling with the UO_2^{2+} ion (Equation 11).

$$BSH + \bullet OH \rightarrow BSH\text{-}OH \quad (10)$$

$$UO_2^{2+} + BSH \rightarrow UO_2^{2+}\text{-}BSH\ (\text{complex}) \quad (11)$$

The reaction in Equation 10 causes the decrease of DMPO-OH competitively with DMPO. Equation 11 impedes the UO_2^{2+} reduction (Equation 1), which decreases DMPO-OH indirectly. Biosubstances, such as saccharides, amino acids, proteins, and microbial cells, decreased the amounts of DMPO-OH through Equations 10 and 11. Polyphenols also decreased DMPO-OH through Equations 10 and 11. However, some polyphenols, such as EGC, MMT, and QBT, enhanced the DMPO-OH formation through the strong reduction of UO_2^{2+} to UO^{2+} (Equation 9).

In the case of vitamins, an anomalous decrease of DMPO-OH was observed, which was caused by the elimination of DMPO-OH (Equation 12) in addition to the elimination of the hydroxyl radical.

$$\text{DMPO-OH} + \text{VC} \rightarrow \text{DMPOH-OH} + \text{VCR} \tag{12}$$

$$\text{DMPO-OH} + \text{VE} \rightarrow \text{DMPOH-OH} + \text{VER} \tag{13}$$

where VCR indicates the ascorbyl radical, VER, the tocopheroxyl radical, and DMPOH-OH, hydroxylamine reduced from DMPO-OH. These results indicate that the true characteristics of the behavior of vitamins in the UO_2^{2+}-H_2O_2 solution could not be obtained by the spin-trapping method.

Conclusion

As the disproportionation of UO_2^+ is negligible, the present pH condition, the hydroxyl radical formation in the UO_2^{2+}-H_2O_2 system, could be described as a stepwise reaction process including the reduction of UO_2^{2+} to UO_2^+ by H_2O_2 and the Fenton-type reaction of UO_2^+ with H_2O_2. In particular, the reduction of UO_2^{2+} should be the most important process. Some biosubstances having anti-oxidant abilities enhanced the hydroxyl radical formation through the reduction of UO_2^{2+}, which suggests that the behavior of biosubstances in the UO_2^{2+}-H_2O_2 system is quite complicated. Some of them suppress the hydroxyl radical formation, and some of them enhance it. To screen anti-oxidants for the ROS formation in the system including DU, the facts reported above should be taken into account.

Acknowledgement

This work has been supported by Grants-in-Aid for Scientific Research (C) (No. 13680592 and 16510038) from the Ministry of Education, Culture, Sports, Science and Technology of Japan, and by the REIMEI Research Resources of Japan Atomic Energy Research Institute.

References

[1] European Commission (2001). *Opinion of the group of experts established according to article 31 of the Euratom Treaty. Depleted Uranium*, European Commission, Directorate-General Environment, Directorate C-Nuclear Safety and Civil Protection, ENV.C4-Radiation Protection.

[2] Friberg, L.; Nordberg, G. F.; Vouk, V. B. (Eds) (1986). *Handbook of the Toxicology of Metals*. Amsterdam, Elsevier, Vol. II, 623-637.

[3] Hodge, H. C.; Stannard, J. N.; Hursh, J. B. (1973). *Handbook of Experimental Pharmacology*, Vol. 36, *Uranium, Plutonium, Transplutonium Elements*. New York, Springer-Verlag, 165-195.

[4] Leach, L. J.; Yuile, C. L.; Hodge, H. C.; Sylvester, G. E.; Wilson, H. B. (1973). A five-year inhalation study with natural uranium dioxide (UO_2) dust-II. Postexposure retention and biologic effects in the monkey, dog and rat. *Health Phys.* **25**, 239-258.

[5] Domingo, J. L. (2001). Reproductive and developmental toxicity of natural and depleted uranium: A review. *Reproductive Toxicology* **15**, 603-609.

[6] Bleise, A.; Danesi, P. R.; Burkart, W. (2003). Properties, use and health effects of depleted uranium (DU): A general overview. *J. Environmental Radioactivity* **64**, 93-112.

[7] Miller, A. C. (Ed.) (2007). *Depleted Uranium. Properties, Uses, and Health Consequences*. Boca Raton, NW, USA, CRC Press.

[8] McQueney, M. S.; Markham, G. D. (1995). Investigation of monovalent cation activation of S-adenosylmethionine synthethase using mutagenesis and uranyl inhibition. *J. Biol. Chem.* **270**, 18277-18284.

[9] Hamilton, M. M.; Ejnik, J. W.; Carmichael, A. J. (1997). Uranium reactions with hydrogen peroxide studied by EPR-spin trapping with DMPO. *J. Chem. Soc., Perkin Transac.* 1997, 2491-2494.

[10] Miller, A. C.; Stewart, M.; Brooks, K.; Shi, L.; Page, N. (2002). Depleted uranium-catalyzed oxidative DNA damage: Absence of significant alpha particle decay. *J. Inorg. Biochem.* **91**, 246-252.

[11] Yazzie, M.; Gamble, S. L.; Civitello, E. R.; Stearns, D. M. (2003). Uranyl acetate cause DNA single strand breaks *in vitro* in the presence of ascorbate (vitamin C). *Chem. Res. Toxicol.* **16**, 524-530.

[12] Nakajima, A.; Ueda, Y.; Endoh, N.; Tajima, K.; Makino, K. (2002). Effect of tannin compounds on metal ion-hydrogen peroxide system. In A. Kawamori, J. Yamauchi, H. Ohta (Eds.), *EPR in the 21st Century* (pp483-487), Amsterdam, Elsevier Science B.V.

[13] Nakajima, A.; Ueda, Y. (2007). Effects of hydroxyl radicals generated from the depleted uranium-hydrogen peroxide systems. *J. Radioanal. Nucl. Chem.* **272**, 251-255.

[14] Nakajima, A.; Ueda, Y.; Tajima, K.; Makino K. Effect of anti-oxidants on the depleted uranium-hydrogen peroxide system. *International Chemical Congress of Pacific Basin Societies* (Pacifichem 2005) (Dec. 16-20, Honolulu, USA).

[15] Nakajima, A.; Matsuda, E.; Ueda, Y. Effect of Polyphenol Compounds on the DNA Damage and ROS Formation by Depleted Uranium. *A Joint Conference of the International Symposium on Electron Spin Science and the 46th Annual Meeting of the Society of Electron Spin Science and Technology* (ISESS-SEST2007), Shizuoka, Japan, November 6-9 (2007).

[16] Savvin, S. B. (1961). Analytical use of Arsenazo III: Determination of thorium, zirconium, uranium and rare earth elements. *Talanta* **8**, 673-685.

[17] Motojima, K.; Yamamoto, T.; Kato, Y. (1969). 8-Quinolinol extraction and spectrophotometric determination of uranium with Arsenazo III. *Bunseki Kagaku* **18**, 208-213.

[18] Weigel, F. (1986). Uranium. In *The Chemistry of the Actinide Elements.* J. J. Katz, G. T. Seaborg, L. R. Morss (Eds.), Vol. 1, (169-442), London, New York, Chapman & Hall.

[19] Makino, K.; Hagiwara, T.; Hagi, A.; Nishi, M.; Murakami, A. (1990). Cautionary note for DMPO spin trapping in the presence of iron ion. *Biochem. Biophys. Res. Commun.* **172**, 1073-1080.

[20] Makino, K.; Hagi, A.; Ide, H.; Murakami, A.; Nishi, M. (1992). Mechanistic study on the formation of aminoxyl radicals from 5,5-dimethyl-1-pyrroline-N-oxide in Fenton systems. Characterization of key precursors giving rise to background ESR signals. *Canadian J. Chem.* **70**, 2818-2827.

[21] Mukai, K.; Nishimura, M.; Kikuchi, S. (1991). Stopped-flow Investigation of the Reaction of Vitamin C with Tocopheroxyl Radical in Aqueous Triton X-100 Micellar Solutions. The structure-activity relationship of the regeneration reaction of tocopherol by vitamin C. *J. Biol. Chem.* **266**, 274-278.

[22] Onishi, H.; Toita, Y. (1965). Separation of uranium from thorium by tri-iso-octylamine extraction and photometric determination of uranium with Arsenazo III. *Bunseki Kagaku* **14**, 1141-1145.

[23] Mizuta, Y.; Masumizu, T.; Kohno, M.; Mori, A.; Packer, L. (1997). Kinetic analysis of the Fenton reaction by ESR-spin trapping. *Biochem. Mol. Biol. Int.* **43**, 1107-1120.

[24] Sakurai, H.; Nakai, M.; Miki, T.; Tsuchiya, K.; Takada, J.; Matsushita, R. (1992). DNA cleavage by hydroxyl radicals generated in a vanadyl ion-hydrogen peroxide system. *Biochem. Biophys. Res. Commun.* **189**, 1090-1095.

[25] Mitsuta, K.; Mizuta, Y.; Kohno, M.; Hiramatsu, M.; Mori, A. (1990). The application of ESR spin-trapping technique to the evaluation of SOD-like activity of biological substances. *Bull. Chem. Soc. Japan* **63**, 187-191.

[26] Finkelstein, E.; Rosen, G. M.; Rauckman, E. J. (1980). Spin Trapping. Kinetics of the Reaction of Superoxide and Hydroxyl Radicals with Nitrones. *J. Am. Chem. Soc.* **102**, 4994-4999.

[27] Madden, K. P.; Taniguchi, H. (2001). The role of the DMPO-hydrated electron spin adduct in DMPO-OH spin trapping. *Free Rad. Biol. Med.* **30**, 1374-1380.

In: Nuclear Chemistry: New Research
Editor: Axel N. Koskinen, pp. 99-154

ISBN: 978-1-60456-957-5

Chapter 4

RAPID AND AUTOMATED ANALYTICAL TECHNOLOGIES FOR RADIOLOGICAL/NUCLEAR EMERGENCY PREPAREDNESS

***Dominic Larivière*[1,*]*, Jeffrey C. Whyte*[2]*, Weihua Zhang*[2]*, Ian Hoffman*[2]*, R. Kurt Ungar*[2]*, Sonia Johnson*[2] *and R. Jack Cornett*[2]

[1]Radioecology Laboratory, Chemistry Department, Laval University, 1045, avenue de la Médecine, Pavillon Alexandre-Vachon, Québec, QC, Canada, G1V 0A6

[2]Radiation Protection Bureau, 775 Brookfield Road, Ottawa, ON, Canada, K1A 1C1

Abstract

The possibility of a radiological/nuclear event and the consequences of its aftermath have led the scientific community to rethink its analytical approach toward environmental and biological radiological monitoring. Rapid procedures leading to precise and unequivocal results are required for timely response, whether dealing with an alpha, beta, or gamma-emitter. This chapter will highlight recent technological developments within the Radiation Protection Bureau, Health Canada (RPB/HC) to cope with the unique requirements associated with an emergency situation. Three types of radioanalyses will be covered in this chapter, ranging from long-lived alpha actinides to radiostrontium and gamma emitters. First, automated methodologies for the rapid detection of ^{89,90}Sr in milk and water will be described. This automated system is developed around a computer-controlled radiochemical procedure followed by on-line counting using stop-flow liquid scintillation. For radiostrontium (^{89,90}Sr), a detection limit of 200 mBq (800 mL sample with a counting time of 50 minutes) is achieved for each nuclide in the absence of a complementary radionuclide. The two stage radiochemical separation of radiostrontium can be performed on eight samples simultaneously and enables the separation of other radionuclides such as ^{226}Ra. Second, a rapid methodology for the on-line detection of plutonium in urine using an inductively coupled plasma mass spectrometer (ICP-MS) will be described. For ^{239,240}Pu urinalysis, a detection limit of 0.58 mBq L^{-1} (^{239}Pu) at an analytical rate of one sample every 15 minutes is possible. The analytical performances

* E-mail address: dominic.lariviere@chm.ulaval.ca. Phone: 418-656-7250. Fax : (418) 656-7916. Radioecology Laboratory, Chemistry Department, Laval University, 1045, avenue de la Médecine, Pavillon Alexandre-Vachon, Québec, QC, Canada, G1V 0A6

of these methodologies have been demonstrated for both reference and inter-comparison samples.

Finally, an automated analysis platform for rapid gamma-spectral analysis used at RPB will be described. It consists of a database server, with storage schema named Linux System for Spectral Information (LINSSI), a high purity germanium gamma spectra analysis software (UniSampo/Shaman), and a communication scripts layer. This platform is used to process daily global Comprehensive Nuclear Test Ban Treaty (CTBT) and weekly Canadian radiological monitoring air filters for airborne radioactivity analysis. Raw gamma-spectral data is emailed to our server from field laboratories for analysis and assessment with results being available in near real-time. This analytical engine can be connected to any radiation spectrometry software, either commercial or in-house. Use of database storage allows the raw data, sample parameters, and final results to serve as a comprehensive radiological information archive. The nuclide detection limit of the system is suitable for both environmental radiation monitoring and emergency preparedness purposes. The performance of this platform including its capabilities and limitations for rapid gamma-peak search, nuclide identification, and radionuclide activity concentration evaluation will be described.

Introduction

In the aftermath of an incident involving a radiological/nuclear (R/N) component, the necessity to analyze a large number of samples in a timely manner is essential for consequence management. For this purpose, analytical methodologies able to unambiguously identify the presence or absence of a specific radionuclide are required[1,2]. This is especially true in the early phase of an event involving unknown radioactive material such as those resulting from radiological terrorism[3]. Once the radiological component(s) of the contamination is known, rapid methodologies that can accurately quantify this type of contamination will play a vital role in the intermediate and late phases of the event where the number of samples to analyze for decontamination and health assessment studies could be overwhelming.

While identifying and quantifying radioactivity may seem to be trivial for an R/N laboratory, additional constraints and limitations may impact the performance of a laboratory normally focussed on routine, non-emergency analysis. The need for high-quality data in as short a time as possible, the need to operate the laboratory continuously and the number of samples to process are just some examples of what laboratories will face in the aftermath of an R/N emergency. The impact of such factors was stressed during a recent radioanalytical emergency response intercomparison exercise hosted by the National Institute of Standards and Technology (NIST). The results from the NIST Radiochemistry Intercomparison Program (NRIP) emergency preparedness exercise demonstrated that even if the activity in the samples is much higher for emergency response than for other environmental intercomparison programs (up to several hundreds of Bq/g), the degree of precision and accuracy of the results provided by the participants within 8 hours of receiving the samples was worse than for exercises with less time-constraint[4]. At post-exercise discussions with all participating laboratories[5], factors such as the insufficient number of staff and equipment, limitation in counting time available to achieve acceptable counting error, and the complexity to perform multiple simultaneous analyses have been described by the participants to explain some of the deviation observed. While it is well established that precision and accuracy might not be the most critical factors in the early phase of an event[2], large discrepancies could have significant consequences on the course of action. One possible avenue to minimize the impact of operating understaffed and underequipped radioanalytical laboratories is via the development of rapid automated radiochemistry protocols. In addition, automated radiochemical analyses can reduce the number of errors associated with the operators, which may be elevated during a stressful situation as those arising from an R/N event, and can minimize the exposure of personnel to radioactivity.

As the development of automated radiochemical procedures relies heavily on current day improvements in traditional radioanalytical chemistry, it is essential to highlight a few key points emerging from this field of research with respect to emergency preparedness. The first point is the need to increase the analytical throughput by reducing the sample preparation and/or counting time[6-8]. The latter can often be done by predetermining the action level requirement necessary for a specific matrix and radionuclide and by optimizing the counting time to be a fraction of this level. The second one is to reduce sample consumption[9,10], especially for in-vivo analyses as the amount of sample received will vary greatly based on availability during the sampling phase. The limited sample size also plays an important role in minimizing the laboratory storage requirement and time requirement for sample loading. The third point is that emergency response methodology should be suitable for a variety of matrices. This can be achieved by using radiochemical procedures that have high selectivity for the radionuclide(s) of interest such as extraction chromatography[11]. Finally, reusability of the system components should be evaluated in order to maximize the use of supplies which might be scarce or hard to obtain in an emergency[7,8,12]. Reusability also limits the quantity of waste generated by the analysis.

Table 1. Radioactive properties of radioisotopes of importance in nuclear emergencies[86].

Element	Isotope	Half-life	Decay mode
Americium	Am-241	432.2 y	α, γ
Cesium	Cs-137	30.17 y	β, γ
Cobalt	Co-60	5.27 y	β, γ
Iodine	I-131	8.04 d	β, γ
Iridium	Ir-192	74.02 d	β, γ
	Pu-238	87.75 y	α
Plutonium	Pu-239	24 100 y	α
	Pu-240	6 560 y	α
Polonium	Po-210	138.38 d	α
Strontium	Sr-90	28.6 y	β
Uranium	U-235	7.03×10^8 y	α
	U-238	4.47×10^9 y	α

Many radiochemical separation and on-line detection systems, with various levels of automation, have been presented and demonstrated over the years. Several thorough reviews[13,14] and book chapters[15-17] have been published with the goal of highlighting the progress made in the field of radioanalytical automation. Therefore, this chapter will not cover the applications discussed in those publications. Nonetheless, many of the methodologies developed with the purpose of assessing nuclear wastes or long-term radiological contamination from nuclear activities have the potential to be implemented for emergency radioanalytical monitoring after the proper re-optimization. Another challenge faced by radiochemists involved in the development of automated radioanalytical procedures for emergency preparedness resides in the number of key potential radionuclides that may require analysis an R/N emergency situation (Table 1). This list, while not exhaustive, stresses the complexity of emergency preparedness, especially if a unique method is required per sample type collected (*e.g.,* air filters, foodstuffs, in-vivo biological samples). Since a significant portion of the radionuclides present in Table 1 are either actinides or gamma-emitters, they represent an excellent research opportunity for radiochemical and data analysis automation. Radiostronitum, the only pure β-emitter in the table, is also of prime interest as its rapid detection is very challenging.

The requirements associated with the development of automated radiochemical systems are largely dictated by counting impediments such as low counting efficiency and/or spectral and non-spectral interference, independent of the detection system used[7,8,18-20]. Consequently, the vast majority of automated radiochemical separations are designed around the concept of minimizing the presence of spectral interferences and improving detection via sample preconcentration. In addition, a portion of the automation development should be dedicated towards simplifying quantification and reporting processes, as data quality objectives (DQOs) would likely change during the various phases of the event[2].

This chapter will highlight the development and performance of three independent automated radioanalytical procedures designed at the Radiation Protection Bureau, Health Canada (RPB/HC) to facilitate the monitoring of population, foodstuffs, and the environment in the aftermath of an R/N event. First, a two-stage automated pre-concentration/separation procedure for the determination of radiostrontium in foodstuffs (*i.e.,* milk and drinking water) by liquid scintillation counting will be presented (*Section 1*). Then an automated methodology for the detection and quantification of ^{239,240}Pu in biological fluid based on mass spectrometric analysis will be described (*Section 2*). Finally, the performance of an automated data analysis sequence for the identification and quantification of γ-emitters will be demonstrated (*Section 3*). Applications of each methodology for an emergency type situation will also be highlighted.

Section 1. Simultaneous Determination of ^{89}Sr and ^{90}Sr in Foodstuffs for Emergency Response

Introduction

Strontium-90 is one of the most radiotoxic components of radioactive fallout[21]. Produced in high yield by the fission of ^{235}U and ^{239}Pu, ^{90}Sr has been dispersed worldwide in the atmosphere by nuclear testing and by accidental releases from nuclear reactors. The high radiotoxicity of ^{90}Sr is the combination of two factors: i) the combined result of its long radiological ($t_{1/2}$ = 28.5y) and biological ($t_{1/2}(B)$ = 13.2y)[22] half-lives and ii) its chemical similarity to other alkaline earth metals, including essential minerals such as calcium and magnesium. This similarity facilitates its deposition in calcium-rich biological structures, including bones and teeth[23], which are subsequently irradiated for years by the high-energy β-particles generated by the decay of ^{90}Sr(E_{max} = 0.554MeV) to ^{90}Y ($t_{1/2}$ = 64h, E_{max} = 2.27MeV).

The determination of ^{90}Sr shortly after the release of radioactive contamination proved to be challenging as its detection is obscured by another radiostrontium isotope (^{89}Sr; $t_{1/2}$ = 50.5d, E_{max} = 1.49MeV) which interferes with radiometric measurements. The results obtained from environmental samples collected after Chinese nuclear tests and the Chernobyl accident have demonstrated that the ^{89}Sr:^{90}Sr activity ratio can vary greatly, ranging from 0.4 to 30, depending on reactor inventories and the time lapsed after the event[24,25]. Since both radiostrontium isotopes cannot be isolated via traditional chemical separation, quantitative discrimination will depend greatly on the characteristics of the analytical instrumentation chosen for the detection. Brun *et al.*[26] have reviewed the various radiometric and mass spectrometric instrumentation available for the detection of radiostrontium in milk and have concluded that liquid scintillation counting (LSC) was by far the most commonly used. Other instrumentation such as gas proportional counting (GPC)[27], Cerenkov counting[28-32], surface barrier detector (SBD)[33,34] and mass spectrometry (MS)[35-37] have been used, to a lesser extent, for the detection of radiostrontium. Unfortunately, not all instrumentation mentioned above can be used for discrete radiostrontium analyses. For example, GPC cannot discriminate between ^{89}Sr, ^{90}Sr and ^{90}Y, limiting its usefulness to a single measurement of gross radiostrontium activity for radioanalytical emergency situations where all radioisotopes would be present. Mass spectrometry, the preferred method for the rapid analysis of many long-lived radionuclides, has great potential for ^{90}Sr as demonstrated by many research groups[19,35,38-40]. However, the detection of ^{90}Sr by mass spectrometry is plagued by issues such as high ^{89}Sr detection limits due to its short half-life, and peak tailing of the most abundant stable strontium isotope (^{88}Sr) at mass-to-charge ratios equal to 89 and 90[19,40]. Cerenkov counting of ^{90}Sr is possible via the ingrowth of ^{90}Y which achieves secular equilibrium after approximately 14 days. This approach for ^{90}Sr detection is however not suited for emergency response due to the required ingrowth period and the need to count the sample twice to assess ^{89}Sr and ^{90}Y contributions to the Cerenkov signal. In addition, this methodology reaches its analytical limit at high ^{89}Sr:^{90}Sr activity ratios where the count rate of ^{89}Sr overshadows the ^{90}Y count rate even after the proper ingrowth period.

LSC has demonstrated great promise for detecting and quantifying radiostrontium, given the different β-spectral distribution of each of the radioisotopes. In addition, LSC has a high counting efficiency which enables shorter counting time per sample for an equivalent activity and background signal, a sought after feature for radioanalytical emergency response methodology[1]. However, in order to use LSC, radiostrontium must be extracted from the sample matrix and separated from possible spectrally-interfering β-emitters such as ^{40}K, ^{87}Rb, ^{134}Cs, ^{137}Cs, and ^{140}Ba. For these reasons, many strontium separation and pre-concentration schemes have been developed for a variety of industrial and environmental matrices using methods encompassing co-precipitation, ion exchange and extraction chromatography[26]. The classical approach for the radiochemical separation and pre-concentration of radiostrontium in milk is based on the greater solubility of calcium nitrate in fuming nitric acid[41]. While effective to separate radiostrontium from most of the β-emitters, the fuming HNO_3 procedure is time-consuming (*i.e.* 23 steps in 7 days), leads to inconsistent chemical recovery yields, and cannot adequately separate Ba from Sr[41,42]. The development and use of procedures based on ion and extraction chromatography for radiostrontium has enabled better and more consistent yields while providing greater selectivity[26], resulting in a lower minimum detectable activity (MDA). In addition, the fact that the separation, pre-concentration, and elution of $^{89,90}Sr$ occur in solution facilitate its coupling with liquid scintillation, especially since the development of flow scintillation analyzers (FSA)[14,43-45]. These methods are regarded as extremely valuable for emergency response as they provide fast and reliable results as well as less cumbersome protocols to follow for non-radiochemists performing the measurements.

This section will highlight the development and analytical performance of a dual stage automated procedure followed by on-line detection of radiostrontium in water and milk, which could be employed during radiological/nuclear emergencies involving ^{89}Sr and ^{90}Sr.

Experimental Section

Automated radiostrontium system. Diagrams of the two sequential separation schemes, hereafter referred to as the primary separation using strong cation exchange (SCX) chromatography and the secondary separation using high performance ion chromatography (HPIC), are shown in Figures 1 and 2, respectively. The SCX system, developed to enable the separation and pre-concentration of 8 samples simultaneously, was automated using Lookout process control software (National Instruments (NI), Houston, TX), with its corresponding data acquisition card (AT 232-4) and FieldPoint relays. Wash solution [WS] selection was performed using a modular valve positioner [MVP] (Hamilton Company, Reno, NV). The wash solutions were distributed to the eight columns [C], each packed with 16mL of Dowex 50W-X8 resin, using a nine port polyether ether ketone (PEEK) manifold flow splitter [SPL] (Alltech, Deerfield, IL). Selection of either sample [S] or wash solution [WS] was performed through a bank of electrically actuated three-way solenoid valves [SV1] (Takasago, Nagoya, Japan). In order to isolate the columns from the system, a by-pass system [BP] was designed using 2 three-way stopcocks [SC-1 and SC-2] (Cole-Parmer, Vernon Hills, IL). Connections between the stopcocks and the 20mL polypropylene (PP) columns (Evergreen Scientific, Los Angeles, CA) were made by fitting the column inlet screw cap with a stainless steel bulkhead hose-barb fitting (Small Parts Inc., Miami Lakes, FL) backed with an O-ring. An IPC-24 multichannel peristaltic pump [MCPP] (Ismatec, Vernon Hills, IL) was used to pull liquids through the system. The second bank of solenoid valves [SV2] allows the user to divert the flow from the column outlets either to waste, or to quartz collection beakers [QCB]. Quartz beakers (Quartz Scientific Inc., Fairport Harbor, OH) were used to collect fractions from the manifold for the subsequent boil down and ashing steps. 1.52mm i.d. Tygon (Cole-Parmer, Vernon Hills, IL) transfer lines were employed throughout the column pre-treatment manifold. Where lines were immersed in wash solutions [WS] or sample [S] reservoirs, polytetrafluoroethylene (PTFE) (Upchurch Scientific, Oak Harbor, WA) was employed.

The HPIC separation was performed using an HPLC chromatography system (Waters, Milford, MA) composed of a fluid handling unit (model 600 EF, maximum flow rate 45mL min^{-1}), a controller (model 600) and an autosampler (model 717+), coupled with a conductivity detector (Pharmacia LBK, Uppsula, Sweden) and a Flow Scintillation Analyzer (Packard 505TR, Downer's Grove, IL). The system was controlled using a BUS/LACE card and SATIN module using Waters Empower chromatography software. The conductivity detector was used to determine stable strontium carrier yields. The flow scintillation analyzer (FSA) used for the simultaneous counting of radiostrontium in 2 windows (Low Energy, High Energy) was purchased with an integrated scintillant pump, electronic diverter valves, additional lead shielding, a large volume (5mL) flow cell, and the spectrum analysis software package with multichannel analyzer (MCA) operating under the Packard Flo-One for Windows software. The detector employs a linear amplifier for the energy scale. A 250 μL cartridge static mixer (Analytical Scientific Instruments, El Sobrante, CA) was placed prior to the detector cell to improve the sample-to-scintillant mixing.

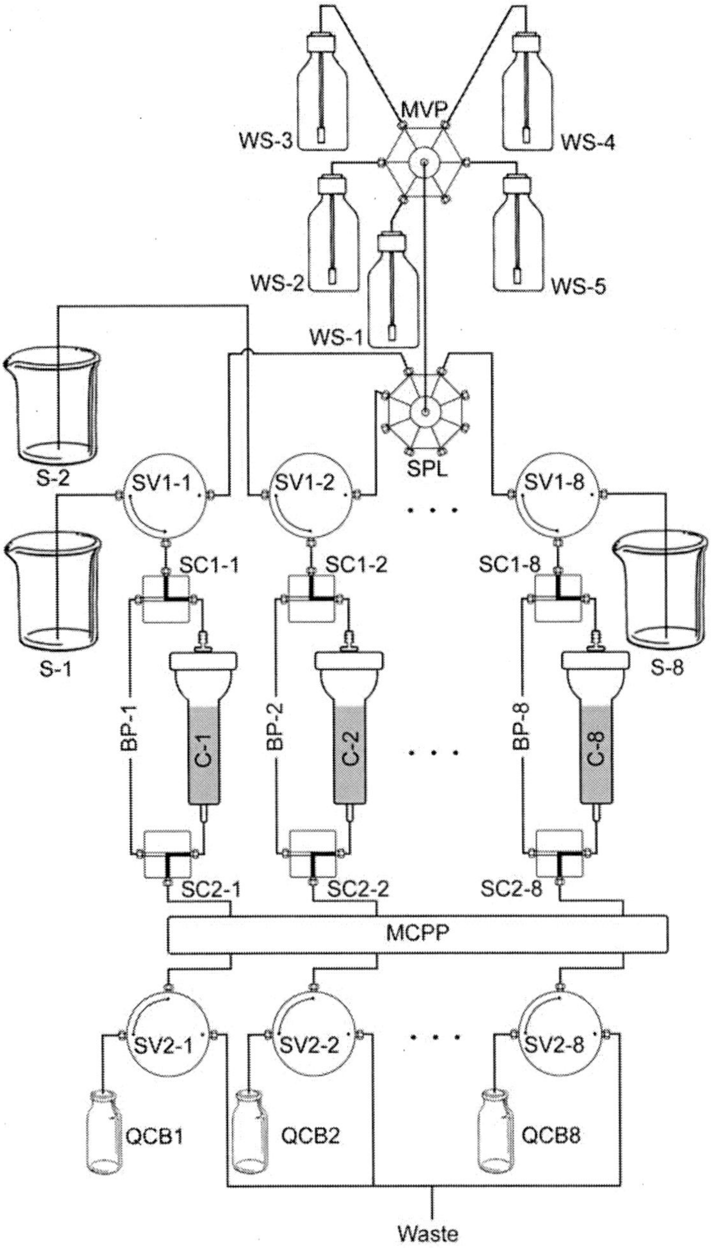

Figure 1. Schematic of the automated strong cation exchange (SCX) manifold for the primary separation. WS: Wash solution, MVP: Modular valve positioner, SPL: 8-port splitter, S: sample, SV1: loading 3-port switching valve, C: Dowex 50X-8 column, MCPP: Multi-channel peristaltic pump, SV2: loading 3-port switching valve, BP: By-pass, QCB: Quartz collection beaker.

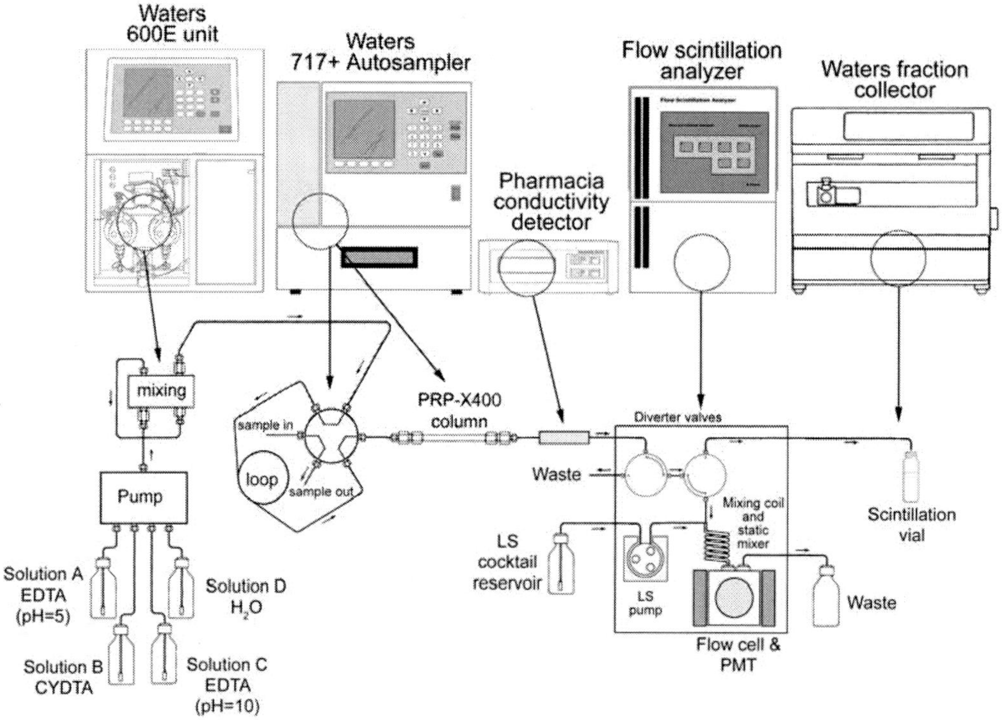

Figure 2. Schematic of the system for the secondary separation by high performance ion chromatography (HPIC) and stop flow counting of radiostrontium.

Primary separation using SCX chromatography. The separation scheme used for SCX chromatography utilizes a pH gradient of diaminocyclohexanetetraacetic acid (CyDTA) and ethylenediaminetetraacetic acid (EDTA) and is a modified version of protocols proposed by Wade and Seim[46], Povondra *et.al.*[47], and Alfaro *et. al.*[48]. First, samples are loaded into pre-conditioned (0.062M $(NH_4)_2EDTA$ at pH=5) 50W-X8 columns at a flow rate of 3.8mL min^{-1} (step 1, Table 2). Following sample loading, columns are washed with 10 column volumes (160mL) of a mixture of CyDTA and acetic acid (step 2), followed by 10 column volumes (160mL) NH_4Cl (step 3). Steps 2 and 3 enable to wash away residual alkaline earth and alkaline ions (Table 2), respectively, improving the purity of the radiostrontium fraction. Radiostrontium is eluted with 125mL $(NH_4)_2EDTA$ at pH=6 (step 4) into a 500 mL quartz beaker. The collected fraction is then boiled down to dryness and ashed at 525°C overnight to remove organics. The ash is dissolved in 2M HCl, transferred to a 50mL quartz beaker, boiled dry, and taken up in a small volume (approximately 2.1mL) of 0.32M CyDTA at pH equal to 5.2. Step 5 of the SCX separation enables the collection of radium and radiobarium fraction, while step 6 is designed to re-equilibrate the resin.

Table 2. Operating sequence of the FI system for the automated separation of ^{89}Sr and ^{90}Sr.

Steps	Description	Solvent[a]	pH	Flow rate ($mL\ min^{-1}$)	Volume (mL)	Elements eluted	Code
Primary separation using SCX							
1	Sample loading	$(NH_4)_2EDTA$ (0.062)	4.8	3.8	800	Ca, Y, Na, K	S
2	Matrix removal-1	CyDTA (0.060) / AcOH (0.040)	5	3.8	160	Mg	W1
3	Matrix removal-2	NH_4Cl (0.374)	5	3.8	160	Cs, Rb	W2
4	Sr elution-1[b]	$(NH_4)_2EDTA$ (0.062)	6	3.8	150	Sr	W3
5	Ra elution	$(NH_4)_2EDTA$ (0.062)	10	3.8	125	Ba, Ra	W4
6	Reequilibration	$(NH_4)_2EDTA$ (0.062)	5	3.8	125		W5
Secondary separation using HPIC							
7	Sample loading	CyDTA (0.32) / AcOH (0.48)	5	2	2		
8	Matrix removal-3[c]	CyDTA (0.04) / AcOH (0.06)	5	2	36		
		$(NH_4)_2EDTA$ (0.062)	5	2	30		
9	Sr elution-2	$(NH_4)_2EDTA$ (0.062)	10	2 - 0.7	16		
10	Column reequilibration	CyDTA (0.04) / AcOH (0.06)	5	2	34		

a. molarity (M) represented in parentheses

b. Fractions collected into a quartz beaker, evaporated to dryness and ashed at 525°C. The residue is subsequently re-dissolved in 2M HCl, re-evaporated to dryness and re-dissolved to be used in step 7.

c. This step is a gradient elution

Secondary Separation using HPIC and radiostrontium detection. The re-dissolved samples resulting from the primary separation are filtered into a conical bottomed 3mL polypropylene (PP) autosampler vial using a polypropylene transfer pipette (Samco Scientific, San Fernando, CA) and a 3mL disposable plastic syringe (Becton Dickinson, Franklin Lakes, NJ) equipped with low hold-up volume (<20μL) hollow fibre syringe filter on the outlet (Dynagard-Spectrum Laboratories Inc., Rancho Dominguez, CA). The filtered samples are injected sequentially into the HPIC using the autosampler (step 7), followed by a CyDTA and EDTA gradient at pH=5 (step 8) to further remove any of the alkaline earth metals remaining from the first SCX separation. Radiostrontium is then eluted in a minimal volume using EDTA at pH=10 which will be used for stop flow counting (step 9). The flow used during step 9 is gradually decreased from 2mL min^{-1} down to 0.7mL min^{-1} to minimize the broadening of the Sr peak and enable its collection into the flow cell.

Chemical yield for radiostrontium is determined via integration of the peak area of the strontium carrier using an on-line conductivity detector positioned prior to the FSA. The eluant conductivity drops as the Sr fraction elutes from the column during the pH=10 step of the gradient. This dip can be integrated tangent-to-tangent, or alternatively, a blank gradient can be run and this chromatogram can be subtracted to yield the more familiar chromatographic peak. Sample vials are weighed prior to and after sample injection, in order to determine the fraction of sample injected.

Once the Sr fraction from the HPIC column reaches the flow scintillation analyzer (FSA) inlet, it is mixed with the Optiphase Supermix scintillation cocktail (Perkin-Elmer, Waltham, MA) at a ratio of 3:2 (cocktail: eluant) via the on-board scintillant pump (LS pump, Figure 2). The mobile phase and cocktail meet initially at a mixing tee, then flow through a mixing coil, followed finally by the static mixer. When the Sr elution peak reaches the cell, the first electronically controlled diverter valve switches toward waste and the LS pump is stopped, isolating the peak within the 5mL detection cell. This is possible by the use of time-triggered events controlled by the Flo-One software. After the flow cell is loaded, the FSA is used as a static counter. Once the end of the chromatographic run is completed, the resin is re-equilibrated (Step 10) and the next sample injection and separation occurs while the first sample is being counted in the FSA flow cell under static conditions.

A 50 minute count time is employed to maintain sample throughput, and the two radionuclides (^{89}Sr:^{90}Sr) are determined simultaneously using two counting windows, namely 2-125 keV for ^{90}Sr, and 135-500 keV for ^{89}Sr. The FSA flow cell is well purged (48 mL) after each stop flow count to minimize possible sample memory effects.

Reagents and materials. A Dowex 50W-X8 strong cation exchange resin, 16 mL of resin, 200-400 mesh, (Sigma-Aldrich Canada, Oakville, ON), and a PRP-X400 poly(styrene–divinylbenzene)-sulfonate cation-exchange (7μm) column (Hamilton Company, Reno, NV) heated at 40°C were employed in the first and second separation stages, respectively. Ethylenediaminetetraacetic acid (EDTA) diammonium salt (≥99.0%), cyclohexanediamine-N,N,N,N'-tetraacetic acid (CyDTA) monohydrate (≥99.0%), and ammonium chloride(≥99.0%), obtained from Fluka (Oakville, ON), were used to produce WS-1 to -5. Glacial acetic acid (AcOH), HPLC grade, concentrated hydrochloric acid (trace metal grade), and ammonium hydroxide 50% v/v, were obtained from Fisher Scientific (Ottawa, ON). Optiphase Supermix biodegradable liquid scintillation cocktail (Wallac Company, Turku, Finland), was used as scintillant throughout the experiment. Stock solutions of stable strontium for chemical yield were prepared from A.C.S. reagent grade salts (J. T. Baker, Phillipsburg, NJ). NIST traceable ^{89}Sr and ^{90}Sr standards (Isotrak-QSA Global, Burlington, MA) were used to prepare quality control spike samples and to determine the FSA detector efficiency. The eluants used for the secondary separation are continually degassed with a Helium sparge.

Milk and water sample preparation for SCX chromatography. Although the automated methodology for radiostrontium is designed to accommodate any aqueous sample, a different sample preparation is required for water and milk. For water samples, 16g of $(NH_4)_2EDTA$ and 5mg of stable strontium (carrier) are added to 800mL of sample. The pH of the solution is then adjusted to 4.8 using a few drops of 50% (v/v) NH_4OH. The sample is then transferred into a conical HPLC bottle to facilitate its complete loading into the resin. Milk sample preparation is performed in multiple stages. First, 5 mg of stable strontium is incorporated into 1L of milk to serve as a yield tracer. The sample is then nebulized at a flow rate of 120-150mL h^{-1} using a peristaltic pump into a 1.5L quartz beaker heated at 400-500°C, similar to the approach proposed by Abbadi *et al.*[49]. This operation, which requires approximately 6-7 hours, enables the evaporation of the milk water content. The beaker containing the milk residue is then placed into a muffle furnace, set at 525°C, and ashed overnight to remove organic material. Upon cooling and ash grinding, wet ashing using 50-75mL of 4M HNO_3 is performed. Following a short reflux (30 min), the sample is evaporated to dryness and dry ashed (2h). Wet ashing is repeated using 50mL of 2M HCl, followed by evaporation near dryness. Finally, 800mL of 2% (m/v) of $(NH_4)_2EDTA$ is added to the sample and the pH of the solution adjusted to 4.8 using 50% (v/v) NH_4OH.

Results and Discussion

Chromatographic separation performance. Optimal column sizing for the primary separation using SCX chromatography was estimated using the average mineral composition of 1L of milk. A 1L milk ash sample contains approximately 1200 (60meq), 120 (9.9meq), 1500 (38.4meq), and 500mg (21.7meq) of calcium, magnesium, potassium, and sodium, respectively[26]. As Ca is effectively complexed in EDTA[50] at pH equal to 4.8, it does not interact with the resin, and can be ignored in a calculation of required column capacity (assuming a sufficient amount of EDTA is present). A 16mL column (27meq) is sufficient to quantitatively retain strontium even though its total capacity is lower than what is required (70meq). This is due essentially to the fact that AG50W-X8 resin has a much higher relative selectivity for strontium (4.95) than for potassium (2.5) and sodium (1.5)[51]. Thus, even though there is a huge excess of monovalent cations present in the sample, 5mg of strontium (0.057meq) added as a carrier is completely adsorbed by the resin under the conditions of sample loading presented (results not shown). Decontamination factors (DFs), as defined in the IUPAC Compendium of Chemical Terminology[52], were experimentally determined to be greater than $1.2x10^4$ and $2.8 \ x10^3$ for Cs and Ba, respectively, after the primary SCX separation. High DF for those two elements are required for LSC analysis, as their respective radioisotopes (*i.e.* ^{137}Cs and ^{140}Ba) spectrally interfere with the radiostrontium spectral signature and are known to be present in fallout contamination[26]. DF greater than 10^5 and 10^4, respectively, were achieved for the overall process (combined primary and secondary separation). These factors compare favourably with typical DF for most elements using the fuming HNO_3 methodology (*e.g.* $DF_{Cs} = 10^5$-10^7)[41]. However, the approach presented here is far more effective for removing alkaline earth metals (*i.e.* Ca, Ba)[41].

As calibration over the 1-20mg strontium carrier range led to linear detector response (results not shown), a simple ratio of peak areas is sufficient to determine the strontium yield. The carrier level adopted, 5mg of stable strontium, was found to be a compromise between the desire to minimize peak volume by having a low carrier level (and hence maximize overall FSA detection efficiency) in the secondary separation and the requirement to keep the carrier level high enough to dwarf the effects of native Sr (milk samples typically contain a maximum of 0.4mg L^{-1} of native Sr)[53] for yield calculation purposes in environmental samples. In order to assess the ruggedness of the HPIC/FSA stop flow method, a pure ^{89}Sr standard (43.70Bq) was separated and analyzed at a carrier level of 4mg (simulating a 80% chemical yield). The sample activity calculated from the experimental result was within 0.2% of the expected activity, indicating that detection efficiency does not vary with chemical yield.

As mentioned previously; ^{90}Sr decays into ^{90}Y, which interferes with radiostrontium LSC measurements. Therefore, it is crucial to investigate the behaviour of ^{90}Y during the dual stage separation procedure. The primary separation leads to a complete removal of yttrium (step 1, Table 2), therefore the main concern remains the ^{90}Y ingrowth following the primary separation. As yttrium is unretained by the PXP-400 resin due to the strength of its metal-chelate complex at the pH chosen[48], it does not pose any issue for on-line radiostrontium analyses. The average ingrowth of ^{90}Y (at midpoint of the stop flow counting interval) is approximately 0.45%, and is therefore insignificant.

For 31 intercomparison and spiked samples analyzed, the reproducibility of chemical yield is excellent ($\overline{X}_{yield} = 100.8\%$, $\sigma=2.1\%$). These chemical recovery yields are far better than those obtained by nitrate co-precipitation ($\overline{X}_{yield} = 61\%$)[41] and extraction chromatography for milk ($\overline{X}_{yield} = 60\%$)[27]. The retention time of the HPIC radiostrontium peak was found to be quite reproducible; a coefficient of variation of 0.1% in the retention time was obtained for 10 injections.

Previous published methodology[48] had performed a similar separation (only EDTA pH gradient) using a two-column HPIC system. The first large column gives an initial Sr enrichment, after which the sample is directed to a second smaller column for separation via a switching valve. The analytical throughput of the automated procedure developed by Alfaro *et al.*[48] is 6 samples per day. This in-series approach limits sample throughput since the HPIC system is entirely committed to a single sample during enrichment and separation steps. The approach chosen by RPB/HC, consisting of a two-step separation based on the SCX chromatographic separation using a manifold in conjunction with a secondary separation by HPIC, overcomes this limitation of the system developed by Alfaro *et al.*[48]. By running the primary separation in parallel (8 samples simultaneously) on the automated ion exchange manifold, multiple samples can be readied for the final separation by HPIC and the subsequent FSA detection. Since the primary separation requires approximately 8 hours while the secondary one lasts about 1 hour, much higher sample throughput can be achieved. If the automated system was to be used without interruption, a sample throughput of 24 samples per two days is achievable. This significant increase in the sample throughput encourages the application of this automated system during R/N emergencies involving radiostrontium isotopes.

On-line flow scintillation analyzer (FSA). Shifting of the beta particle spectral continuum due to variability in the quench can lead to variability in the proportion of sample counts registered (counting efficiency) in the chosen energy window[54]. Therefore, it is necessary to consider how changes in matrix and eluant composition might affect the performance of the FSA. Consistency in quench of radiostrontium fractions (step 9, Table 2) after sequential separation of environmental samples was determined off-line using a Packard 2770TR liquid scintillation counter. Fractions collected were analyzed, after addition of an Instagel liquid scintillation cocktail, to determine the quench parameter (Table 3). It is evident from these results that the strontium fraction is highly quenched ($tSIE_{average}$=270.2), but that the quench level is quite constant (relative standard deviation (RSD)=1.03%). Therefore, quenching can be excluded as a variable from the efficiency calculation, eliminating the need to perform a quenching curve. This finding is important for emergency preparedness as the production of such curve is time-consuming.

Table 3. Variation in quenching for solution collected from the HPIC System (6.3 and 10 mL volume for the radiostrontium fraction and the Optiphase Supermix cocktail, respectively)

Sample	^{90}Sr acitivity (Bq L^{-1})	Transformed Spectral Index of External Standard (tSIE)
EML-QAP 9709	0.64	267.49
Water	10.0	273.08
Water	1.0	270.04

EML-QAP – Environmental Monitoring Laboratory – Quality Assurance Program Intercomparison sample.

Table 4. Miscibility of various liquid scintillation cocktails with a solution for 2% (v/v) $(NH_4)_2$EDTA at pH =10.

Cocktail Name (supplier)	Cocktail : Mobile Phase Ratio	
	1:2	1:1
Ultima Flo AF (Perkin-Elmer)	M	I
Ultima Flo M (Perkin-Elmer)	I	NA
Ultima Flo AP (Perkin-Elmer)	I	NA
Hionic Fluor (Packard Bioscience)	I	NA
Ultima Gold AB (Perkin-Elmer)	I	NA
Ultima Gold LLT(Perkin-Elmer)	I	NA
Ultima Gold XR(Perkin-Elmer)	I	NA
Scintisafe Plus 50% (Fisher Scientific)	M	I
Optiphase Supermix (Wallac)	M	M

M = Miscible, I = Immiscible, NA = Not Analyzed

In order to attain low minimum detectable activities (MDA) in the on-line flow system, a cocktail suitable for chromatographic flow and miscible in comparable proportions with the

HPIC eluant is desirable[55]. Nine LSC cocktails were evaluated (Table 4), but most were abandoned since they did not meet our desired miscibility ratio of 1:2 (cocktail:mobile phase, C:MP). Optiphase Supermix was the only cocktail tested that was miscible at the lower C:MP ratio of 1:1. However, Optiphase Supermix is not a flow cocktail. At the optimum HPIC flow condition for minimal strontium peak width (0.7mL min^{-1}), a ratio of approximately 3:2 (C:MP) was the leanest ratio practically attainable since 1mL min^{-1} is the lowest flow rate that the scintillant pump will deliver under factory set conditions.

A series of viscosity determinations were made using an Ostwald capillary viscometer mounted in a temperature controlled water bath. Measurements of $(NH_4)_2EDTA$, Optiphase Supermix cocktail, and a mixture with a 3:2 (C:MP) ratio were made. Reynolds numbers (k) were then computed based on the following equation:

$$k = \frac{\rho v_s D}{\eta} \tag{1.1}$$

where ρ is the solution density, v_s is the mean fluid velocity, D is the tubing diameter, and η is the viscosity. Given representative tubing diameters and the measured viscosity listed in Table 5, it was possible to ascertain the nature of the flow profile in various mixing zones. Values (k) calculated was below 10 for all tubing diameters tested for a solution with 3:2 (C:MP) ratio at room temperature and with a total flow rate equal to 2mL min^{-1}. These values are well below the critical values ($k \approx 2200$)[56] implying that poor mixing will be occurring as the sample flows through the FSA unit. Despite the higher viscosity of Optiphase Supermix relative to flow cocktails, it is not expected that a turbulent flow regime could be achieved with any cocktail given the flow rates and tubing diameters employed in this separation. To overcome such an issue, an additional mixing element (static mixer) was incorporated after the mixing tee and coil arrangement in the flow system to facilitate the mixing of the scintillant and the mobile phase.

Table 5. Impact of the solution temperature on the solution computed viscosity (η in cP).

Fluid	Temperature (°C)				
	15	**20**	**25**	**30**	**40**
Optiphase Supermix (OS)	81.2	60.6	45.7	34.8	20.7
2% $(NH_4)_2EDTA$ [pH = 10]	1.5	1.3	1.1	1.0	0.78
3:2 (OS: $(NH_4)_2EDTA$ [pH=10])	64.8	51.6	41.4	33.4	22.3

Flow counting. In flow scintillation analysis, the sample can be counted in two different modes: i) normal continuous flow and ii) stopped-flow. The latter is generally preferred as it produces the highest count rate of the two approaches mentioned[57]. As stated by Grate and Egorov[13] in their article describing the theory of stopped-flow detection, achieving the highest fraction of the sample zone collected in the detector cell (D_m) is critical to achieve the maximum count rate. Even with the system constraints mentioned previously such as the cocktail:mobile phase ratio dictated by the on-board scintillant pump, it was possible to achieve a D_m of 0.8. Another important parameter to improve stopped-flow detection is the sample residence time (t_R)[57]. Longer t_R will lead to better MDA. The residence time in the flow cell was set to 50 minutes, almost the entire time required to complete the secondary separation of Sr (57min). This t_R enabled to maintain an acceptable sample throughput (0.81 sample h^{-1}, if the 17min for re-equilibration is included) while obtaining a reasonable MDA for each radiostrontium isotope (0.16 and 0.2Bq L^{-1} for ^{89}Sr and ^{90}Sr, respectively).

The use of the time resolved LSC (TRLSC) noise rejection, another feature of flow scintillation analysis[57], was also tested but did not lead to improved detection limits when coupled with the 5mL TRLSC cell (a scintillator material is impregnated in the cell windows) purchased with the detector.

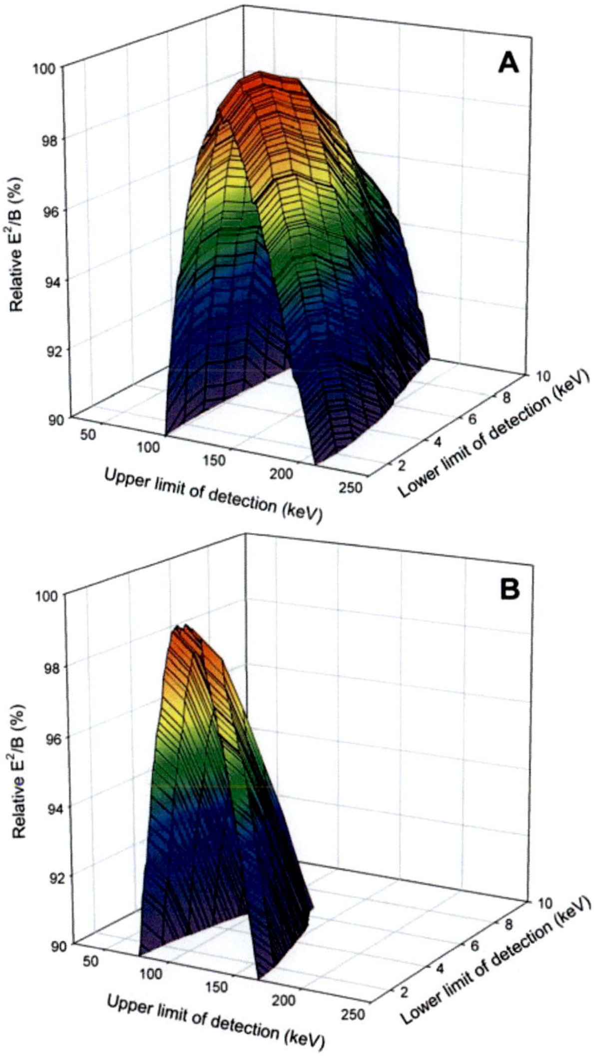

Figure 3. Optimization of the method Figures of merit (E^2/B (%)) for a ^{90}Sr solution containing A) no ^{89}Sr and B) 50-500 Bq of ^{89}Sr.

Optimization of figures of merit for stop flow radiostrontium analysis. In liquid scintillation counting, a widely accepted figure of merit (FOM) used to assess sensitivity of detection, is the term E^2/B, where E is the detection efficiency (usually expressed in percent), and B is the background, in counts per minute (CPM)[58]. Varying the counting region of the lower discriminator limit (LDL) and upper discriminator limit (UDL) settings will govern the counting efficiency and background for any particular radionuclide to be measured and flow cell used[57]. The relative E^2/B for each radionuclide for each pair of lower discriminator level (LDL) and upper discriminator level (UDL) values was computed. Based on a certain background of the other radionuclide, the E^2/B was obtained by applying the desired scalar to the respective pure spectrum net counts. Once the highest E^2/B value is located in the matrix, it was assigned a value of 100%. All other values are then normalized to this maximum. Subsequently, a 3-D surface plot can be constructed of LDL, UDL, and relative E^2/B, whereby the optimum of the two discriminator settings can be evaluated (Figure 3). The goal is to pick lower and upper window boundaries where the FOM is high, and also where it does not change quickly with any slight shift (due to instrument response or quench) in LDL/UDL. ^{89}Sr was optimized for a 5Bq ^{90}Sr background, and ^{90}Sr was optimized for a 50-500 Bq background of ^{89}Sr, to respect the emergency response requirements of the method. Figure 3 illustrates the results of the relative E^2/B window optimizations for ^{90}Sr at two levels of ^{89}Sr. Note that the relative E^2/B values for ^{90}Sr in the 2-125 keV range are high and do not change drastically as the background of ^{89}Sr increases from 0 to 500 Bq. The optimum ^{90}Sr window was set at 2-125 keV (low energy region, LE), and analogously, the optimum ^{89}Sr window was determined to be 135-500 keV (high energy region, HE). The spectra of pure ^{89}Sr, pure ^{90}Sr, and a mixture of the two with these windows are presented in Figure 4. Note that the counting windows chosen were a compromise between the desire to achieve low detection limits for each radionuclide, and the need to maximize the accuracy of determining mixtures of the two.

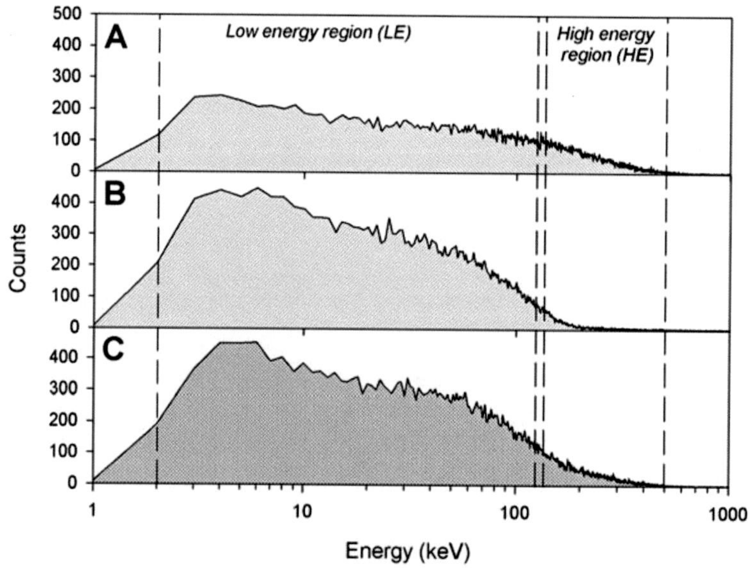

Figure 4. Stop-Flow Liquid Scintillation Counting Spectra (count time = 50 min) of A) ^{89}Sr (10.8 Bq), B) ^{90}Sr (9.9 Bq) and C) a mixture (^{89}Sr(26.1 Bq) and ^{90}Sr (24.8 Bq)).

Calculation of the respective radiostrontium activities. Based on the 2 counting windows, calibration is possible using the contribution of each radionuclide to each counting window. By determining the 4 efficiency values from injections of pure ^{89}Sr and pure ^{90}Sr, the following two equations can be solved to yield the respective activities of each radiostrontium isotope (A_{89} and A_{90} for ^{89}Sr and ^{90}Sr, respectively):

$$\text{Net CPS}_{LE} = \varepsilon_{89,LE} \times A_{89} + \varepsilon_{90,LE} \times A_{90} \tag{1.2}$$

$$\text{Net CPS}_{HE} = \varepsilon_{89,HE} \times A_{89} + \varepsilon_{90,HE} \times A_{90} \tag{1.3}$$

where Net CPS_{LE} and Net CPS_{HE} are the net counts per second in the low energy (2-125 keV) and high energy (135-500 keV) windows, $\varepsilon_{89,LE}$ and $\varepsilon_{89,HE}$ are the counting efficiencies for ^{89}Sr in the low and high energy windows, and $\varepsilon_{90,LE}$ and $\varepsilon_{90,HE}$ are the counting efficiencies for ^{90}Sr in the low and high energy windows. Solving the previous equation to determine the activity of the each radiostronium isotope leads to:

$$A_{90} = \frac{\varepsilon_{89,HE} \times \text{Net CPS}_{LE} - \varepsilon_{89,LE} \times \text{Net CPS}_{HE}}{\varepsilon_{89,HE} \times \varepsilon_{90,LE} - \varepsilon_{89,LE} \times \varepsilon_{90,HE}} \tag{1.4}$$

and

$$A_{89} = \frac{\varepsilon_{90,LE} \times \text{Net CPS}_{HE} - \varepsilon_{90,HE} \times \text{Net CPS}_{LE}}{\varepsilon_{90,LE} \times \varepsilon_{89,HE} - \varepsilon_{90,HE} \times \varepsilon_{89,LE}} \tag{1.5}$$

These equations were developed into a spreadsheet, along with the corresponding propagation of errors in quadrature, to enable one to rapidly quantify ^{89}Sr and ^{90}Sr during emergency response situations. The initial calibration curve was created to derive the efficiencies employed in a multi-point weighted fit based on both pure components and mixtures. These values were assumed to be constant with the instrumental conditions used.

However, subsequent ^{90}Sr results based on these original efficiencies were not as accurate (results not shown), highlighting the necessity to determine the most appropriate efficiency calibration strategy. After re-calibration, it was obvious that the 4 detection efficiencies had changed (Figure 5). These changes in efficiencies lead to errors in the calculation of the two radionuclide activities. Determinations of ^{89}Sr and ^{90}Sr activity were subsequently made based on 4 efficiency values measured in conjunction with the samples on the day of the analysis (hereafter called "daily efficiencies") using a single point standard methodology for each radionuclide.

As stated previously, from a practical and emergency-based point-of-view, it would be convenient to calibrate for both radionuclides on an infrequent basis, especially considering that frequent purchase of ^{89}Sr standards will be necessary due to its relatively short half-life.

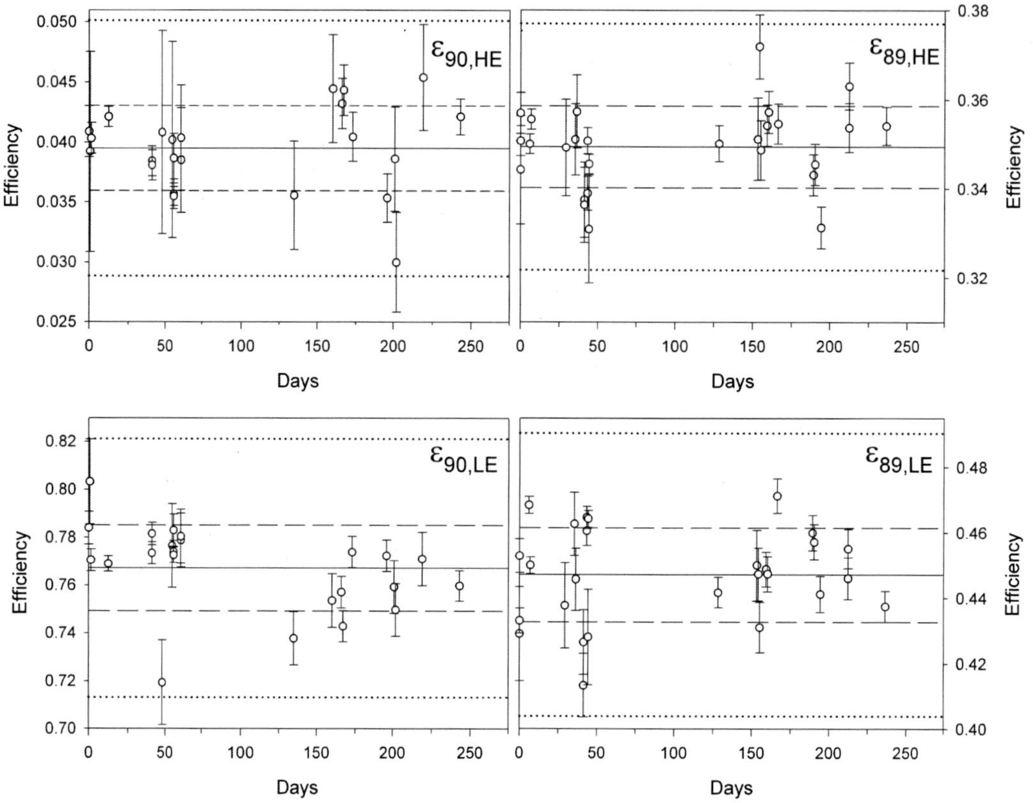

Figure 5. Temporal variation of the low (LE) and high energy (HE) efficiencies (e) for ^{89}Sr and ^{90}Sr. Solid, dashed, and dotted lines represent the nominal values, 1σ, and 3σ error, respectively.

For this reason, a comparison exercise assessing the variation in measured activities of ^{89}Sr and ^{90}Sr based on calculations performed using nominal (solid line, Figure 5) and daily efficiencies was undertaken. When nominal efficiencies were employed in the calculations, the result for a particular day was proportionally biased by the deviation of the nominal efficiency from the actual daily efficiencies. These deviations might be the result of variation in experimental conditions such as scintillation detector response, mobile phase composition, flows and laboratory temperature. Nevertheless, the calculation approach did not lead to any consistent bias in the calculated ^{90}Sr value as demonstrated in Figure 6 and reported by Walter *et al.*[59] using a similar method based on 2-window liquid scintillation determination of ^{89}Sr:^{90}Sr. Linear regression plots of measured versus expected activities for a solution composed of a mixture of ^{89}Sr and ^{90}Sr using daily and time averaged efficiencies yield essentially a linear correlation (dashed line, Figure 6). However, from that figure it is apparent that daily efficiency determination yields much better results, especially for ^{90}Sr activities below 5 Bq. Therefore, if an accurate assay is desired, daily efficiency determination should be performed prior to sample analysis as it will lead to better precision.

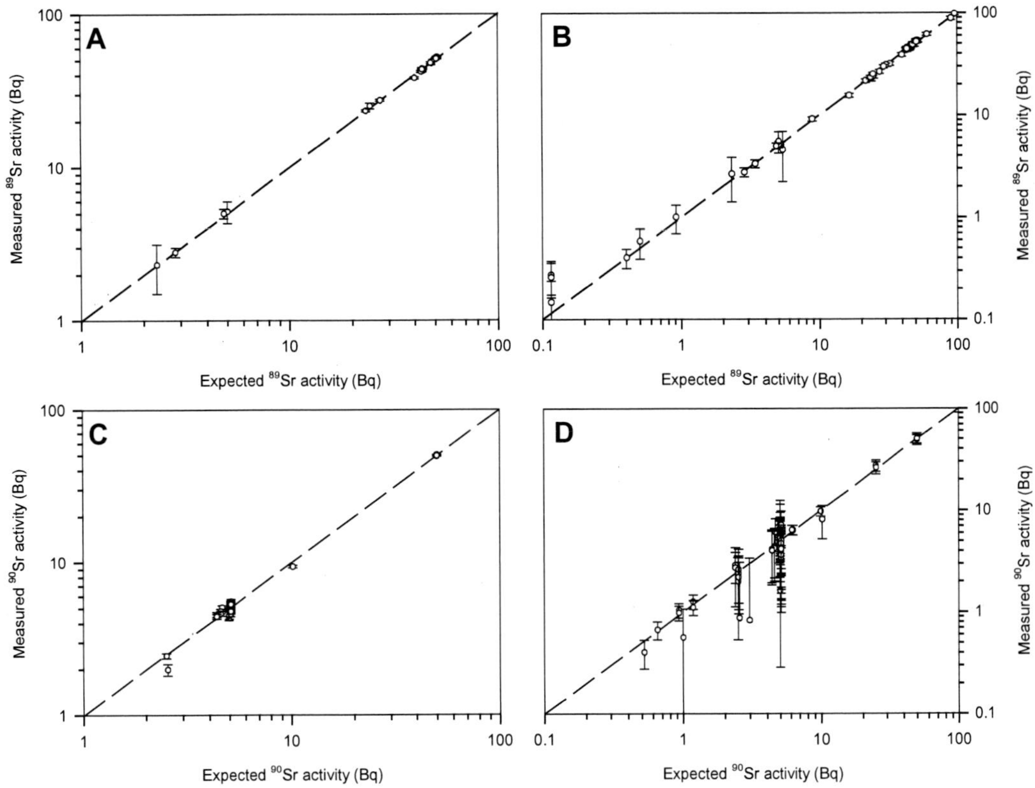

Figure 6. Measured versus expected activity for ^{89}Sr (A,B) and ^{90}Sr (C,D) using daily and nominal efficiencies, respectively. Dashed lines represent an ideal correlation.

Quantification of radiostrontium in water and milk. Table 6 presents some results obtained for spiked samples, certified reference materials and intercomparison samples. Since the method has been developed to be used for emergency response, samples with significant activities of both radionuclides were measured. Excellent agreement between measured and expected activities was obtained for all ^{89}Sr measurements (Table 6). In addition, an excellent correlation was found between measured and expected ^{90}Sr results performed on ^{89}Sr-free samples even at activities as low as 0.3Bq L^{-1}. When both radiostrontium isotopes were present, overestimation of ^{90}Sr activity occurred, mainly for samples with ^{89}Sr:^{90}Sr > 1.

The radiostrontium minimum detectable activities (MDAs) achieved by the automated methodology are, 0.16 and 0.2Bq L^{-1} for ^{89}Sr and ^{90}Sr, respectively. These MDAs are well below national and international guideline levels for milk and water for this radioelement (Table 7), enabling accurate determination of ^{89}Sr and ^{90}Sr at various activity ratios, including those expected during nuclear emergency situations (*e.g.* 0.4-30)[24,25].

Table 6. Analysis of radiostrontium activity (Bq L^{-1}) in water and milk certified reference materials (CRM) inter-comparison programs (ICP) samples, and spiked solutions (SS).

Sample identification	Sample matrix	Sample type	^{89}Sr measured[a]	^{89}Sr expected	^{90}Sr measured	^{90}Sr expected	^{89}Sr:^{90}Sr
Milk	Milk	SS	28.4 ± 0.31	30.0	4.35 ± 0.32	3.52	8.52
IAEA-152	Powder Milk	CRM	0.10 ± 0.05	N.P.	0.33 ± 0.33	0.24	NA
WHO/IRC-64L	Powder Milk	CRM	-0.076 ± 0.05	N.P.	0.38 ± 0.05	0.30	NA
Drinking water-1	Water	SS	16.67 ± 0.42	16.97	3.07 ± 0.48	2.48	6.84
Drinking water-2	Water	SS	33.83 ± 0.83	33.94	2.87 ± 0.87	2.48	13.69
EML-QAP 9709	Water	ICP	-0.01	N.P.	0.65 ± 0.06	0.64	NA
EML-QAP 9803	Water	ICP	0.01	N.P.	1.02 ± 0.06	0.96	NA
EML-QAP 9809	Water	ICP	-0.01 ± 0.05	N.P.	0.45 ± 0.05	0.40	NA
EML-QAP 9903	Water	ICP	-0.04 ± 0.05	N.P.	0.86 ± 0.06	0.79	NA
EML-QAP 9909	Water	ICP	-0.02 ± 0.05	N.P.	0.36 ± 0.05	0.34	NA
MAPEP-97-W5	Water	ICP	-0.06	N.P.	6.48 ± 0.12	6.34	NA
MAPEP-98-W6	Water	ICP	0.03 ± 0.08	N.P.	7.48 ± 0.26	7.54	BA
USEPA blind PE-B	Water	ICP	0.35 ± 0.05	0.38	0.36 ± 0.05	0.29	1.31
USEPA Fall 97	Water	ICP	0.36 ± 0.06	0.41	0.49 ± 0.06	0.56	0.73
USEPA Spring 98	Water	ICP	0.25 ± 0.06	0.12	1.16 ± 0.06	1.18	0.10

N.P. – Not Present; NA – Not Applicable, USEPA – United States Environmental Protection Agency; EML-QAP – Environmental Monitoring Laboratory – Quality Assurance Program; MAPEP – Mixed Analyte Performance Evaluation Program; WHO/IRC – World Health Organization/International Reference Centre.

a- error presented is equal to 1σ.

Table 7. National and international guidelines for activity of ^{89}Sr and ^{90}Sr for water and foodstuffs (Bq L^{-1} or kg^{-1})

<table>
<tr><th>Sample type</th><th>^{89}Sr</th><th>^{90}Sr</th></tr>
<tr><td>Drinking water, Milk[87]</td><td>300</td><td>30</td></tr>
<tr><td>Milk and Infant Foods[88]</td><td>NA</td><td>100</td></tr>
<tr><td>Milk diet for infants[89]</td><td>NA</td><td>160</td></tr>
<tr><td>Foods for General Consumption[90]</td><td>1000</td><td>100</td></tr>
<tr><td>Dairy Products[91]</td><td colspan="2">125</td></tr>
<tr><td>Total Diet[92]</td><td>NA</td><td>160</td></tr>
<tr><td>Methodology MDA[a]</td><td>0.16</td><td>0.2</td></tr>
</table>

MDA – Minimum detectable activity, NA – not specified
a. MDA if only one of the radiostrontium isotopes is present.

Summary

This new automated method is capable of determining environmental levels of ^{89}Sr and ^{90}Sr simultaneously, at ratios expected in a nuclear emergency. The method has led to a considerable improvement in turnaround time (1-2 days) compared to the separation technique previously employed in our laboratory which took approximately 14 days[42]. The high degree of automation of the two separation stages frees the analyst to perform other laboratory tasks, and assures a high degree of reproducibility in the results. Both the SCX chromatography and HPIC resin can be used repeatedly without concerns about decrease in chemical yield or risk of cross-contamination via carry-over. To date, our system has handled over one thousand injections. The combination of liquid scintillation detection in a large volume flow cell under stop flow conditions leads to MDAs that have been unattainable previously with on-line LSC detection systems.

Section 2. Automated Plutonium Urinalysis by Mass Spectrometry

Introduction

The omnipresence of plutonium in the nuclear fuel cycle makes it a prime candidate for internal exposure in the case of an event involving nuclear materials[60]. Events such as the explosion of the 4th reactor at the Chernobyl nuclear power plant, the 2000 Cerro Grande Fire, and the possibility of a radiological/nuclear terrorist attack have highlighted the necessity of large-scale population monitoring strategies[1,23,61]. As noted in the introduction of this chapter, methodologies designed for emergency response biomonitoring have very different requirements from those used in routine monitoring. Amongst the various long-lived actinides, plutonium isotopes are considered one of the most radiotoxic as the result of the relatively short half-life ($t_{1/2}$) of its most predominant isotopes (Table 1) and the high energy α-emission associated with their decay. Various radiometric analytical methodologies including fission-track analysis, alpha energy emission, and liquid scintillation, are employed for the detection of plutonium in urine[60]. However, none so far rivals the extreme sensitivity and multi-isotopic capability of mass spectrometry for the detection of plutonium[62]. While accelerator mass spectrometry (AMS)[63,64], thermal ionization mass spectrometry (TIMS)[10,65,66] and resonance ionization mass spectrometry (RIMS)[67] have shown exceptional sensitivity for plutonium detection, the necessity to modify the physical nature of the urine via co-precipitation makes the automation of methodologies using such instrumentation less than ideal. On the other hand, inductively coupled plasma mass spectrometry (ICP-MS) which is designed to enable the detection of aqueous ions is already well suited, with respect to sample introduction, for urinalysis.

While current high-sensitivity ICP-MS instruments have the capacity to detect Pu in the sub-Bq L^{-1} range, the complexity of the urine matrix prohibits its direct determination. Therefore, Pu urinalysis is only possible after sample preparation. Several research groups[7,8,18,68] have already demonstrated that the quantification of Pu in urine is possible using separation and on-line detection. Using extraction chromatographic (*e.g.* TRU®, TEVA®) or ion-exchange (*e.g.* MetPac™ CC-1) resin as the stationary phase, separation and pre-concentration of plutonium was possible in urine. However, uranium-plutonium overlapping peaks[68] and resin reusability[7] issues arose from the use of MetPac™ and TRU, respectively. The poor selectivity of TRU resin for uranium over plutonium[69] also prevents ^{238}Pu detection, a challenging isotope for inorganic mass spectrometry due to its short half-life (Table 1) and the omnipresence of uranium. Many studies[8,9] have shown that TEVA resin produces an effective separation between uranium and plutonium, reducing the problems associated with peak tailing and molecular interferences. In addition, the resin can be reused, a practical feature for rapid on-line detection system. In this study, the development of a completely automated plutonium urinalysis system based on improvements from a previously published methodology[8] will be described. Investigation of various data and quantification processes for emergency preparedness will also be investigated. Finally, improvement in detectability through new chromatographic processes will also be presented.

Experimental Section

Automated flow injection system. Figure 7 is a schematic diagram of the improved automated flow injection system (AFIS) developed at RPB/HC, which is composed of four sections. The first section is a multi-solvent delivery system (single pump ICS-3000, Dionex, Sunnyvale, CA) which is used to deliver up to four different solvents in either an isocratic or gradient mode. The second section is an autosampler (AS unit, Dionex, Sunnyvale, CA) which is used to automatically load acidified urine samples into a 5 mL sample loop attached to the unit. The third section is composed of two biocompatible analytical-scale two-position, six-port switching modules (MX9900-000, Upchurch Scientific, Oak Habor, WA) to direct sample and eluant flow and a peristaltic pump (Isotech, Plymouth Meetings, PA) to introduce 2% HNO_3 (v/v) for the rinsing and cleaning of the nebulization system. The modules and the pump are controlled by the Chromeleon software (Version 6.80, Dionex, Milford, MA). The last section of the AFIS consists of the Apex-HF unit and the Element sector field inductively coupled plasma mass spectrometer (ICP-SFMS).

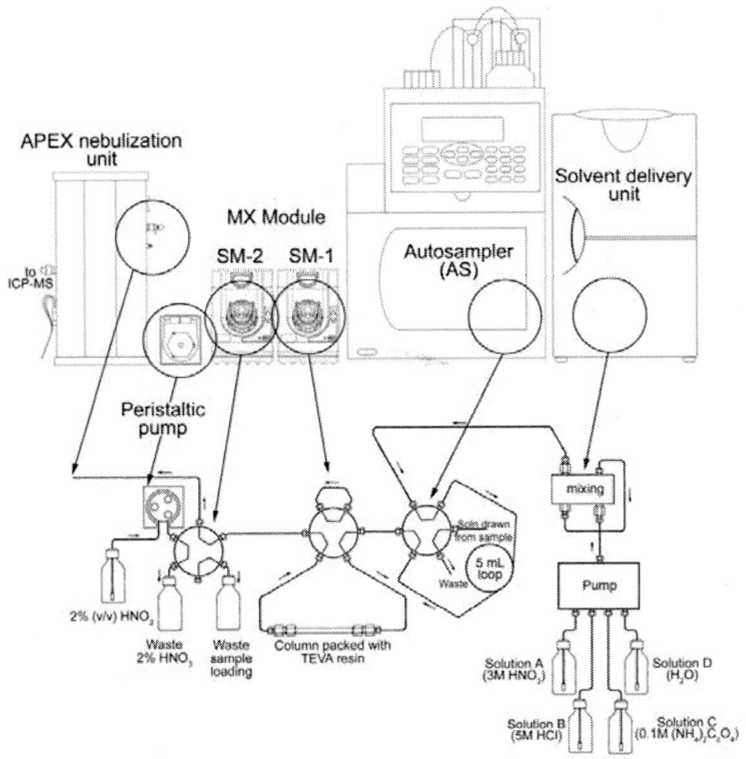

Figure 7. Schematic of the Flow Injection Automated System.

A stainless steel analytical grade column (Alltech, Columbia, MD), 4.6 mm i.d. × 50 mm long, coated inside with PEEK was filled with TEVA resin (50-100 μm particle size, Eichrom Technologies Inc., Darien, IL). All transport lines used in the AFIS were made of 0.762 mm i.d. PEEK tubing (Alltech, Columbia, MD) with the exception of the transfer line between the second switching module (SM-2) and the PFA nebulizer which was assembled using 1.1 mm i.d. polytetrafluoroethylene (PTFE) tubing (Alltech, Columbia, IL). 10-32 PEEK high pressure fittings with PEEK ferrules (Upchurch Scientific, Oakhabor, WA) for coned ports were used to connect the switching modules, multi-solvent delivery system, and the analytical-grade column together. A 5 mL sample loop (Rheodyne, Rohnert Park, CA) was connected to the injection valve of the autosampler to ensure a constant volume during the sample loading phase.

Detailed operations of the AFIS are described in Table 8. The entire protocol lasts 889 seconds, 289 seconds of which are dedicated to autosampler rinsing and loading the sample loop. Ideally, these actions could be performed immediately after the sample elution (step 2, Table 8), allowing overlap between the current sample pre-concentration/elution (step 3 and 4) and the subsequent sample loop loading (step 1). Such a strategy was used by Lariviere *et al.*[8] to maximize sample throughput of their manual injection system. This overlap would reduce the overall analytical time per sample down to 600 seconds with the AFIS with automated injection, increasing the daily sample throughput to 144 samples. Unfortunately, such an approach is not possible with the autosampler used in the development of the AFIS.

Table 8. Operating sequence of the AFIS for the on-line preconcentration and detection of $^{239,240}Pu$ by ICP-MS

Step	Time (s)	Medium	Flow rate (mL/min)	Switching modules (SM) position		Peristaltic pump	Injection valve[a]	Flow direction in RFC
				SM-1[b]	SM-2[b]			
1	289	3M HNO_3	1.0	On	Off	Off	load	Direct
2	180	3M HNO_3	3.0	On	Off	Off	elute	Direct
3	120	5M HCl	3.0	On	Off	On	load	Direct
4[c]	270	0.01M $(NH_4)_2C_2O_4$	1.0	On	On (60[d])	Off (60[d])	load	Reversed[e]
5	30	0.1M $(NH_4)_2C_2O_4$	4.0	On	Off	Off	load	Direct
Total	889							

Step	Step description
1	3M HNO_3 is pumped through the column while the sample is loading the sample loop. The time requirement of this step is defined by the
2	3M HNO_3 is pumped through the sample loop to load the sample and rinse the residual elements from the resin
3	5M HCl is pumped through TEVA to elute Th
4	0.01M $(NH_4)_2C_2O_4$ is pumped through TEVA to elute Pu
5	0.1M $(NH_4)_2C_2O_4$ is used to rinse resin from any residual elements

a - Injection valve position, *elute* signifies that the medium is passing through the sample loop while *load* is the opposite.

b - SM-1 and SM-2 refer to Switching Modules shown in Figure 7.

c - Acquisition of data by the ICP-MS is triggered by the ICS-3000 module 60 seconds after the start of step 3

d - The value in parentheses represents the time elapsed after the beginning of the step at which the process occurs

e - In DFC, the flow is maintained in the direct direction during this step.

Table 9. Impact of the Transient signal processing on the quality of the analytical results

	^{239}Pu (mBq L^{-1})	^{240}Pu (mBq L^{-1})	$^{240}Pu/^{239}Pu$ (mass)
Maximum signal (20)[a]	39 ± 4	68 ± 6	0.47 ± 0.05
Peak summation (20)	45.3 ± 1.8	70.0 ± 3.0	0.42 ± 0.01
Peak summation (60)	48.7 ± 1.4	73.9 ± 2.3	0.408 ± 0.006
Peak fitting and integration (20)	45.2 ± 2.1	69.9 ± 3.4	0.42 ± 0.01
Certified values[66]	49.9 ± 0.3	75.1 ± 0.5	0.410 ± 0.003

a. value in parentheses represent the dwell time used in msec

Table 10. Impact of calibration on analytical results

		Calibration standard activity (mBq L^{-1})		Activity (mBq L^{-1})			
		^{239}Pu	^{240}Pu	^{239}Pu	Δ (%)	^{240}Pu	Δ (%)
Single point calibration	A	9.47	10.50	8.44 ± 0.20	-15.6	4.78 ± 0.56	-2.5
	B	55.1	53.9	10.52 ± 0.25	5.2	5.05 ± 0.59	3.0
	C	111	109	10.15 ± 0.24	1.5	5.10 ± 0.60	4.0
	D	563	551	10.09 ± 0.24	0.9	5.13 ± 0.60	4.7
Double point calibration		A+B		10.69 ± 0.25	6.9	5.07 ± 0.58	3.4
		A+C		10.23 ± 0.23	2.3	5.11 ± 0.59	4.3
		A+D		10.10 ± 0.24	1.0	5.13 ± 0.60	4.8
		B+C		10.15 ± 0.22	1.5	5.10 ± 0.59	4.0
		B+D		10.07 ± 0.24	0.7	5.13 ± 0.60	4.8
		C+D		10.08 ± 0.24	0.8	5.13 ± 0.60	4.7
Triple point calibration		A+B+C		10.27 ± 0.22	2.7	5.11 ± 0.59	4.3
		A+C+D		10.10 ± 0.24	1.0	5.13 ± 0.60	4.8
		A+B+D		10.09 ± 0.24	0.9	5.13 ± 0.60	4.8
		B+C+D		10.07 ± 0.24	0.7	5.13 ± 0.60	4.8
Quadruple point calibration				10.09 ± 0.24	0.9	5.13 ± 0.60	4.8
Certified values				*10.00 ± 0.07*	----	*4.90 ± 0.03*	----

Table 11. Impact of cross-calibration on analytical results

		Calibration standard activity (mBq L^{-1})		Activity (mBq L^{-1}) from cross-calibration			
		^{239}Pu	^{240}Pu	^{239}Pu	Δ (%)	^{240}Pu	Δ (%)
Single point calibration	A	9.47	10.50	9.12 ± 0.22	-8.8	4.42 ± 0.52	-9.7
	B	55.1	53.9	9.63 ± 0.23	-3.7	5.51 ± 0.65	12.5
	C	111	109	9.73 ± 0.23	-2.7	5.32 ± 0.62	8.5
	D	563	551	9.79 ± 0.23	-2.1	5.28 ± 0.62	7.8
Double point calibration	A+B			9.67 ± 0.22	-3.3	5.60 ± 0.65	14.3
	A+C			9.76 ± 0.23	-2.4	5.36 ± 0.60	9.4
	A+D			9.80 ± 0.23	-2.0	5.29 ± 0.62	8.0
	B+C			9.73 ± 0.23	-2.7	5.32 ± 0.59	8.5
	B+D			9.80 ± 0.23	-2.0	5.27 ± 0.61	7.6
	C+D			9.80 ± 0.23	-2.0	5.28 ± 0.62	7.8
Triple point calibration	A+B+C			9.75 ± 0.23	-2.5	5.38 ± 0.60	9.8
	A+C+D			9.80 ± 0.23	-2.0	5.29 ± 0.62	8.0
	A+B+D			9.80 ± 0.23	-2.0	5.28 ± 0.61	7.9
	B+C+D			9.80 ± 0.23	-2.0	5.27 ± 0.61	7.6
Quadruple point calibration				9.80 ± 0.23	-2.0	5.28 ± 0.61	7.8
Certified values				*10.00 ± 0.07*	----	*4.90 ± 0.03*	----

Pu detection and quantification. All measurements were performed on an Element2 ICP-SFMS (ThermoFinnigan, Bremen, Germany). A high sensitivity inlet system (Apex-HF, Elemental Scientific Inc., Omaha, NE) coupled with a Microflow perfluoroalkoxy (PFA) nebulizer was used as the sample introduction system. Instrumental conditions for the Element2 ICP-SFMS and the Apex-HF were set to ensure high instrumental sensitivity and are identical to those published elsewhere[8].

Reagents and materials. High-purity water (18 MΩ-cm) prepared by a Milli-Q® ultrapure water purification system (Millipore, Bedford, MA) was used for all reagent dilutions. Optima grade concentrated HNO_3 (Fisher Scientific, Ottawa, ON) was used to prepare diluted solutions. ACS-certified ammonium oxalate monohydrate salt and concentrated optima grade HCl (both from Fisher Scientific, Ottawa, ON) were used to prepare 0.1M $(NH_4)_2C_2O_4$ and 5M HCl eluent solutions, respectively. Standard solutions of $^{239,240,242}Pu$ (SRM-4330B, -4338A, and -4334G, respectively) were purchased from the NIST (Gaithersburg, MD). Urine samples containing various levels of $^{239,240}Pu$, provided as part of an intercomparison exercise[66], were analyzed to assess the efficiency of the automated protocol for radiobioassay.

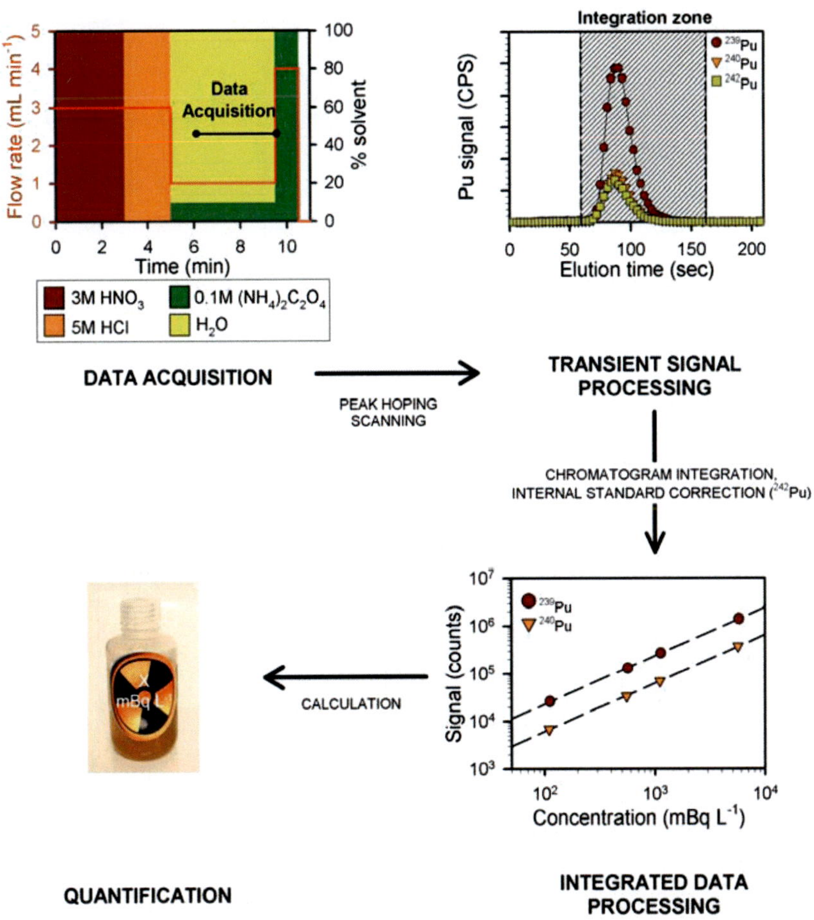

Figure 8. Data acquisition and processing for the quantification of transient plutonium signal.

Sample preparation. Sample preparation was performed as follows: 1.875 mL of concentrated nitric acid and 134 pg of ^{242}Pu were added to approximately 5 mL of human urine. The solution was then made to 10 mL using urine. The sample, now in 3M HNO_3, was shaken to homogenize and analyze. Prior to injection, the samples were filtered through a 0.45 μm nylon filter to remove any particulate that could clog the system and loaded directly into 10-mL polystyrene sample vial (Dionex, Sunnyvale, CA).

Data treatment and isotopic ratio. While the sample preparation and measurements are performed automatically by the AFIS, the data treatment process still has to be done manually by the user. This process illustrated in Figure 8, consists of obtaining the transient signal generated by ^{239}Pu, ^{240}Pu, and ^{242}Pu during step 4 (Table 8) from an export file generated by the Element2 ICP-MS software. ^{238}Pu signal was not acquired as it was unresolved from background ^{238}U signal. Signals from each analyte are integrated into a peak area via summation using a pre-defined integration zone. Activity in the sample is then determined based on comparison of peak area between samples and standards, after proper correction of background signal and of instrumental drift and chemical yield using ^{242}Pu as an internal standard.

Results and Discussion

Transient data acquisition and processing. While instrumental sensitivity of ICP-MS has significantly improved since its introduction in the nuclear science field in the mid-1980's, separation and pre-concentration of many long-lived radionuclides is still required to obtain figures of merit that are usable for occupational and emergency bioassay testing[19]. These chemical procedures, if performed in parallel with on-line detection, will generate a signal that is time dependent. Chromatographic data such as those generated by the AFIS are converted from a transient signal into activities and isotopic ratios using calculations. Longerich *et al.*[70] have highlighted the importance of data signal acquisition parameters including settling time, dwell time, sweep per reading and point per spectral peak on the figures of merit of transient signal generated from laser ablation experiments. As flow injection analysis and laser ablation share similarities with respect to the type of signals generated, investigation of these parameters is essential. As measurements of plutonium in urine were performed using a sector-field instrument with a unique magnet settling mass (m/z =239), the impact of settling time is very limited and therefore this parameter was not included in the investigation. Both transient data acquisition and processing were investigated. First, the impact of transient signal processing on the analytical accuracy and precision was assessed. Second, the effect of acquisition parameter (*i.e.* detector dwell time) was evaluated.

In order to find the most appropriate transient signal processing, three approaches were evaluated: i) maximum intensity determined among data generated by the ICP-SFMS, ii) peak summation, a summing of every signal generated within the integration zone, and iii) peak fitting and integration. The first two signal processing approaches were performed in a Microsoft Excel® spreadsheet, while the latter was done using Peakfit software (Systat Software Inc, San Jose, CA). Data treatment for all the three approaches was investigated using the same data from triplicates and the same integration zone (60 - 160 sec). Table 9 presents the results obtained for ^{239}Pu and ^{240}Pu quantification for each approach tested. Based on these results, peak summation and peak fitting and integration provide the most accurate and precise analytical results of the three data processing methodology tested with respect to those provided with the certified reference material. The main advantage of the signal summation technique is that it is simple to perform. The Element2 Software enables rapid transfer of chromatographic data into ASC-type file, which can be opened using Microsoft Excel®. A pre-defined template can rapidly be developed to accelerate data processing. For these reasons, peak summation was chosen as the preferred methodology for handling data generated by the AFIS.

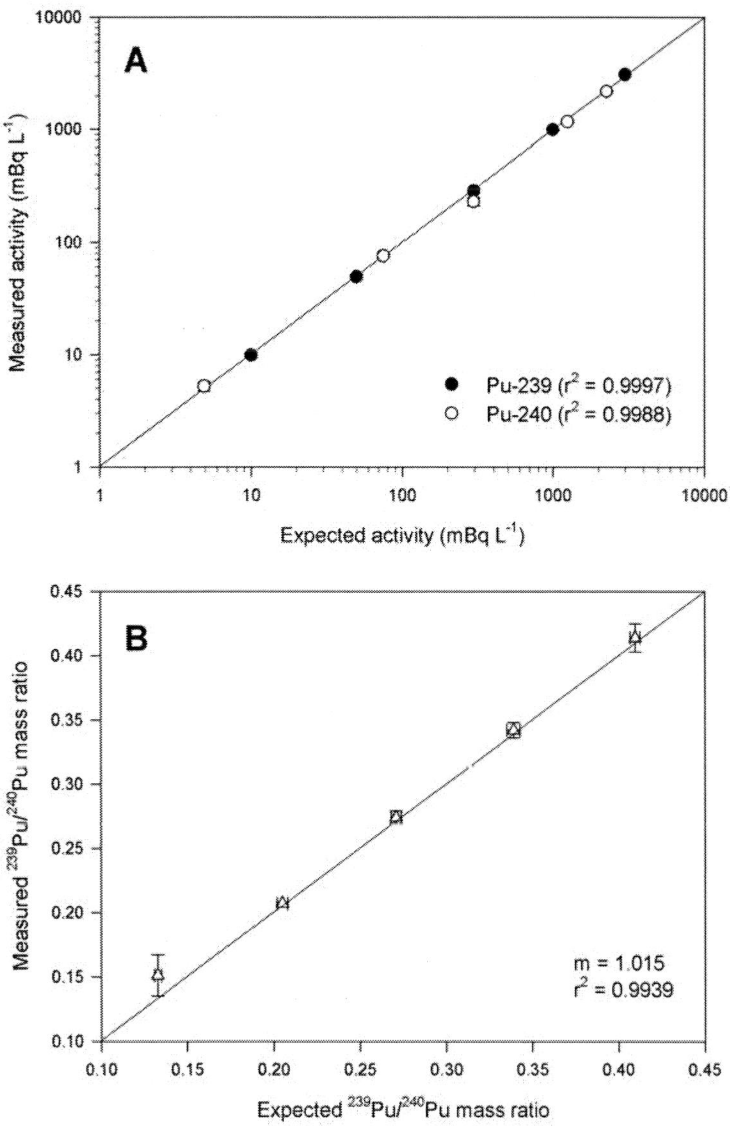

Figure 9. Measured versus expected values for A) activity and B) isotopic ratio for ^{239}Pu and ^{240}Pu urinalysis.

As stated earlier in this chapter, mass spectrometric analyses of plutonium can be used to obtain isotopic information such as the $^{239}Pu/^{240}Pu$ ratio. To determine the accuracy and precision of isotopic ratio measured on time dependent signal, data generated during the signal processing investigation were used. As shown in Table 9, signal summation and peak integration provided again more accurate and precise isotopic ratio than maximum intensity. The ratio found using the two data processing approaches also demonstrates that the uncertainties on the measurement is relatively small (RSD < 5%) which is sufficient to detect isotopic changes but fail to meet the International Atomic Energy Agency (IAEA) requirements for isotopic signature uncertainties in safeguarding nuclear material[71]. However, considering that the measurements are performed on samples with low levels of contamination and not on nuclear fuel, such a finding is not surprising. Nonetheless, the AFIS and the peak summation processing approach are providing enough precision to determine changes in the percent range (Figure 9).

Begley and Sharp[72] have demonstrated that analytical precision can be improved by increasing the dwell time and/or the number of sweeps. Increasing the dwell time is beneficial to minimize background fluctuations and hence improve detection limits and isotopic ratio precision. Laborda *et al.*[73] found that the standard deviation of the baseline should be reduced in proportion to the square root of the dwell time and that detection limits should improve accordingly. The impact of the dwell time on the accuracy of ^{239}Pu, ^{240}Pu, and isotopic ratio was assessed by measuring urine samples of known isotopic composition (n=3) at dwell time varying from 20 msec to 100 msec. For simplicity, number of sweeps per reading and points per peak were kept constant. The transient signal generated was then evaluated using the signal summation approach using a constant time integration window for all the sample times tested. Since the total measurement time was kept constant, the number of replicates had to be decreased with the increase in sample time. Between 60 and 235 sweeps were performed when the dwell time was varied from 20 to 100 msec dwell time per peak. The most accurate isotopic ratio ($^{239/240}Pu$) was obtained for a dwell time of 60 msec (Table 9). Since the isotopic ratio in the solution prepared was expected to be 0.410 ± 0.003 (and confirmed by TIMS and alpha-spectrometry analysis)[66], 60 msec dwell time results leads to better precision and accuracy. The shorter or longer sample dwell times tested deteriorated the isotopic ratio accuracy, probably due to background fluctuation influences and an insufficient number of acquisition points collected within the integration zone, respectively. These finding are in agreement with other studies on transient signals[73].

Simplification of the external calibration procedure. In a radiological/nuclear emergency, qualitative and quantitative values generated by radiobioassay analysis are critical to medical teams to assess which individuals are internally contaminated and to what extent. This type of information is essential to providing the proper treatment. Lariviere *et al.* used a combination of 2 blank correction, 6 standards calibration, and an internal standard (IS) method (^{242}Pu) to quantify the samples provided as part of an intercomparison exercise[66]. While this type of procedure ensures the highest degree of reliability in the results, it is disadvantageous with respect to sample counting time. Due to the large variability in urine matrix composition, the use of an internal standard is critical to assess the plutonium chemical yield and far more appropriate than matrix-matched calibration which is time-consuming with respect to the number of samples to be counted. ^{242}Pu and ^{244}Pu were tested and led to equivalent chemical yield (results not shown), but ^{242}Pu was preferred as the IS since it is not significantly affected by thorium carbide ($^{232}Th^{12}C$) molecular interference that might result from high natural thorium in urine.

The possibility to reduce the number of standards used during calibration without significantly affecting precision and accuracy was investigated using data generated from 4 urine sample replicates of known plutonium activity. Single, double, triple, and quadruple point calibrations were tested. Calibrating with any of the calibration methods led to quantitative results that lay within ±16% of the certified value (Table 10). As expected, calibration based on 4 points provides the most accurate activity for ^{239}Pu and ^{240}Pu. Single point calibrations, although providing acceptable results are not usable for emergency purposes as any unexpected changes in the signal would greatly impact the quantification process, a risk unacceptable when an erroneous result could be detrimental to the patient's health. Double point calibrations, especially if performed with standards of higher activity produce results with accuracy below 5%. Using higher activity standards is important to minimize the impact of background fluctuation and molecular interference contributions, which are substantial for lower count rate. Using such calibration methodology reduces the number of blanks and standards to be run to three, which represent approximately 45 minutes of calibration before a sample can be analyzed, a significant reduction in QA/QC time requirements (approximately one hour). However, it is important to clearly identify the linearity range of the methodology before using such a calibration approach. Using the AFIS proposed in this chapter, linearity was observed between 0.58 and 5 000 mBq L^{-1} (Figure 9).

Table 12. Figures of merit for $^{239,240}Pu$ urinalysis for the proposed protocol.

	Emergency protocol	
Sample volume, mL	5 (4.06)	
Time of analysis, min	14.8	
Sample frequency, sample d^{-1}	97	
Precision for 10 mBq L^{-1} (% RSD, 10 replicates)	2.4	
Chemical recovery (%, 24 samples)	84 ± 4	
	^{239}Pu	**^{240}Pu**
Absolute sensitivity (10^4 counts mBq^{-1})	4.60 ± 0.04	1.25 ± 0.01
Method detection limits ($3\sigma_B$, mBq L^{-1})	0.58	0.84
Method absolute detection limits (μBq)	2.9	4.2
Method quantification limits ($15\sigma_B$, mBq L^{-1})	2.92	4.21

Another strategy tested to minimize the impact of calibration on actual sample throughput was the use of cross calibration. As ionization energies for all isotopes of an element are virtually identical[19], the calibrated signal from one isotope can be used to quantify another. Using experimental analytical results such as absolute sensitivity, it is possible to demonstrate such a concept. However, conversion of absolute sensitivity (Table 12) from activity unit to mass unit is required. By rearranging the mass-activity equation found in Lariviere *et al.*[19], it is possible to determine a mass-activity conversion ratio for ^{239}Pu and ^{240}Pu using the following equation:

$$\frac{m(^{239}Pu)}{m(^{240}Pu)} = \frac{A(^{239}Pu) \times MM(^{239}Pu) \times t_{1/2}(^{239}Pu)}{A(^{240}Pu) \times MM(^{240}Pu) \times t_{1/2}(^{240}Pu)} = \frac{A(^{239}Pu) \times 239.0522(g\ mol^{-1}) \times 24\,100 \pm 30y}{A(^{240}Pu) \times 240.0538(g\ mol^{-1}) \times 6\,560 \pm 10y} = 3.66 \pm 0.01 \tag{1.6}$$

where m, A, MM and $t_{1/2}$ represent the mass of the radioisotope, its activity, its molar mass, and its half-life, respectively. By solving this equation for ^{239}Pu and ^{240}Pu, a value of 3.66±0.01 is obtained, which indicates that for the same sample activity, ^{239}Pu will have 3.66 times more atoms than ^{240}Pu. Dividing the absolute sensitivity of ^{239}Pu by that of ^{240}Pu (Table 12), a value of 3.68±0.03 can be obtained. This value is in excellent agreement with the calculated one, indicating that cross-calibration between plutonium isotopes is possible.

Table 11 shows results for ^{239}Pu activity when ^{240}Pu standards were used for calibration and vice-versa. The same calibration strategies as those mentioned above were also tested for cross-calibration. ^{239}Pu calibration for ^{240}Pu analysis led to overestimation in most cases (7.6 to 14.3%) while ^{240}Pu calibration led to underestimation for ^{239}Pu (-2.0 to -8.8%). The results from this exercise highlight the better precision achieved by simply calibrating with ^{240}Pu, independent of the plutonium isotope studied. This conclusion is probably a consequence of the higher contribution of weak spectral and non-spectral interferences at m/z=239, which impact the accuracy. Contrary to the multi-point calibration approach, cross-calibration does not result in a reduction of the number of samples to analyse. However, this approach can be beneficial to reduce sample contamination originating from previously run standards. As reported by Lariviere *et al.*[8], TEVA resin can retain up to 1.3% of the plutonium from the previous samples. This cross-sample contamination can be reduced below 0.06% if two or more rinses are run between samples. This *modus operanti* is however very time-consuming and will significantly slow down the analytical pace. Assuming that calibration for the detection of ^{239}Pu in urine would be performed using ^{240}Pu standards, cross-contamination between standards and samples would be completely eliminated. This would enable the insertion of standards throughout the sample sequence without worrying about sample cross-contamination. Note that this strategy is not effective for analysis requiring ^{240}Pu activitiy or plutonium isotopic signature, often needed as part of a radioanalytical emergency response program.

Possible automation improvements. For an analytical system generating data in a time dependent format, peak shape will influence many analytical parameters including detection and quantification limits, and precision. At equivalent integration area, a narrower peak would have a more limited contribution of the background signal. As a portion of the variation in precision is the result of background fluctuations is from detector dark noise, slight fluctuations in plasma, and instrumental conditions[19], a narrower peak compromise in a narrower integration zone, would have a higher degree of precision associated with it. All other analytical figures were maintained constant. Detection (c_L) and quantification (c_D) limits are defined as:

$$c_L = \frac{3\sigma_B}{S} \tag{1.7}$$

$$c_D = 5c_L \tag{1.8}$$

where σ_B and S are the standard deviation associated with 10 replicate measurements of the blank and the normalized sensitivity, respectively[74]. An increase in precision, which translates into the reduction in the blank and sample standard deviations, would result in lower limits of detection.

Narrower elution peaks are also beneficial with respect to practical considerations for both on-line and off-line analysis. In an on-line system, narrower chromatographic peaks reduce the size of integration zone, allowing higher sample throughput. For off-line analysis, narrower elution peaks lead to smaller volume collection, facilitating subsequent radiochemical steps such as evaporation or precipitation in the radiometric and mass spectrometric determination. For both types of analysis, narrower peaks mean less eluant volume requirement, a possible limiting factor in an emergency response situation where shortage of storage space and chemical or other supplies can be expected. For all of these reasons, the development of radiochemical separation systems should include an investigation regarding the performances of the elution process. The following paragraphs summarize the finding of a novel strategy to improve peak shape: reversed-flow chromatography.

Peak broadening in chromatography has been linked to numerous experimental factors including eluant velocity, column length, and Eddy diffusion. As demonstrated by Horwitz and Bloomquist[75], extraction chromatography is not exempt from such considerations. These factors are directly linked with the contact time between the mobile phase and the stationary phase (extractant). Therefore, optimizing the analytical conditions to create favorable interaction conditions could provide improvements in peak shape. One interesting concept to minimize contact time during the elution is called reversed-flow chromatography (RFC). Originally proposed by Bailly and Tondeur[76] for the separation of Na^+ and K^+ by ion-exchange chromatography, RFC is now used in a wide variety of applications[77]. As extraction chromatography has similarities with affinity chromatography with respect to its great selectivity and high distribution ratio (k'), it is imaginable that it could be used in nuclear chemistry. The operation principle of RFC is simple. The extraction of the analyte is performed in the direction in which the resin has been packed (Figure 10A) while the re-extraction is performed in the opposite direction (Figure 10B). This is in contrast with conventional extraction chromatography (referred hereafter as direct-flow chromatography (DFC)) where extraction and re-extraction are performed using a constant and unidirectional flow. Performing re-extraction in the opposite direction limits the interaction time between the analyte and stationary phase during this process. This in turn leads to minimizing diffusion, which should translate into a narrower elution peak measurable using parameters such as full-width half-maximum (FWHM), peak area, and peak maximum signal[77].

In order to test the applicability of RFC in radiochemistry procedures based on extraction chromatography, the first 6-port MX module of the AFIS (Figure 7) was replaced by a 10-ports MX module. This change enables the sample loading and rinsing steps (1, 2, and 3 in Table 8) to occur in direct mode and the elution one (Step 4 in Table 8) to occur in reversed-flow mode. DFC results used to compare the analytical performances of RFC were obtained by maintaining the flow of mobile phase constant throughout all the steps of the AFIS. To facilitate the comparison of DFC and RFC peak parameters, shift in peak elution time must be corrected. This shift is the result of changes in tubing volume associated with a modification of direction of the mobile phase flow. The tubing volume difference between RFC and DFC was determined by measuring the time elapsed between the injection and the detection of a non-extracted analyte, in this case ^{133}Cs, in direct and reversed configuration. The time shift was determined to be 7 seconds, which represents a difference in tubing volume of 116 μL. Peaks were shifted accordingly in subsequent experiments to correct for this difference. Based on measurements performed in triplicate using 50 mBq L^{-1} (^{239}Pu) urine samples, improvements in peak area (4.5%), peak maximum signal (4.1%), and peak narrowness (8.0%) were obtained for RFC over DFC using identical loading and elution conditions (Figure 10). These improvements, while modest, indicate a better elution of Pu from TEVA. The rest of the experimental results were therefore collected using RFC.

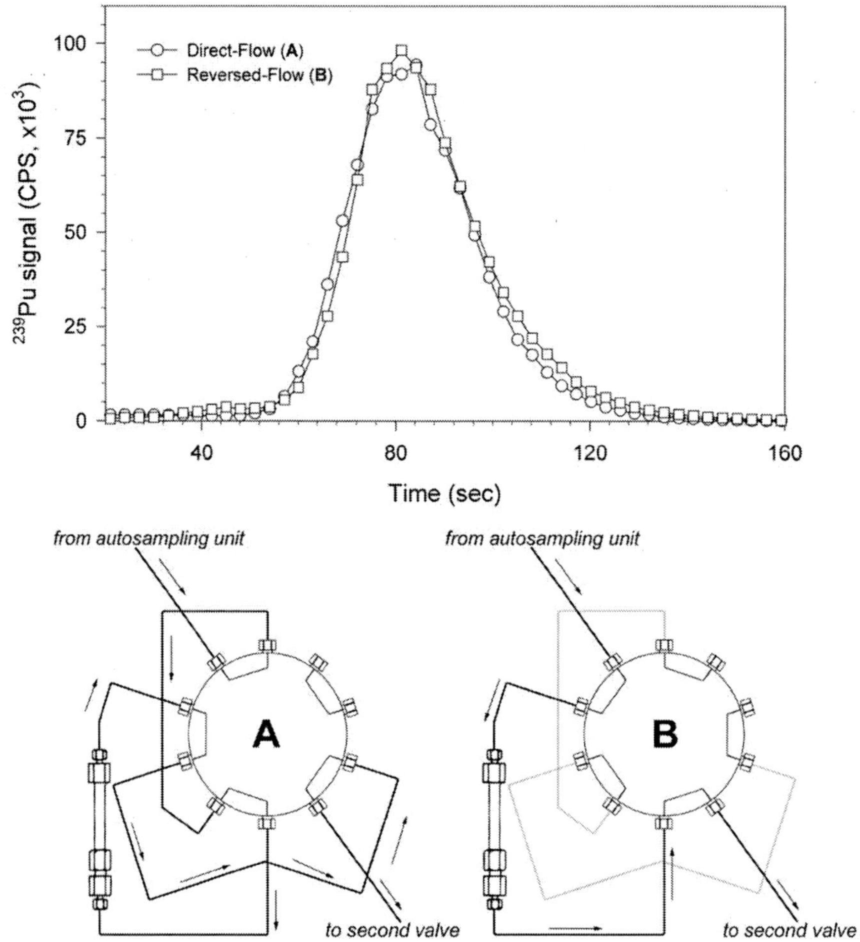

Figure 10. Transient monitoring of plutonium eluant using A) direct- and B) reversed-flow chromatography.

Figures of merit of automated flow injection system (AFIS). As stated previously, the method presented in this chapter is a modified version of an automated flow injection system (AFIS) already described in detail elsewhere[19]. The complete automation of the system, achieved through the addition of an autosampler and an external triggering mechanism to the mass spectrometric system, enables the system to perform separation, preconcentration, and quantification of Pu in urine without prior sample preparation other than acidification. The volume limitations associated with the addition of an autosampling component in the system led to a complete reassessment of eluant volumes (Table 8). Note that the investigation of optimal loading, rinsing, and elution conditions (results not shown) led to the use of eluants with identical chemical nature and concentration as those presented elsewhere[19]. This stresses the specificity of the eluants used for the separation/removal of matrix and molecular interferences. The only modification in the operating sequence is a reduction in eluant volume used during steps 2 and 3. Although the sample loading volume of the new system (5 mL) is half of the original methodology[8], the sample analysis time remained unchanged. The expected gain in analysis time per sample resulting from the smaller sample volume loaded is overshadowed by the system's incapability to perform simultaneous steps. In the manual injection method, sample loop loading occurred during the elution of the previous sample, an action impossible with the current autosampler. Therefore all the time reduction typically associated with a minimization of the sample and eluant volumes is lost as part of the subsequent loading. An autosampler unit that would allow sample loop loading during the chromatographic elution steps (3 to 5 in Table 8) would reduce the time of analysis from 889 to 600 seconds. This in turn would increase the daily sample throughput to approximately 144 as opposed to 97.

Although no gain was obtained on the analytical throughput with respect to sample loading automation, small improvements in precision and chemical recovery were observed with the new methodology. The use of an autosampling unit led to a decrease of 2.6% in precision at the 10mBq L^{-1} range. The lower standard deviation associated with the measurements is the result of two factors. First, sample loading automation reduced the presence of air in the sample loop, a phenomenon which occurs frequently with manual injection, leading to discrepancies in the sample volume injected. This is also supported by the improvement in standard deviation obtained for pressure readings collected by the solvent delivery unit (SDU) during the sample injection (results not shown). Second, the enhanced interactivity between the SDU and the mass spectrometer via triggering led to far more repeatable chromatograms. The higher degree of reliability of the automated external triggering versus manual triggering eliminates fluctuations in the acquisition time windows, ensuring identical peak positioning between each measurement, improving analytical precision. The increase in chemical recovery can be linked to the reduction in sample volume necessary for the implementation of an autosampling unit. As demonstrated by Torres *et al.*[31], elution peak shape in extraction chromatography is linked to the amount of sample loaded onto a predefined volume of resin. At higher sample loads approaching the resin maximum capacity, bleeding and tailing are observed, leading to lower chemical yield. As technical limitations lead to a decrease in the sample load for an equivalent volume of resin, the measurement of a higher recovery yield for plutonium is expected.

Detection limits of 0.58 and 0.84mBq L^{-1} for ^{239}Pu and ^{240}Pu, respectively, were obtained by measuring the signal generated at m/z = 239 and 240 for 10 urine samples free of plutonium. Accounting for the reduction in sample loading volume, these values are slightly better then those previously published for the manual injection methodology[8]. The method detection limit of 0.58mBq L^{-1} (^{239}Pu), below the 4.8mBq L^{-1} action level guideline proposed by Pappas *et al.*[9], demonstrates its usefulness during radiological/nuclear emergency response efforts. The automated methodology figures of merit also compare advantageously to other extraction based methods coupled with mass spectrometry detection[7,9], especially in light of the small volume of urine required for the analysis. Currently, only one manual plutonium bioassay procedure based on ICP-MS measurements[9] provides results with a smaller volume (1 mL).

In order to test the accuracy and precision of the proposed automated methodology, samples prepared for a plutonium urinalysis method intercomparison exercise[66] were measured based on a 2-point calibration in RFC mode. Note that cross-calibration was not used in this part of the experiment. The results, presented in Figure 9, demonstrate the excellent correlation between measured and expected activities. The wide range of plutonium activities tested (over three orders of magnitude) highlights the great linearity of the system in the mBq and Bq L^{-1} range (r^2 = 0.9997 and 0.9988 for ^{239}Pu and ^{240}Pu, respectively), a required feature in emergency response methodologies. Analytical bias of the method can be determined by calculating the slope between the expected and measured values for each plutonium isotope tested. Slope analysis demonstrated that ^{239}Pu and ^{240}Pu activities were overestimated and underestimated, respectively, by approximately 3%. Nonetheless, the degree of precision and accuracy obtained for these results undoubtedly demonstrates that the improvements presented herein improve the analytical detection capability of plutonium by using this on-line mass spectrometric setup.

In addition, to facilitating the detection of plutonium isotopes, mass spectrometry instrumentation can also provide valuable information on plutonium isotopic ratio. Figure 9 highlights the system capability to measure accurate isotopic ratios based on transient signal quantification. Note that the largest deviation from the expected measurement in Figure 9 is associated with the sample having the lowest plutonium activity being affected by background fluctuation and molecular interferences the most. Therefore, in order to obtain an accurate plutonium isotopic ratio, urine samples with plutonium activities higher than 50mBq L^{-1} should be used.

Summary

The plutonium urinalysis system presented in this chapter proposes a complete automated solution for emergency response bioassay. Theoretical and practical considerations to increase sample throughput have lead to the development of a system that can analyze over 95 samples daily with Pu activity in the mBq L^{-1} range. Reduction in sample volume consumption compared to the original methodology, a feature essential for a radiological/nuclear emergency response methodology, was also achieved by the introduction of an autosampling unit. The level of automation achieved is also another desired feature making it user friendly, minimizing human error and possible exposure to high doses of ionizing radiation. The use of simple calibration approach and data treatment facilitate data reporting, a bottle neck in laboratories subjected to short deadlines[4]. Finally, the use of reversed-flow chromatography helps decrease peak broadening and elution volume which can impact detection limits.

Section 3. Rapid Gamma Spectra Processing Using an Automated Analysis Platform

Introduction

Health Canada operates a variety of aerosol samplers for purposes such as routine environmental monitoring, nuclear emergency monitoring, counter-terrorism, and Comprehensive Nuclear Test Ban Treaty (CTBT) verification. Health Canada maintains approximately 30 samplers nationally, spread throughout the country as shown in Figure 11A.

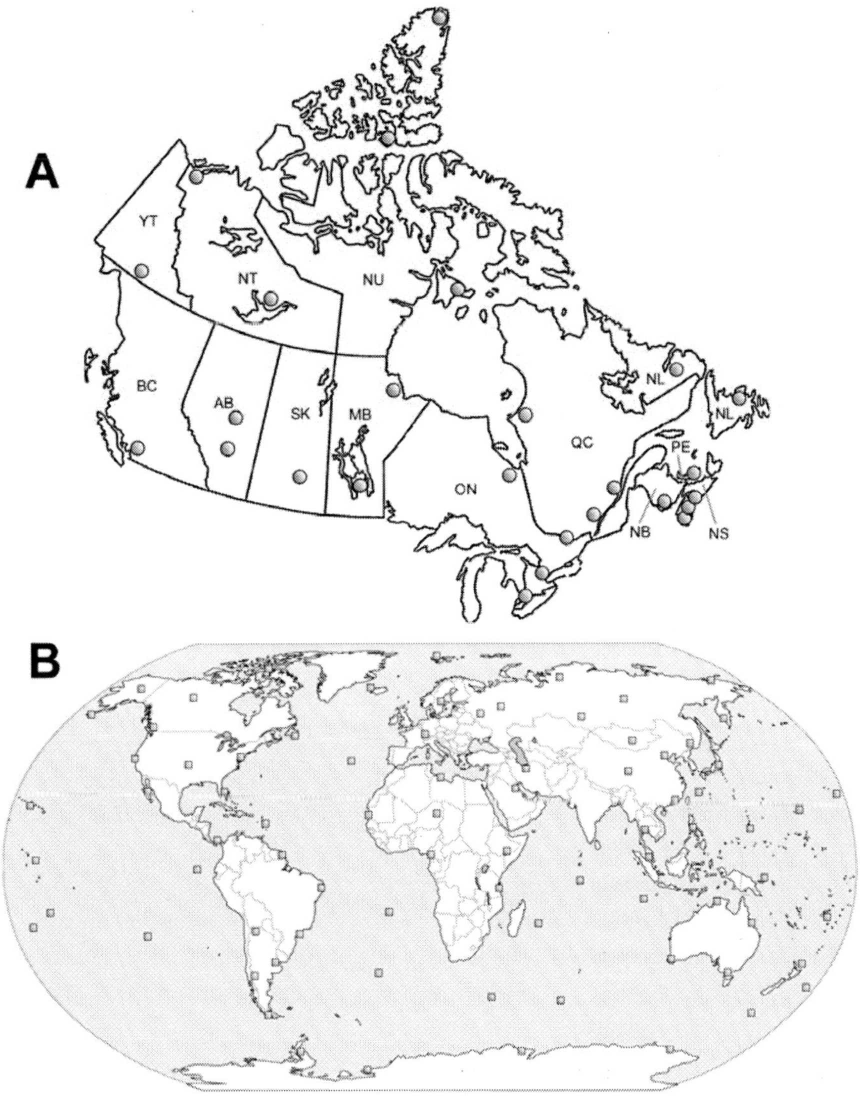

Figure 11. Map of A) Canadian Radiological Monitoring Network (CRMN) Facilities B) Comprehensive Nuclear Test Ban Treaty (CTBT) Radionuclide Monitoring Facilities.

The majority of the nationally managed samplers operate on a weekly sampling cycle. In the event of an emergency or situation of national interest, there are provisions to operate all samplers daily. The samplers used as the Canadian contribution to the international network to verify the CTBT and sites of special scientific interest are used in a daily sampling cycle. The Canadian CTBT nuclear network is complemented with raw data obtained from daily samples collected worldwide as part of the CTBT International Monitoring System (IMS) shown in Figure 11B. During high frequency sampling periods when large numbers of samples must be analyzed, it becomes an even greater operational challenge to maintain high-quality analyses while providing expedient results for decision makers. Health Canada's National Data Centre (NDC) must receive, process, report and archive sample information, sample raw spectral data, measurement parameters, and final results of the spectra from a large number of airborne particulate samplers utilizing different technologies, software and data formats. To accomplish these objectives in an efficient, timely and cost-effective manner, a high degree of automation is required during every stage of the monitoring process from sampling through to generation of final results. Key system components that have been automated or optimized are the: sampling, acquisition, analysis (rapid nuclide identification), radiological event categorization, centralized reporting, and electronic data sharing.

System Descriptions

Radiological monitoring networks. The CTBT IMS includes a network of 80 fully automated and manual radionuclide stations as shown in Figure 11B. These stations are equipped with samplers for airborne particulate monitoring and a field laboratory for measurement of samples as shown in Figure 12. The components are labeled as follows: a) air intake, b) barcode and timing tape, c) HPGe detector (note shielding has been removed in this photo), d) filter stacking and wrapping mechanism, e) polypropylene filters (on 6 arms), and f) exhaust fan. Intake air from the exterior environment is sucked through the filter by the exhaust fan. Airborne particulates are deposited in the filter on each of the sampling arms. The filters on each of the six arms are then stacked to form a single sample which is wrapped in mylar to prevent contamination and marked with a barcode and sample collection parameters. Once the sample is wrapped and marked, it is measured by the detector. A calibration source (not shown) is available to perform calibration of the detector which can be inserted remotely into the shield. Normally, these stations operate in a "short cycle" mode. This mode includes 24 hour air sampling by filtering air with a flow rate of 500 to 1000 m^3 h^{-1}, a 24 hour decay period to reduce interfering activity from ^{214}Pb and ^{214}Bi, and the acquisition of the sample gamma spectrum over approximately 20 hours in a compact geometry close to a HPGe high resolution detector (relative efficiency $\geq$ 40%; resolution $<$ 2.5 keV at 1332 keV; baseline sensitivity range from 10 to 30 $\mu Bq\ m^{-3}$ for ^{140}Ba).

Figure 12. IMS Radionuclide Aerosol Sampler/Analyzer (RASA) fully automated sampling system.

The Canadian Radiological Monitoring Network (CRMN) consists of 26 air sampling stations distributed throughout Canada as shown in Figure 11A with laboratory facilities located in Ottawa, Ontario. A polypropylene filter, 20 x 25 cm, is used for aerosol collection. Particulate collection efficiency of the filter is better than 85% at Ø = 0.1 μm and 98% at Ø = 0.4 μm. 12,600 m^3 of air is collected over a 7 day sampling period. During emergency operations, the samples are collected on a daily basis. Gamma spectrum and gross beta measurements are carried out on these filters. Gamma-ray spectrometry is performed using an HPGe high resolution detector. Figure 13 shows a typical high volume air sampler manufactured by Hi-Q Environmental Products (San Diego, CA).

Figure 13. A typical CRMN sampler for weekly sample collection.

Laboratory Facilities. Health Canada has a full service laboratory able to measure gamma spectra in a fully automated robotic measurement system as shown in Figure 14. These systems are relatively common for applications requiring high-throughput analyses. This system is fully integrated into a Laboratory Information Management System (LIMS) developed by LabWare (Wilmington, DE). The LIMS ensures samples can be tracked at all times by using barcodes for chain of custody purposes. All parameters associated with the sample including measurements and analyses are stored in the LIMS. While the robotic measurement systems can easily perform automated sample measurements, interactive spectral analysis are typically carried out by analysts using software packages such as Genie2000. However, should further analysis be desired, the LIMS is integrated with other high-performance automated gamma spectroscopy software platforms that use the raw spectral data stored in the LIMS and are described in the next section.

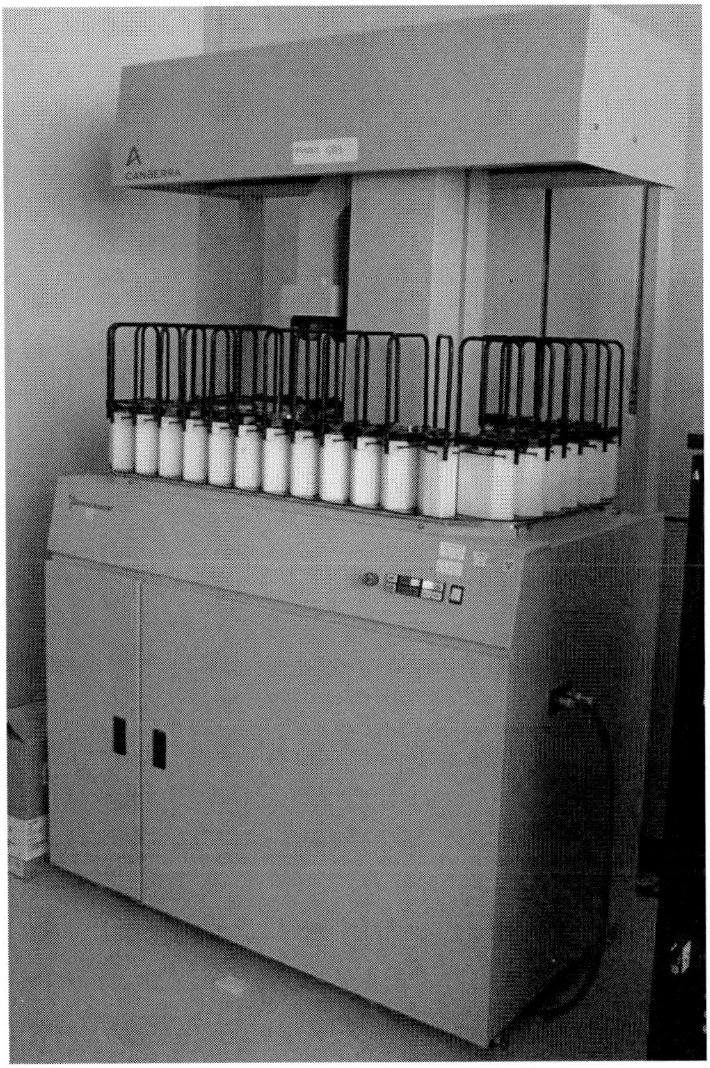

Figure 14. Robotic gamma spectroscopy system for high throughput Laboratories.

Software for gamma spectra analysis. With the large number samples that can be generated by the various sampling platforms and networks, automated software is used to process all the collected spectra, and perform as many tasks as possible automatically during the analysis. For samples from the CTBT network and samples needing further processing from the LIMS, there is a fully automated or interactive analysis software platform available. The software package consists of three linked components: Unisampo gamma spectrum analysis software for peak finding, peak energy and intensity determination, the Shaman expert system for nuclide identification, classification, and activity calculation, and Aatami for radionuclide analysis and evaluation software to perform both peak quantitative analysis and gamma line explanation. With this level of integration and automation, analyst review is not normally needed as most samples contain only natural background radionuclides that the software platform can easily identify. Further efficiencies and power arise when the results of all analyses are written to a database where an analyst can focus on cases of interest. This will be illustrated fully in the database section. For now, a closer look at two of the important or unique features of the Unisampo/Shaman analysis system will illustrate the power of this platform.

The first important feature is the automatic calibration done in Unisampo. Peak energy and shape calibrations are updated for each spectrum analysis. In this approach, large, well-shaped, known background peaks in the spectrum, such as 46.5keV of ^{210}Pb, 477.6keV of ^{7}Be, 238.6keV of ^{212}Pb, 583.2 of ^{208}Tl and so on, are used for energy and peak shape re-calibrations. The updated energy and shape calibrations are then reapplied to each spectrum. This is particularly important for daily automatic analysis to correct electronic gain drift causing channel-to-energy parameterization error and poor peak fitting. In Shaman, algorithms have been implemented for estimation of coincidence sum peaks, double and single escape as well as cosmogenic peaks. Based on the photopeak efficiency and the total efficiency of the detector, coincidence summing correction factors for each energy peak are calculated and used for activity correction automatically.

The second important feature is intelligent nuclide identification. Shaman uses the peak quantitative analysis results from Unisampo together with an internal comprehensive nuclide library of 3648 nuclides and 80062 gamma-ray and X-ray lines to identify, or name, each spectrum peak quantitatively within a specified energy tolerance. This library has been tuned for the purposes of environmental radiation monitoring, and in this case utilizes an "aerosol hypothesis" in choosing a nuclide. Therefore, identification decisions made by the software reflect the environmental monitoring design criteria; however the library is also designed to flag anthropogenic nuclides as being "suspicious".

The entire process from raw data delivery to final results being stored in a database from the automated system is under 5 seconds on a modern PC. The peaks that have been flagged as "suspicious" can be reviewed interactively by reprocessing the raw data, or by retrieving the spectral data from the database. Once retrieved, the analyst is free to review or change any of the analysis parameters such as the calibration and interpolation equation for energy, efficiency, and peak shape. The analyst can add or remove peaks and modify the chosen nuclide identification by looking at a list of possible nuclides. The entire automated analysis platform is further augmented by connection to a well-designed relational database.

The database system. Linux System for Spectral Information (LINSSI)[78] is an open-source environmental monitoring database schema implemented on a MySQL platform developed in partnership with Radiation Physics Group of Helsinki University of Technology (RPG/HUT)[79], Finnish National Data Centre (FiNDC), the Aerosol Laboratory of the Finnish Radiation and Nuclear Safety Authority (ASL/STUK), and RPB/HC.

The LINSSI database system consists of the database and the necessary scripts to do reporting and database administration through a combination of web pages and command line scripts. The database is designed for mobile measurements, laboratory measurements, and field measurements of environmental radiation. It can provide data for scientific studies and operational requirements by interfacing different analysis tools in a common platform. The website[80], describes the structure and contents of the database in a PDF format.

There are many advantages of using a database for storage of environmental radiation monitoring data. During routine operations, analysts can rely on the automated software possessing and use the database to determine which samples need further investigation. Due to the design of the relational database, one can view all parameters associated by station, sample, measurement, analysis or any combination of these. For example, samples that are missing ever-present atmospheric nuclides such as ^{7}Be, ^{220}Rn and ^{222}Rn progeny indicate a potential measurement, calibration or analysis problem. All measurements and analysis results are stored in a central repository and can be retrieved for further analysis, review, and study. This allows analysts to quickly pick out "suspicious" events, and to focus on them for an in-depth analysis. The ability to focus on unusual phenomena rather than the more mundane typical sample processing is one of the advantages of the system, which has led to direct improvements in the nuclide identification algorithm. Since the processing software platform is linked to the database, any results from other software that can be stored in the database can be reviewed and reprocessed with Unisampo, Shaman, and Aatami or web-based tools. This capability also allows intercomparisons amongst different analysis platforms. By using the database, a central repository, generation of sample reports, nuclide trends, or any other parameter can be made on yearly, monthly, or any other time-scale desired as soon as measurements are complete.

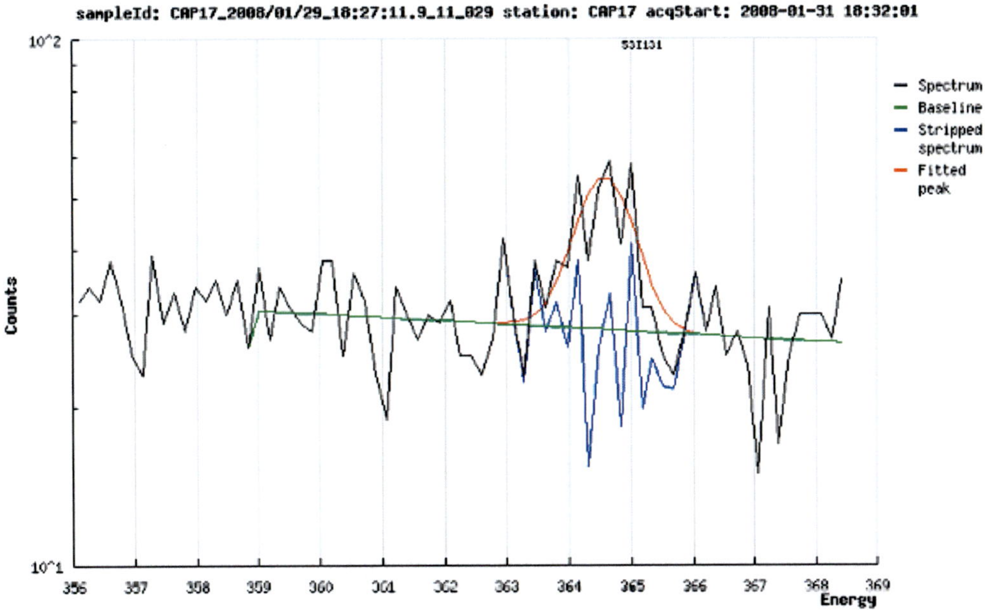

Figure 15. Close up view of ^{131}I Peak.

To illustrate the power and utility of a database for storing and reviewing spectroscopy results, it may help to look at an example. Suppose we are interested in observations of ^{131}I, a common medical isotope used for thyroid treatment and diagnosis that is not typically found in aerosols, at the monitoring station in St. John's, Newfoundland. A query of the database through a web application developed specifically for LINSSI will rapidly show that the most recent observation of ^{131}I at St. John's was detected in the sample from January 29, 2008. In our ^{131}I example, the existence of a peak at approximately 364.5 keV could indicate the presence of Iodine in the sample. After displaying the spectra in the web application, one can zoom in on the region of interest as shown in Figure 15 and decide on the validity of the peak. Note the figure shows the channel counts, peak fit, baseline, and stripped spectra to aid in the review process. The peak is well defined, thus confirming the presence of the peak, and supporting the hypothesis of ^{131}I being the cause. Additionally, the web application allows the analyst other important options besides the visual review; various textual reports can be generated in PDF format, and finally, the results or raw data can be transformed into either the original or a different data format for further processing or reporting.

In an emergency situation, it is the ability to quickly: inventory, process, and analyze samples to focus on specific parameters that can differentiate between a positive and negative outcome in emergency response. The ^{131}I query serves to illustrate how easy it is to retrieve from the database any possible parameter or combination of parameters to assist decision-makers in planning their emergency response operations.

Results and Discussions

The automatic pipeline has been operational at the Health Canada radionuclide NDC since April 2004. By issuing a simple query to the LINSSI database a graph such as the one in Figure 16 or 17 can be created in seconds. Furthermore, it is also equally trivial to generate histograms of nuclide observation frequency in order to examine the types of nuclides observed and focus on specific nuclides of interest.

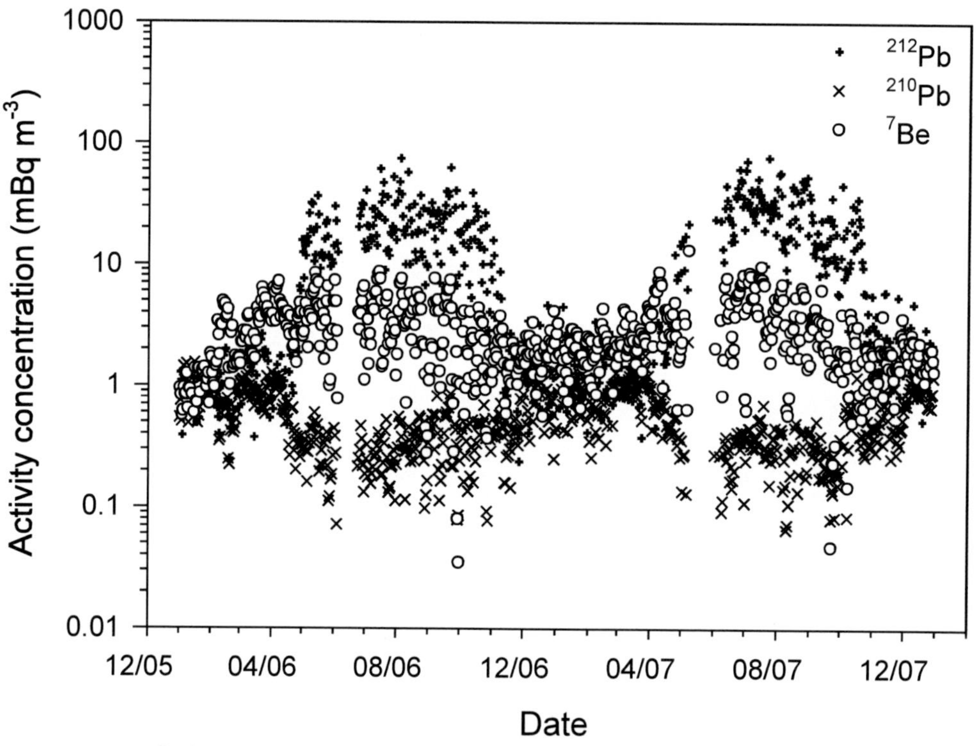

Figure 16. Trend plots of ^{7}Be, ^{210}Pb and ^{212}Pb at Yellowknife station from 01/01/2006 to 12/31/2007.

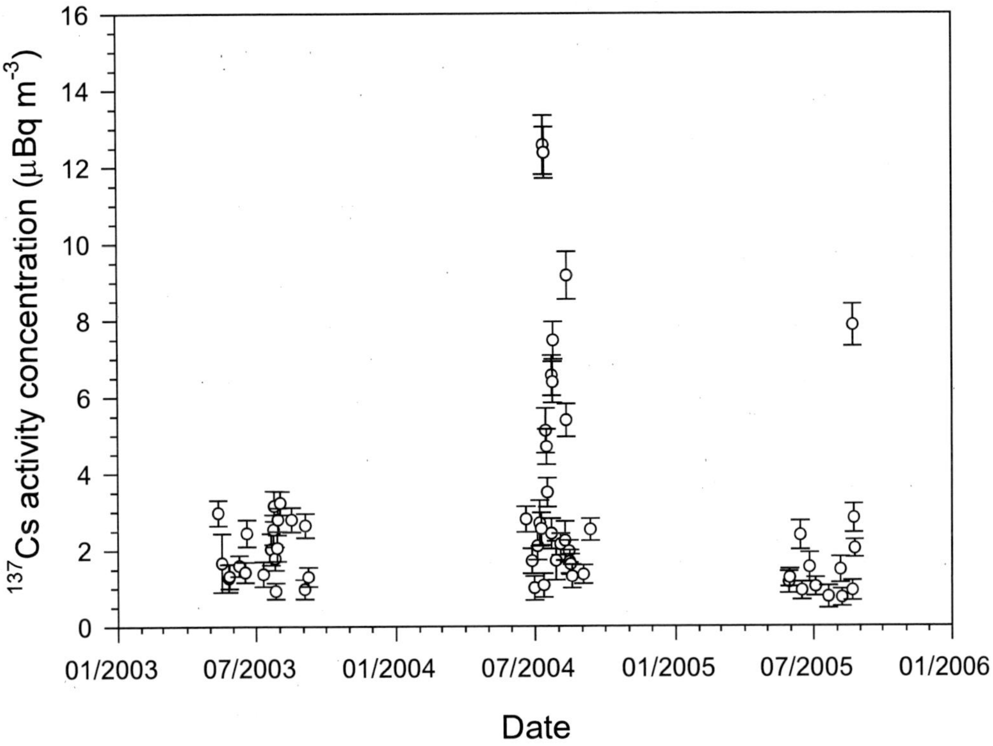

Figure 17. Daily ^{137}Cs observed at Yellowknife monitoring station from 01/01/2003 to 01/01/2006.

The CTBT and CRMN networks are both very sensitive systems giving clear signals for natural radioactivity, including cosmogenic ^{7}Be, and ^{220}Rn and ^{222}Rn progeny. Figure 16 gives monitoring results for two years of ^{7}Be, ^{210}Pb and ^{212}Pb from Yellowknife station showing nearly continuous environmental monitoring. When the environment is monitored continuously and data quality can be verified, analysts can focus on unusual events such as the ^{131}I example above. Storing naturally occurring radioactive nuclides in the database also serves a further purpose. By examining the activity of natural nuclides it is possible to map air circulation. This assists atmospheric modelling specialists in the development of atmospheric transport models (ATM) which are used in an emergency situation to predict or "backtrack" aerosol transport and to estimate nuclide deposition and washout. To illustrate the need for to combine ATM and environmental monitoring, we look at the following example which clearly illustrates how unusual observations and ATM can identify the source location and show the transport pathway.

Radiocaesium is an environmental contaminant, primarily resulting from nuclear weapons testing and nuclear accident releases. Gustafson *et al.*[81] estimated the cumulative ^{137}Cs deposition on the Earth's surface (on soil and vegetation) and concluded that the fallout by the end of 1960s ranged between 2.2 and 3.7 GBq km^{-2} over most of the USA. Daily observations from Yellowknife station for ^{137}Cs are plotted in Figure 17. In winter, the ^{137}Cs concentration in airborne particulate is lower than the detection limit, thus unobservable. In summer, the observed ^{137}Cs concentration typically ranges from 0.002 to 0.003 (mBq m^{-3}).

It is worth noting that unusually high ^{137}Cs concentrations were observed sequentially during the summer of 2004. Concentrations ranged from 0.006 to 0.014 (mBq m^{-3}), exceeding the typical mean value by a factor between 3 to 7. These results strongly indicate local re-suspension of ^{137}Cs. Satellite imaging and meteorological plume modeling both suggest the high levels of ^{137}Cs were associated with forest fires occurring 10 to 200 km distant from Yellowknife. These observations demonstrate that long-range transport of radionuclides that were originally deposited in forests after nuclear weapons tests were released into the atmosphere, decades later, by forest fires.[82] Although this example was ultimately traced back to natural causes, the same process would be used in an emergency situation to improve emergency response by providing a more accurate assessment of the contaminated area.

System validation and inter-comparison. A comparison study was carried out between the automated gamma spectra pipeline processing capability in the domains of peak finding, nuclide identification and activity calculation and the corresponding results from manual review by analysts. Radiocesium was used for this purpose because of its frequent observation and low activity concentrations. In the study, about 700 spectra from Yellowknife and Ottawa stations were processed manually with the Genie2000 software in parallel with Unisampo/Shaman automatic processing. For each manually processed spectrum, ^{137}Cs was assumed to be present in the sample, and a peak was inserted at 661.8 keV. Interactive review was then employed to determine if the peak insertion can be statistically justified or is indistinguishable from the statistical variability of the baseline. The ^{137}Cs activity concentrations obtained by the automatic pipeline were queried directly from LINSSI. Obtained ^{137}Cs activities in the same spectra are plotted in Figure 18. It is apparent that the measured activities of the two methods are evenly distributed on both sides of a one-to-one correlation line. Therefore, within measurement uncertainty, agreement between the two methods is good.

As shown in Figure 18, when activities are greater than 1.2:Bq m^{-3}, ^{137}Cs identification between the two methods was in full agreement. For activities on the order of Currie's L_c value[83], six disagreements were reported between automatic Unisampo/Shaman processing and interactively reviewed Genie2000 analysis. In-depth investigation reveals the six discrepancies involved missing four real peaks and the reporting of two false positive peaks; these "discrepancy" points are shown on the y-axis and the x-axis, respectively, of Figure 18. Ideally, spectrum for a ^{137}Cs standard should be used for this comparison, but a real scenario study was deemed acceptable. If we assume human interactive review is more reliable, we would conclude that the automatic pipeline missed four real peaks and reported two false positives, *i.e.* took spurious peaks as real ones. Out of 700 processed spectra, the false positive rate of the automatic pipeline was only 0.3%. Missed peaks are due to the default settings of the peak search sensitivity algorithm. For example, when a small peak lies near a large one, the small peak may be missed by the peak search algorithm. If a robust baseline fitting algorithm is applied to spectrum analysis, these types of errors can be eliminated by checking the residual between the fitted data and actual channel counts.

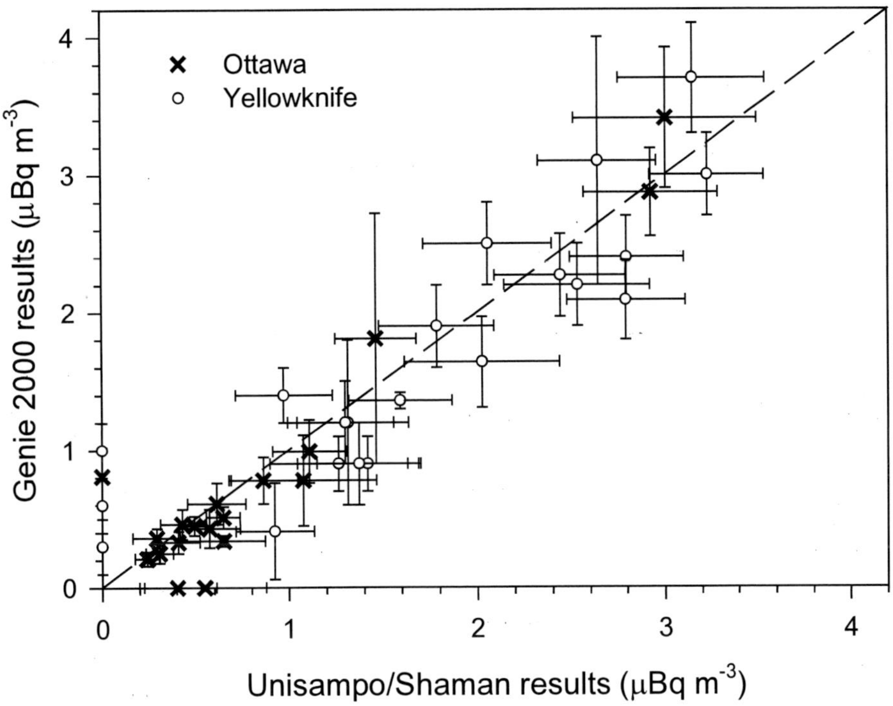

Figure 18. Correlations between automatic Unisampo/Shaman pipeline and manually processed Genie 2000 ^{137}Cs activity concentration.

A two phase radionuclide inter-comparison exercise was organized by CTBT to evaluate their data processing capabilities in relation to those of the NDCs. In the first phase, participating NDCs processed real-world spectra from daily measurements at radionuclide stations in the IMS. In the second phase, using Monte Carlo generated artificial spectra, a total of 430 anthropogenic nuclides were artificially "spiked" into 100 real background spectra at varying, but known concentrations. The spectra were then delivered to the participants for performance assessment of their radionuclide analysis software and procedures. These artificial spectra were designed to challenge the capability of the participants to identify peak signals near the background level.[84]

In the first phase, spectral sets provided excellent data to perform a nuclide-finding benchmark study. All these real-world spectra were automatically and interactively analyzed with the platform. This test demonstrated that the automated analysis process has high accuracy in the nuclide identification. One hundred interactive and automated analysis results are collected together and plotted in Figure 19. The listed nuclides are most frequently observed in these real-world spectra. This study clearly indicates that when nuclide activity concentration is more than 4 times the MDC, the nuclide can be identified automatically with 100% accuracy. The missed events were mainly due to signals too small to be distinguishable from the statistical variability of the baseline or because of difficulties in X-peak deconvolution, such as ^{234}Th.

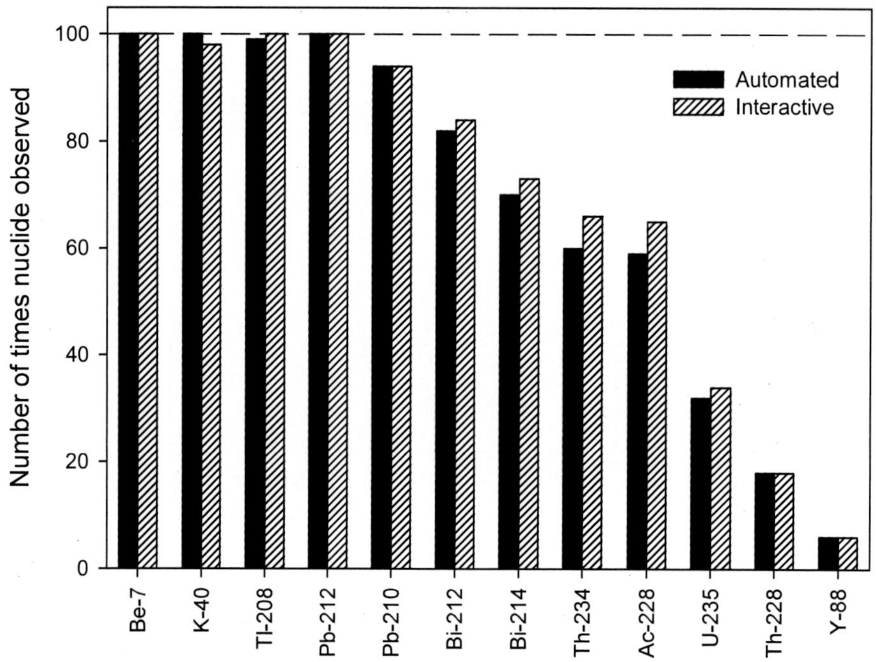

Figure 19. Nuclide identification capability comparisons between interactive and automated analysis for selected nuclides in 100 spectra.

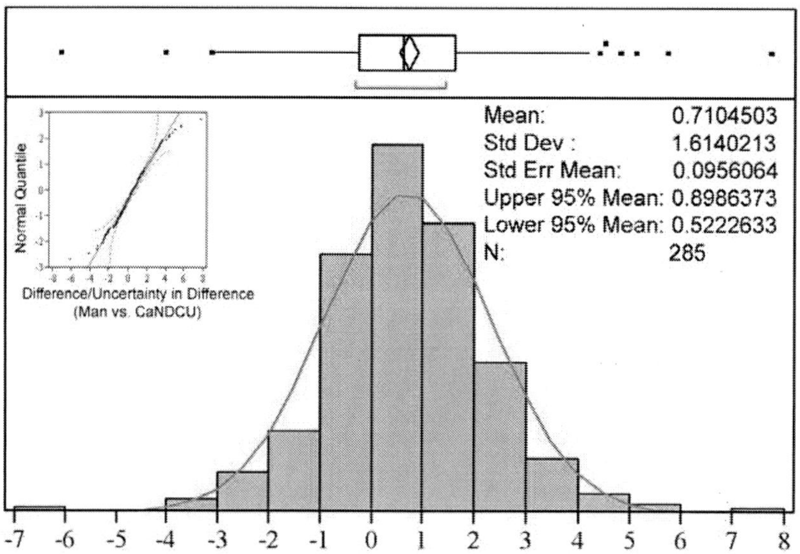

Figure 20. Normal distribution of activity difference between observed and reference value.

Of the 430 "spiked" nuclides used during the second phase, 285 were identified by Health Canada's NDC. One study analysed the standardised residuals (*i.e.* difference between measured and true value divided by calculated uncertainty). The mean and empirical standard deviation of the standardised residuals were computed to estimate the normal distribution of the residuals, the existence of a biais, or an unknown uncertainty (see Figure 20). It is found that these residuals are normally distributed. However, a small average bias has been observed, and there is about 60% more uncertainty than is predicted by the individual uncertainty. Similar results were observed by the IDC and other NDCs. The bias may be explained by the imposition of a decision threshold applied to small peaks.

Summary

The automatic analysis platform, Unisampo/Shaman, Aatami and LINSSI, used by Health Canada has been processing gamma spectra for over four years. The implementation of large-scale automation in this system represents a significant improvement over the standard approach of performing manual interactive analysis with unconnected systems. The efficiency improvement as a result of a fully automated analysis procedure would greatly accelerate the processing of environmental monitoring data in the event of nuclear accidents, nuclear explosions, unintended releases of radioactive materials, and terrorist attacks, in turn, affecting the effectiveness of emergency responses. Over 90 000 spectra, an overwhelming number that would be impossible to analyze interactively, have been processed since the system began operation. The large amount of radionuclide information saved in the database provides a valuable archive for scientific research activities including non-nuclear scientific domains such as climate and atmospheric studies. Inter-comparison exercises have demonstrated that the automated platform has a level of performance nearly comparable to that of analyst-reviewed in areas of rapid peak search, nuclide identification and event reporting and is well suited to its applications of treaty verification, nuclear emergency response and environmental gamma spectroscopy. The large amount of gamma spectra analysis parameters stored in the LINSSI database and ease of data access provide a foundation to create useful tools for statistical studies. For example, prediction limits of false-positive peak and false-negative identification rates at different peak intensities are a common useful study. The database-compiled results can be used as input for gamma spectrum analysis and signal categorization of different radioactivity levels as well as a guideline on deciding whether an analytical procedure satisfies requirements and thus is suitable for measurement.

Conclusion

As demonstrated throughout this chapter, the analytical performances of automated systems are effective to assess radiological contamination found in a variety of matrices including urine, milk, and air filters, at levels that are sufficiently low to meet action level requirements. Using instrumentation as varied as ICP-MS, FSA, and automated γ-spectrometry, it is possible to develop rapid methodologies that cover a large variety of radioemitters (α, β, γ) likely to be uncovered during R/N events. During the development of automated methodologies for emergency readiness, considerations with respect to multi-analyte capability should be evaluated, as those methodologies would be more beneficial with respect to analytical throughput. For example, the methodology presented in section 1 also enables the separation of radium without significantly affecting sample throughput, a valuable feature especially for samples only available in limited quantity. Ideally, the automated methodologies should be usable for non-emergency monitoring, supporting their implementation and sustainability. For instance, the methodologies described in sections 1 and 2 have been used with little modification for the routine measurement of radiostrontium in milk and water[85] and plutonium in urine[8], respectively.

Finally, the development of certified reference materials (CRMs) and intercomparison exercises aimed at assessing the analytical performances of radioanalytical laboratories for emergency readiness are critical to enable the testing and development of new rapid methodologies (automated or not). While a general requirement for QA/QC, the availability of CRMs is essential to assess the analytical performances of emergency response methodologies as samples collected during real R/N emergency situation are difficult to obtain and actual emergencies are, thankfully, rare.

Acknowledgements

The authors would like to thank Amila DeSilva, Thomas A. Cumming, Keith T. Henderson, Vera Kochermin, Stephen Kiser, Dr. Chunsheng Li, and Ann Porada for their help in the development of the automated systems, STUK (Radiation and Nuclear Authority, Finland) and Helsinki University of Technology for their work on LINSSI and Unisampo/Shaman, and Dr. N. S. St-Amant and D. Beaton for their thorough reviewing of this manuscript. Funding for sections 1 and 2 was provided by Health Canada, and the CBRNE Research and Technology Initiative (CRTI), respectively.

References

[1] Green, N., An evaluation of rapid methods of radionuclide analysis for use in the aftermath of an accident. *Science of the Total Environment* 1993, 130, 207-218.

[2] Wilber, D.; Daigler, D.; Nielsen, E. C.; Riedhauser, S. R.; Shanks, A.; Thompson, R. C.; Nasstrom, J. S., Nuclear/Radiological emergency response in the USA. *International Journal of Emergency Management* 2007, 4, (3), 339-355.

[3] Karam, P. A., Radiological Terrorism. *Human and Ecological Risk Assessment* 2005, 11, (3), 501-523.

[4] Inn, K. G. W.; Outola, I.; Nour, S.; Kurosaki, H.; Albin, L.; Berne, A., Radioanalytical emergency response exercise. *Journal of Radioanalytical and Nuclear Chemistry* 2006, 269, (2), 351-360.

[5] Inn, K. G. W.; Outola, I.; Nour, S.; Kurosaki, H.; La Rosa, J. J., NRIP'06 Workshop. In *52nd Annual Radiobioassay and Radiochemical Measurements Conference*, Chicago, IL, 2006.

[6] Ohtsuka, Y.; Takaku, Y.; Nishimura, K.; Kimura, J.; Hisamatsu, S.; Inaba, J., Rapid method for the analysis of plutonium isotopes in a soil sample within 60 min. *Analytical Sciences* 2006, 22, (2), 309-311.

[7] Epov, V. N.; Benkhedda, K.; Cornett, R. J.; Evans, R. D., Rapid determination of plutonium in urine using flow injection on-line preconcentration and inductively coupled plasma mass spectrometry. *Journal of Analytical Atomic Spectrometry* 2005, 20, (5), 424-430.

[8] Lariviere, D.; Cumming, T. A.; Kiser, S.; Li, C.; Cornett, R. J., Automated flow injection system using extraction chromatography for the determination of plutonium in urine by inductively coupled plasma mass spectrometry. *Journal of Analytical and Atomic Spectrometry* 2008, 23, 352-360.

[9] Pappas, R. S.; Ting, B. G.; Paschal, D. C., Rapid analysis for plutonium-239 in 1 ml of urine by magnetic sector inductively coupled plasma mass spectrometry with a desolvating introduction system. *Journal of Analytical Atomic Spectrometry* 2004, 19, (6), 762-766.

[10] Elliot, N. L.; Bickel, G. A.; Linauskas, S. H.; Paterson, L. M., Determination of femtogram quantities of Pu-239 and Pu-240 in bioassay samples by thermal ionization mass spectrometry. *Journal of Radioanalytical and Nuclear Chemistry* 2006, 267, (3), 637-650.

[11] Horwitz, P.; McAlister, D.; Dietz, M., Extraction Chromatography Versus Solvent Extraction: How Similar are They? *Separation Science and Technology* 2006, 41, (10), 2163-2182.

[12] Kim, C. S.; Kim, C. K., Determination of Pu isotopes in seawater by an on-line sequential injection technique with sector field inductively coupled plasma mass spectrometry. *Analytical Chemistry* 2002, 74, (15), 3824-3832.

[13] Grate, J. W.; Egorov, O. B., Automating analytical separations in radiochemistry. *Analytical Chemistry* 1998, 70, (23), 779A-788A.

[14] Grate, J. W.; Strebin, R.; Janata, J.; Egorov, O.; Ruzicka, J., Automated analysis of radionuclides in nuclear waste: Rapid determination of Sr-90 by sequential injection analysis. *Analytical Chemistry* 1996, 68, (2), 333-340.

[15] Egorov, O. B.; O'Hara, M. J.; Addelman, R. S.; Grate, J. W., Automation of radiochemical analysis: from groundwater monitoring to nuclear waste analysis. In *Radioanalytical Methods in Interdisciplinary Research: Fundamentals in Cutting-Edge Applications*, Laue, C. A.; Nash, K. L., Eds. ACS Symposium Series: American Chemical Society Publication 2004; pp 246-270.

[16] Grate, J. W.; Egorov, O. B., Chapter 14 - Automated radiochemical separation, analysis, and sensing. In *Handbook of Radioactivity Analysis*, 2nd ed.; L'Annunziata, M. F., Ed. Elsevier Science: San Diego, CA, 2003; pp 1129-1164.

[17] Miley, H. S.; Aalseth, C. E., Automated Laboratory and Field Radionuclide Analysis System. In *Radioanalytical Chemistry*, Kahn, B., Ed. Springler Science + Business Media, LLC: New York, NY, 2007; pp 318-361.

[18] Epov, V. N.; Benkhedda, K.; Evans, R. D., Determination of Pu isotopes in vegetation using a new on-line FI-ICP-DRC-MS protocol after microwave digestion. *Journal of Analytical Atomic Spectrometry* 2005, 20, (9), 990-992.

[19] Lariviere, D.; Taylor, V. F.; Evans, R. D.; Cornett, R. J., Radionuclide determination in environmental samples by inductively coupled plasma mass spectrometry. *Spectrochimica Acta - Part B* 2006, 61, (8), 877-904.

[20] Maxwell, S., Rapid analysis of emergency urine and water samples. *Journal of Radioanalytical and Nuclear Chemistry* 2008, 275, (3), 497-502.

[21] Klusek, C. S., Sr-90 in food and bone from fallout. *Journal of Environmental Quality* 1987, 16, (3), 195-199.

[22] Muck, K.; Sinojmeri, M.; Whilidal, H.; Steger, F., The long-term decrease of Sr-90 availability in the environment and its transfer to man after a nuclear fallout. *Radiation Protection Dosimetry* 2001, 94, (3), 251-259.

[23] Hodgson, S. A.; Ham, G. J.; Youngman, M. J.; Etherington, G.; Stradling, G. N., A review of measurements of radionuclides in members of the public in the UK. *Journal of Radiological Protection* 2004, 24, (4), 369-389.

[24] Denschlag, H. O.; Diel, A.; Glasel, K. H.; Heimann, R.; Kaffrell, N.; Knitz, U.; Menke, H.; Trautmann, N.; Weber, M.; Herrmann, G., Fallout in the Mainz Area from the Chernobyl Reactor Accident. *Radiochimica Acta* 1987, 41, (4), 163-172.

[25] Swindle, D. L.; Kuroda, P. K., Variation of Sr89/Sr90 Ratio in Rain Caused by Chinese Nuclear Explosions of December 28,1966 and June 17,1967. *Journal of Geophysical Research* 1969, 74, (8), 2136-2140.

[26] Brun, S.; Kergadallan, Y.; Boursier, B.; Fremy, J. M.; Janin, F., Methodology for determination of radiostrontium in milk: a review. *Lait* 2003, 83, (1), 1-15.

[27] Brun, S.; Bessac, S.; Uridat, D.; Boursier, B., Rapid method for the determination of radiostrontium in milk. *Journal of Radioanalytical and Nuclear Chemistry* 2002, 253, (2), 191-197.

[28] AlMasri, M. S., Cerenkov counting technique. *Journal of Radioanalytical and Nuclear Chemistry-Articles* 1996, 207, (1), 205-213.

[29] Chang, T. M.; Chen, S. C.; King, J. Y.; Wang, S. J., Rapid and accurate determination of Sr-89/90 in radioactive samples by Cerenkov counting. *Journal of Radioanalytical and Nuclear Chemistry-Articles* 1996, 204, (2), 339-347.

[30] Poletiko, C.; Lemercier, R.; Leiba, B.; Llug, P., Measurement of Sr-90 by the Cerenkov counting technique. *Journal of Radioanalytical and Nuclear Chemistry-Articles* 1994, 178, (1), 131-141.

[31] Torres, J. M.; Garcia, J. F.; Llaurado, M.; Rauret, G., Rapid determination of strontium-90 in environmental samples by single Cerenkov counting using two different colour quench curves. *Analyst* 1996, 121, (11), 1737-1742.

[32] Walter, K.; Trautmann, N.; Herrmann, G., Simultaneous determination of Sr-90 and Sr-89 by Cerenkov-radiation and liquid-scintillation counting - a reevaluation for low-level counting. *Radiochimica Acta* 1993, 62, (4), 207-212.

[33] Celiktas, C., Beta spectrometers with surface barrier detector and plastic scintillator: applications to 90Sr, 204Tl, 210Pb and 14C. *Turkish Journal of Physics* 2001, 25, (2), 97-107.

[34] Dabrowski, A.; Jagusztyn, M.; Roman, J., *Strontium-90-yttrium-90 measurements using silicon counters*; Centralne Lab. Ochrony Radiologicznej,: Warsaw, Poland, 1966; p 18.

[35] Baumgartner, F.; Kim, M. A.; Probst, T., Comparative study of radiometric and mass-spectrometric detection limits of rapid Sr-90 Y-90 determination. *Radiochimica Acta* 1993, 61, (3-4), 235-240.

[36] Taylor, V. F.; Evans, R. D.; Cornett, R. J., Determination of Sr-90 in contaminated environmental samples by tuneable bandpass dynamic reaction cell ICP-MS. *Analytical and Bioanalytical Chemistry* 2007, 387, (1), 343-350.

[37] Vonderheide, A. P.; Zoriy, M. V.; Izmer, A. V.; Pickhardt, C.; Caruso, J. A.; Ostapczuk, P.; Hille, R.; Becker, J. S., Determination of Sr-90 at ultratrace levels in urine by ICP-MS. *Journal of Analytical and Atomic Spectrometry* 2004, 19, (5), 675-680.

[38] Berryman, N.; Probst, T., Rapid determination of Sr-90 by electrothermal vaporization inductively coupled plasma mass spectrometry (ETV-ICP-MS). *Radiochimica Acta* 1997, 76, (4), 191-195.

[39] Grinberg, P.; Willie, S.; Sturgeon, R. E., Determination of natural Sr and Sr-90 in environmental samples by ETV-ICP-MS. *Journal of Analytical and Atomic Spectrometry* 2007, 22, (11), 1409-1414.

[40] Hou, X. L.; Roos, P., Critical comparison of radiometric and mass spectrometric methods for the determination of radionuclides in environmental, biological and nuclear waste samples. *Analytica Chimica Acta* 2008, 608, (2), 105-139.

[41] Sunderman, D. N.; Townley, C. W., *The radiochemistry of Barium, Calcium, and Strontium*; NAS-NS-3010; National Academy of Sciences: Washington, DC, 1960.

[42] Hunt, W. C., *Personal Communication*, 2008

[43] Grahek, Z.; Eskinja, I.; Kosutic, K.; Lulic, S.; Kvastek, K., Isolation of radioactive strontium from natural samples: A semi-automatic procedure. *Journal of Radioanalytical and Nuclear Chemistry* 1999, 241, (3), 617-626.

[44] Grate, J. W.; Fadeff, S. K.; Egorov, O., Separation-optimized sequential injection method for rapid automated analytical separation of Sr-90 in nuclear waste. *Analyst* 1999, 124, (2), 203-210.

[45] Hastie, K. H.; DeVol, T. A.; Fjeld, R. A., Development of a homogeneous pulse shape discriminating flow-cell radiation detection system. *Nuclear Instrumentation Methods and Physics Research* 1999, 422, (1-3), 133-138.

[46] Wade, M. A.; Seim, H. J., Ion Exchange Separation of Calcium and Strontium - Application to Determination of Total Strontium in Bone. *Analytical Chemistry* 1961, 33, (6), 793-795.

[47] Povondra, P.; Sulcek, Z., Determination of calcium in strontium salts and strontium in calcium-rich materials. *Chemist-Analyst* 1961, 50, 79, 93-94.

[48] Alfaro, J.; Apfel, T.; Diercks, H.; Knochel, A.; Sengupta, R.; Todter, K., Trace Analysis of the Radionuclides Sr-90 and Sr-89 in Environmental-Samples .3. Development of a Fast Analytical Method. *Angewandte Chemie International Edition* 1995, 34, (2), 186-189.

[49] Abbadi, S.; Diercks, H.; Knochel, A.; Sen Gupta, R.; Todter, K., Rapid procedures for the determination of radioactive strontium isotopes in food and environmental samples. *Kerntechnik* 1997, 62, 91-95.

[50] Strelow, F. W. E.; Weinert, C. H. S., Comparative Distribution Coefficients and Cation-Exchange Behaviour of Alkaline Earth Elements with Various Complexing Agents. *Talanta* 1970, 17, (1), 1-12.

[51] Bio-Rad Laboratories, *AG50W and AGMP-50 Cation Exchange Resins - Instruction Manual*; LIT203 Rev. B; Bio-Rad Laboratories: Hercules, CA, p 28.

[52] McNaught, A. D.; Wilkinson , A., *International Union of Pure and Applied Chemistry (IUPAC) - Compendium of Chemical Terminology*. 2nd ed.; Blackwell Science: Oxford, UK, 1997; p 464.

[53] Bureau of Chemical Safety / Health Products and Food Branch, *Average concentrations (ng/g) of trace elements in foods for Total Diet Study from 1993 to 1999*. http://www.hc-sc.gc.ca/fn-an/alt_formats/hpfb-dgpsa/pdf/surveill/metal_conc_plomb_93-99_e.pdf (*last accessed 11-04-2008*),

[54] L'Annunziata, M. F., Chapter 5 - Liquid Scintillation Analysis: Principles and Practice. In *Handbook of Radioactivity Analysis*, L'Annunziata, M. F., Ed. Elsevier Science: San Diego, CA, 2003; p 1326.

[55] PerkinElmer, *Appendix B: Principles of Radio-HPLC Flow Detection*; Publication No. 1694243 Rev. C; PerkinElmer Inc: Downers Grove, IL, 2006, p 30.

[56] Schuster, P.; Wagner, J., Holographic velocimetry for flow diagnostics. *Experimental Mechanics* 1988, 28, (4), 402-408.

[57] L'Annunziata, M. F., Chapter 12 - Flow Scintillation Analysis. In *Handbook of Radioactivity Analysis*, 2nd ed.; L'Annunziata, M. F., Ed. Elsevier Science: San Diego, CA, 2003; pp 989-1054.

[58] Kessler, M. J., *Liquid scintillation Analysis: Science and Technology*, ; 169-3052, Rev. G; Packard Instrument Co,: Meriden, CT, 1989; pp 5-16.

[59] Walter, K.; Trautmann, N.; Herrmann, G., Simultaneous Determination of Sr-90 and Sr-89 by Cerenkov-Radiation and Liquid-Scintillation Counting - a Reevaluation for Low-Level Counting. 1993, 62, (4), 207-212.

[60] McAninch, J. E.; Hamilton, T. F., *Measurement of Plutonium and other Actinide Elements at the Center for Accelerator Mass Spectrometry: A Comparative Assessment of Competing Techniques*; LANL Technical Report UCRL-ID-133118; Lawrence Livermore National Laboratory: Livermore, CA, 1999; p 28.

[61] Stricklin, D. L.; Tjarnhage, A.; Nygren, U., Application of low energy gamma-spectrometry in rapid actinide analysis for emergency preparedness. *Journal of Radioanalytical and Nuclear Chemistry* 2002, 251, (1), 69-74.

[62] Kim, C.-S.; Kim, C.-K.; Martin, P.; Sansone, U., Determination of Pu isotope concentrations and isotope ratio by inductively coupled plasma mass spectrometry: a review of analytical methodology. *Journal of Analytical Atomic Spectrometry* 2007, 22, 827-841.

[63] Marchetti, A. A.; Brown, T. A.; Cox, C. C.; Hamilton, T. F.; Martinelli, R. E., Accelerator mass spectrometry of actinides. *Journal of Radioanalytical and Nuclear Chemistry* 2005, 263, (2), 483-487.

[64] Priest, N. D.; Pich, G. M.; Fifield, L. K.; Cresswell, R. G., Accelerator mass spectrometry for the detection of ultra-low levels of plutonium in urine, including that excreted after the ingestion of Irish sea sediments. *Radiation Research* 1999, 152, (6), S16-S18.

[65] Lewis, D.; Miller, G.; Duffy, C. J.; Efurd, D. W.; Inkret, W. C.; Wagner, S. E., Los Alamos National Laboratory thermal ionization mass spectrometry results from intercomparison study of inductively coupled plasma mass spectrometry, thermal ionization mass spectrometry, and fission track analysis of mu Bq quantities of Pu-239 in synthetic urine (LA-UR-001698). *Journal of Radioanalytical and Nuclear Chemistry* 2001, 249, (1), 115-120.

[66] Li, C.; Lariviere, D.; Kiser, S.; Moodie, G.; Falcomer, R.; Elliot, N.; Burchart, L.; Paterson, L.; Epov, V.; Evans, R. D.; Pappas, R. S.; Smith, J.; Cornett, R. J., Method intercomparison for the analysis of $^{239/240}$Pu in human urine. *Journal of Analytical Atomic Spectrometry* 2008, 23, (4), 521 - 526.

[67] Erdmann, N.; Herrmann, G.; Huber, G.; Kohler, S.; Kratz, J. V.; Mansel, A.; Nunnemann, M.; Passler, G.; Trautmann, N.; Turchin, A.; Waldek, A., Resonance ionization mass spectroscopy for trace determination of plutonium in environmental samples. *Fresenius Journal of Analytical Chemistry* 1997, 359, (4-5), 378-381.

[68] Hang, W.; Zhu, L. W.; Zhong, W. W.; Mahan, C., Separation of actinides at ultra-trace level from urine matrix using extraction chromatography-inductively coupled plasma mass spectrometry. *Journal of Analytical Atomic Spectrometry* 2004, 19, (8), 966-972.

[69] Horwitz, E. P.; Dietz, M. L.; Chiarizia, R.; Diamond, H.; Maxwell, S. L.; Nelson, M. R., Separation and preconcentration of actinides by extraction chromatography using a supported liquid anion exchanger: application to the characterization of high-level nuclear waste solutions. *Analytica Chimica Acta* 1995, 310, (1), 63-78.

[70] Longerich, H. P.; Jackson, S. E.; Gunther, D., Laser ablation inductively coupled plasma mass spectrometric transient signal data acquisition and analyte concentration calculation. *Journal of analytical atomic spectrometry(Print)* 1996, 11, (9), 899-904.

[71] International Atomic Energy Agency (IAEA), *International Target Values 2000 for Measurements Uncertainties in Safeguarding Nuclear Materials*; STR-327; International Atomic Energy Agency (IAEA): Vienna, 2001.

[72] Begley, I. S.; Sharp, B. L., Occurrence and reduction of noise in inductively coupled plasma mass spectrometry for enhanced precision in isotope ratio measurement. *Journal of Analytical Atomic Spectrometry* 1994, 9, (3), 171-176.

[73] Laborda, F.; Medrano, J.; Castillo, J. R., Data acquisition of transient signals in inductively coupled plasma mass spectrometry. *Analytica Chimica Acta* 2000, 407, (1-2), 301-309.

[74] Zander, A. T., Spectral Interferences and Line Selection in Plasma Emission Spectrometry. In *Inductively Coupled Plasma in Analytical Atomic Spectrometry*, 2 ed.; Montaser, A.; Golightly, D. W., Eds. VCH Publishers, Inc.: New York, NY, 1992; pp 299-335.

[75] Horwitz, E. P.; Bloomquist, C. A. A., High speed-high efficiency separation of the transplutonium elements by extraction chromatography. *J. Inorg. Nucl. Chem* 1973, 35, (1), 271–284.

[76] Bailly, M.; Tondeur, D., Two-way chromatography: flow reversal in non-linear preparative liquid chromatography. *Chemical Engineering Science* 1981, 36, (2), 455-469.

[77] Kim, S. U.; Berninger, J. A.; Yu, Q.; Wang, N. H. L., Peak compression in stepwise pH elution with flow reversal in ion exchange chromatography. *Industrial & Engineering Chemistry Research* 1992, 31, (7), 1717-1730.

[78] Aarnio, P., *LINSSI SQL Database for Gamma-Ray Spectrometry - Part I: Database*; TKK-F-A841; Helsinki University of Technology: Espoo, 2006; p 96.

[79] Aarnio, P. A.; Ala-Heikkilä, J.; Isolankila, A.; Kuusi, A.; Moring, M.; Nikkinen, M.; Siiskonen, T.; Toivonen, H.; Ungar, R. K.; W., Z., LINSSI - Database for gamma-ray spectrometry. In *Methods And Applications of Radioanalytical Chemistry– MARC VII*, American Nuclear Society, International Topical Conference: Kailua-Kona, Hawaii, Usa, 2006.

[80] Aarnio, P. A., *LINSSI - Database for Gamma-Ray Spectrometry*. http://www.hut.fi/Units/AES/projects/radphys/linssi/ (*last accessed 10-04-2008*),

[81] Gustafson, P. F.; Nelson, D. M.; Brar, S. S.; Muniak, S. E., *Recent Trends in Radioactive Fallout*; Argonne National Laboratory: Argonne, IL, 1970; p 246.

[82] Wotawa, G.; De Geer, L. E.; Becker, A.; D'Amours, R.; Jean, M.; Servranckx, R.; Ungar, K., Inter- and intra-continental transport of radioactive cesium released by boreal forest fires. *Geophysical Research Letters* 2006, 33, (12), L12806.1-L12806.4.

[83] Currie, L. A., Limits for Qualitative Detection and Quantitative Determination - Application to Radiochemistry. *Analytical Chemistry* 1968, 40, (3), 586-593.

[84] Toivonen, H.; Pelikan, A., *Generation of Artificial Spectra for SPT1 Radionuclide Inter-comparison Exercise*; STUK - Radiation and Nuclear Safety Authority: Helsinki, 12 September, 2005; p 9.

[85] Whyte, J. C.; Ungar, R. K.; DeSilva, A.; Henderson, K. T., Radiostrontium and radium analysis in low-level environmental samples after a multi-stage semi-automated chromatographic sequential separation. *Journal of Radioanalytical and Nuclear Chemistry*, in preparation.

[86] Centers for Disease Control and Prevention, *Emergency Preparedness & Response - Radiation Emergencies - Radioactive Isotopes*. http://emergency.cdc.gov/radiation/isotopes/ (*last accessed 10-04-2008*),

[87] Health Canada (HC), *Canadian Guidelines for the Restriction of Radioactively Contaminated Food and Water Following a Nuclear Emergency*; H46-2/01-254E; Health Canada: Ottawa, 2000.

[88] Food and Agriculture Organization/World Health Organization (FAO/WHO), *Codex Alimentarius Volume 1A: Guideline Levels for Radionuclides in Foods following Accidental*

Nuclear Contamination for use in International Trade, Section 6.1 CAC/GL 5-1989; Joint FAO/WHO Food Standards Programme: Geneva, Switzerland, 1989.

[89] World Health Organization (WHO), *Derived intervention levels for radionuclides in food - Guidelines for application after widespread radioactive contamination resulting from a major radiation accident.* ; 92-4-154233-0 World Health Organization: Geneva, Switzerland, 1988; p 60.

[90] International Atomic Energy Agency (IAEA), *Intervention criteria in a nuclear or radiation emergency*; Safety Series No. 109; IAEA: Vienna, 1994.

[91] Council Regulation (CEC), *On the special conditions for exporting foodstuffs and feeding stuffs following a nuclear accident or any other case of radiological emergency.*; No. 2219/89 Council Regulation (Euratom): 1989.

[92] United States Food and Drug Administration (USFDA), *Accidental radioactive contamination of human food and animal feeds: Recommendations for State and local agencies.*; United States Department of Health and Human Services, Food and Drug Administration, Center for Devices and Radiological Health,: Rockville, MD., 1998.

In: Nuclear Chemistry: New Research
Editor: Axel N. Koskinen, pp. 155-180

ISBN: 978-1-60456-957-5

Chapter 5

FIBER BRAGG GRATINGS AND THEIR APPLICATIONS AS TEMPERATURE AND HUMIDITY SENSORS

Qiying Chen* and Ping Lu
Department of Physics and Physical Oceanography,
Memorial University of Newfoundland
St. John's, Newfoundland A1B 3X7, Canada

Abstract

As an important waveguiding medium, optical fiber plays significant roles in optical communications, optoelectronics, and sensors. A new type of microstructure inscribed in the optical fibers, i.e., fiber Bragg gratings (FBGs), has received considerable attention in recent years. A FBG is a type of distributed Bragg reflector constructed in a short segment of optical fiber that reflects specific wavelengths of light and transmits all the other components. In this chapter, optical properties of FBGs will be reviewed first with the underlying physical mechanisms. Different techniques to fabricate FBGs will be illustrated with the comparison of their advantages and drawbacks. For their important sensing applications, FBGs as temperature and humidity sensors will be discussed. The FBG sensors exceed other conventional electric sensors in many aspects, for instance, immunity to electromagnetic interference, compact size, light weight, flexibility, stability, high temperature tolerance, and resistive to harsh environment. A novel approach to realize individual temperature or humidity measurement, and their simultaneous measurement will be demonstrated by the use of FBGs coated with different polymers. The polymer-coated FBGs indicate linear shifts in the Bragg resonance wavelengths of the gratings with the temperature changes. A polyimide-coated FBG is sensitive to humidity due to the unique hygroscopic properties of polyimide while an acrylate-coated FBG shows insensitivity to humidity. The experimental results are in good agreement with the theoretical analysis.

Keywords: Fiber Bragg grating (FBG), humidity sensing, temperature sensing, polymers.

---

* E-mail address: qiyingc@mun.ca; Phone: 1 709 737-8878; Fax: 1 709 737-8739.Corresponding author.

1. Introduction

Guided wave optics is the forefront of research in optics nowadays. For modern applications, such as optical telecommunications and optical sensors, optical waveguides are the key components in which generation, modulation, propagation, and detection of light are governed by the principles of guided wave optics. An optical waveguide is a dielectric structure that transports energy at a wavelength in the infrared or visible portions of the electromagnetic spectrum [1]. Common waveguides used for optical communications are highly flexible fibers composed of nearly transparent dielectric materials. Charles Kao first suggested the possibility that low-loss optical fibers could be competitive with coaxial cable and metal waveguides for telecommunications applications in 1966 [2]. The commercial applications of optical fibers were not possible until 1970 when Corning Glass Work discovered an optical fiber with a loss less than the benchmark level of 10 dB/km [3,4].

With the increasing interests in the studies of all-fiber systems, fiber Bragg gratings (FBGs) have been applied in many photonic devices. A FBG is a type of distributed Bragg reflector constructed in a short segment of optical fiber that reflects specific wavelengths of light and transmits all the other components. A FBG can also be regarded as a fiber device with a periodic variation of the refractive index of the fiber core along the length of the fiber. The physical mechanism of inscribing the Bragg grating in a fiber is the photosensitivity of the fiber core. When a germanium-doped (GeO_2-doped) fiber is exposed to a high-intensity ultraviolet light, the refractive index of the fiber core is permanently changed [5]. The amount of change in the refractive index varies from 10^{-5} to 10^{-3}. The phenomenon of the change in the refractive index of the fiber upon light irradiation is usually referred as photosensitivity. In 1978, Hill *et al.* first reported the photosensitivity phenomenon in optical fibers achieved by the interference between counter-propagating waves inside the fiber core [6]. A decade later, Meltz *et al.* obtained the first FBG which was imprinted in Ge-doped silica single mode fiber by transverse coherent 244 nm UV beams from a tunable excimer-pumped dye laser with a frequency-doubled crystal [7]. Since then, FBGs have been applied in optical communications and optical fiber sensor networks owing to their unique advantages and versatility as in-fiber devices. FBGs are now widely used in a variety of lightwave communication applications such as fiber laser [8, 9], fiber amplifier [10], fiber Bragg filter [11,12], wavelength division multiplexers/demultiplexers [13,14], and dispersion compensation [15]. On the other hand, FBGs have been used as excellent sensor elements for measuring many environmental parameters, including temperature, strain, bending, refractive index, pressure, and flow rate [16].

In the following part of this chapter, basic optical properties of FBGs will be briefly reviewed first together with their fabrication techniques followed by an overview of FBG sensors. The third part of this chapter will discuss FBG sensors for temperature and humidity measurements.

2. Properties of Fiber Bragg Gratings

2.1. Optical Properties

The fundamental properties of optical fibers and FBGs inscribed on the fibers have been discussed in several books and monographs [5, 17-21]. In this section, a brief review on the optical properties of FBGs will be given and details can be found from these references.

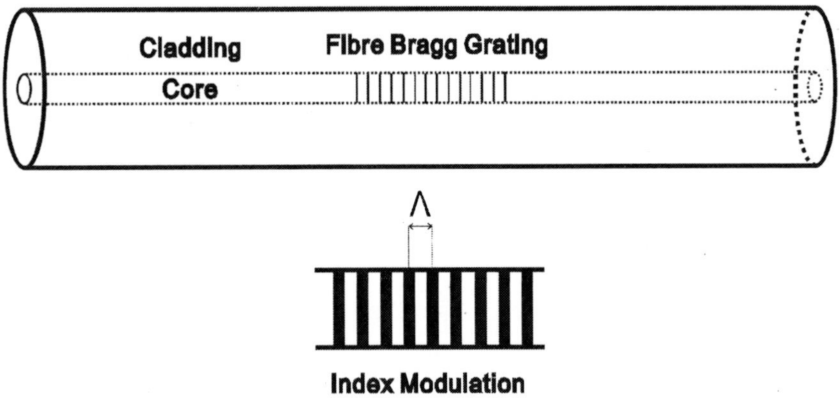

Figure 1. Illustration of a uniform fiber Bragg grating.

For a FBG consisting of a periodic modulation of the refractive index in the core of an optical fiber, the phase fronts are perpendicular to the fiber longitudinal axis and the grating planes are of a constant period, as illustrated in Fig. 1. Light propagating along the fiber core will interact with each grating plane, in which the Bragg condition is used for the discussion of the light propagation,

$$2\Lambda \sin\theta = n\lambda, \tag{1}$$

where Λ is the spacing between the grating planes, θ is the angle between the incident light and the scattering planes, λ is the light wavelength, and n is an integer. In the case when the Bragg condition is not satisfied, the light reflected from each of the subsequent planes becomes progressively out of phase and will eventually disappear. Only when the Bragg condition is met, the contributions of reflected light from each grating plane add constructively in the backward direction to form a back-reflected peak with a center wavelength defined by the grating parameters, i.e., the Bragg wavelength.

The Bragg grating resonance condition is the requirement to satisfy both energy and momentum conservation, in which the energy conservation ($\hbar\omega_f = \hbar\omega_i$) requests that the frequency of the reflected radiation should be the same as that of the incident radiation while the momentum conservation implies that the incident wave vector, $\vec{k}_i$, plus the grating wave vector, $\vec{K}$, equal the wave vector of the scattered radiation $\vec{k}_f$,

$$\vec{k}_i + \vec{K} = \vec{k}_f \tag{2}$$

where the grating wave vector $\vec{K}$ has a direction normal to the grating plane with a magnitude $2\pi/\Lambda$.

The diffracted wave vector is equal in magnitude with opposite direction with regard to the incident wave vector. The momentum conservation condition can be simplified as the first –order Bragg condition

$$\lambda_B = 2n_{eff}\Lambda \tag{3}$$

where λ_B is the Bragg grating wavelength, which is the center wavelength of the input light in the free space that will be back-reflected from the Bragg grating. n_{eff} is the effective refractive index of the fiber core at the free-space centre wavelength.

For a uniform FBG inscribed in the core of an optical fiber with a refractive index n_0. The refractive index profile is [22]

$$n(x) = n_0 + \Delta n \cos\left(\frac{2\pi x}{\Lambda}\right) \tag{4}$$

where Δn and x are the amplitude of the induced refractive index perturbation and the distance along the fiber longitudinal axis, respectively. The reflectivity of a grating with constant modulation amplitude and period can be described as [22].

$$R(l,\lambda) = \frac{\Omega^2 \sinh^2(sl)}{\Delta k^2 \sinh^2(sl) + s^2 \cosh^2(sl)} \tag{5}$$

where $R(l,\lambda)$ is the reflectivity which is a function of the grating length l and wavelength λ, Ω is the coupling coefficient, $\Delta k = k - \pi/\lambda$ is the detuning wave vector, $k = 2\pi n_0/\lambda$ is the propagation constant, and $s = \sqrt{\Omega^2 - \Delta k^2}$. The coupling coefficient Ω for the sinusoidal variation of index perturbation along the fiber axis has been found as

$$\Omega = \frac{\pi \Delta n \eta(V)}{\lambda} \tag{6}$$

where $\eta(V)$ is a function of the normalized frequency V of the fiber that represents the fraction of the fiber mode power contained in the core, $\eta(V) \approx 1 - V^{-2}$. The normalized frequency V can be expressed as [22]

$$V = \left(2\pi/\lambda\right) a \left(n_{co}^2 - n_{cl}^2\right)^{1/2} \tag{7}$$

where a is the core radius, n_{co} and n_{cl} are the refractive indices of the core and cladding, respectively.

At the Bragg grating center wavelength, there is no wave vector detuning and Δk equals zero, the expression for the reflectivity becomes

$$R(l,\lambda) = \tanh^2(\Omega l). \tag{8}$$

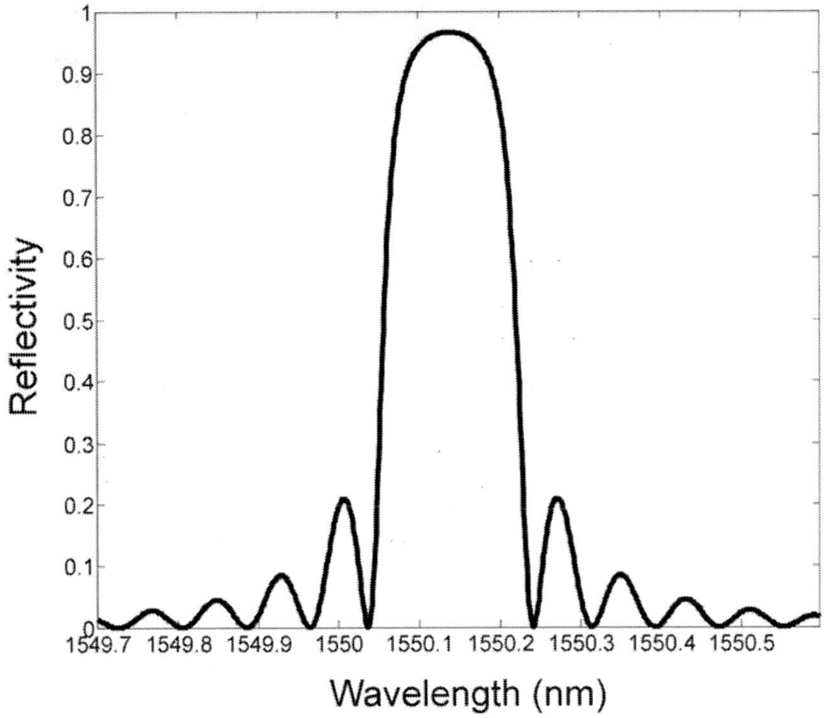

Figure 2. Simulation of the reflection spectrum of a FBG as a function of wavelength detuning.

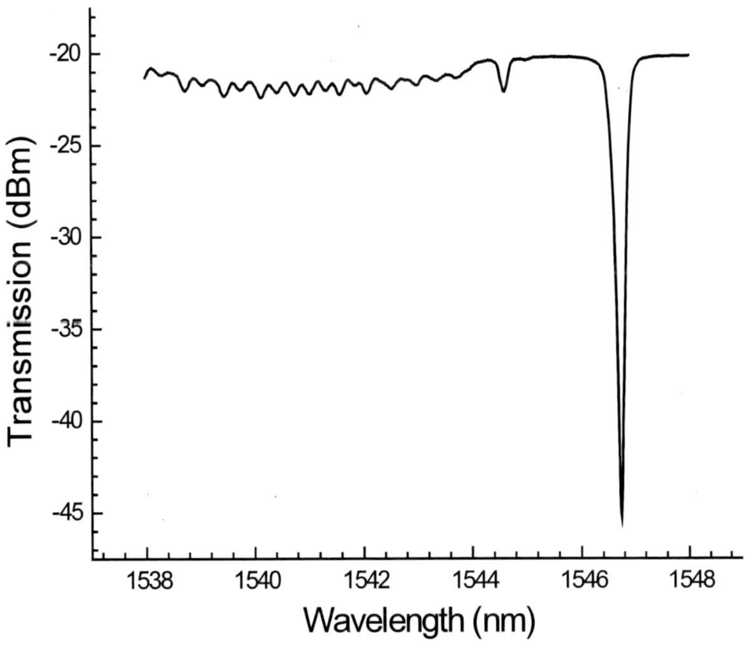

Figure 3. Transmission spectrum of a typical fiber Bragg grating exhibiting loss to radiation modes at the short-wavelength side.

The reflectivity increases with the change in the induced refractive index. It can also be found that the reflectivity increases with the increase in the length of the grating. Figure 2 gives a simulated reflection spectrum as a function of the wavelength detuning, in which the side lobes of the resonance are due to multiple reflections to and from opposite ends of the grating region.

A Bragg grating written in a highly photosensitive fiber exhibits a pronounced transmission feature on the short-wavelength side of the Bragg peak (Fig. 3). This feature is only observable in the transmission spectrum and only the main peak is visible in the reflection spectrum, which is due to the light leaving the fiber from the side. Radiation-mode coupling as a loss mechanism on core-mode transmission has been suggested to explain this phenomenon [23, 24]. For the cylindrical cladding-air interface, the transmission spectrum of the Bragg grating consists of multiple sharp peaks that modulate the radiation-mode coupling. The light energy in the grating will be coupled to other modes at shorter wavelengths: some will be reflected or absorbed, and the others will be radiated away from the fiber. These interactions are seen as a series of many transmission dips in the spectrum at wavelengths that are shorter than the Bragg wavelength.

2.2. Photosensitivity and Fabrication Techniques

As discussed in the previous section, photosensitivity in an optical fiber refers to a permanent change in the index of refraction of the fiber core when exposed to light of specific wavelength and intensity. Photosensitivity in optical fibers has significant importance in the fabrication of FBGs for applications in telecommunications and sensors.

After the first observation of weak changes ($\sim 10^{-6}$) in the index of refraction in germanosilica fibers by Hill *et al.* [6] in 1978, Meltz *et al.* showed that a strong index of refraction change occurred when a germanium-doped fiber was exposed to UV light close to the absorption peak of a germanium-related defect at a wavelength range of 240–250 nm [7]. Photosensitivity of optical fibers can be regarded as a measure of the amount of refractive index change in a fiber core. It is desirable to fabricate photoinduced devices in standard optical fibers for compatibility with existing systems, however, the standard single mode telecommunication fibers with a doping concentration of germanium around 3%, typically display weak index changes of 10^{-5}, far below the expected value on the order of 10^{-4}. Hydrogen loading (hydrogenation) [25-27], flame brushing [28], boron co-doping [29,30], and short wavelength light source [31,32] have been used for enhancing the photosensitivity in silica optical fibers.

Bragg gratings have been written in many types of optical fibers using various methods. However, the mechanism of index change is not fully understood. Several models have been proposed for these photoinduced changes in the refractive index, for instance, the color center model [33,34], the dipole model [35], the compaction model [36], the stress-relief model [37-39]. The general idea from these theories is that the germanium–oxygen vacancy defects, Ge–Si or Ge–Ge, are responsible for the photoinduced index changes.

Direct optical inscription of high quality gratings into the cores of optical fibers has been actively pursued by many research laboratories and various techniques have been reported. Different FBG fabrication techniques can be classified as internal inscription and external inscription techniques. Adopted primarily during the earlier years [6], internally writing technique uses relatively simple experimental setup in which the standing wave inside the optical fiber photoimprints a Bragg grating with the same pattern as the standing wave. In recent years, the internal inscription technique has been superseded by the external inscription technique due to the intrinsic limitation of the internally written gratings. The reported externally written fabrication techniques can be categorized into four groups, i.e., interferometric technique, point-by-point exposure technique, phase mask technique, and most recently ultrashort pulse lasers technique.

The interferometric fabrication technique, the first external writing technique to inscribe Bragg gratings in photosensitive fibers, was demonstrated by Meltz *et al.* [7], in which two beams split from one UV light recombined to form an interference pattern and expose a photosensitive fiber inducing a refractive index modulation in the fiber core. Either amplitude-splitting or the wave-front-splitting interferometer has been used in the fabrication of FBGs.

The point-by-point exposure technique [40], which produces grating fringe one-by-one, is accomplished through UV light irradiation over consecutive steps. After each irradiation, either the fiber or the laser beam itself is translated in the axial direction of the fiber by a computer controlled translation stage. The advantage of this technique is that a laser beam of high intensity is obtained due to the focus of the laser beam with a flexibility to alter the Bragg grating parameters. The drawback of the point-by-point technique is that it is a time-consuming process. During the slow fabrication process, errors in the grating spacing due to thermal effects and/or small variations in the fiber's strain can occur, which limits the gratings to a very short length.

One of the most effective methods for inscribing Bragg gratings in photosensitive fiber is the phase-mask technique [41, 42]. This method employs a diffractive optical element, i.e., phase mask, to spatially modulate the UV writing beam. The phase-mask grating has a one-dimension surface-relief structure fabricated in a high-quality fused silica block transparent to the UV writing beam. The profile of the periodic gratings is chosen such that, when an UV beam is incident on the phase mask, the zero-order diffracted beam is suppressed to less than a few percent of the transmitted power. In addition, the diffracted plus and minus first orders are maximized to contain, typically, more than 35% of the transmitted power. A near-field fringe pattern is produced by the interference of the plus and minus first-order diffracted beams. The period of the fringes is one-half that of the mask. The interference pattern photoimprints a refractive-index modulation in the core of a photosensitive optical fiber placed in contact with or in close proximity immediately behind the phase mask. A cylindrical lens may be used to focus the fringe pattern along the fiber core. The phase mask greatly reduces the complexity of the fiber grating fabrication system. The simplicity of using only one phase mask provides a robust and an inherently stable method for reproducing FBGs.

Recently, with the development of ultrafast laser technology, change of refractive index in transparent dielectrics with ultrashort laser pulses for photonic devices has been an interesting topic. Some published papers reported the use of femtosecond lasers to fabricate FBGs in commercial, non-photosensitized, and unhydrogenated fibers. These reported techniques include FBG inscription with standard or special phase masks [43,44] and direct point-by-point writing [45]. The wavelengths in the near-infrared region around 800 nm and ultraviolet (UV) around 248 nm have been used for FBG fabrication [45,46]. On the other hand, femtosecond lasers have also been demonstrated as a powerful tool in trimming FBGs to realize continuously tuning of the grating performance [47]. The ultrafast laser enhancement of photosensitivity response and modification of anisotropic index profile in silica fiber is a powerful technique to precise control of the performance of FBG devices for applications in optical filtering and polarization mode dispersion management.

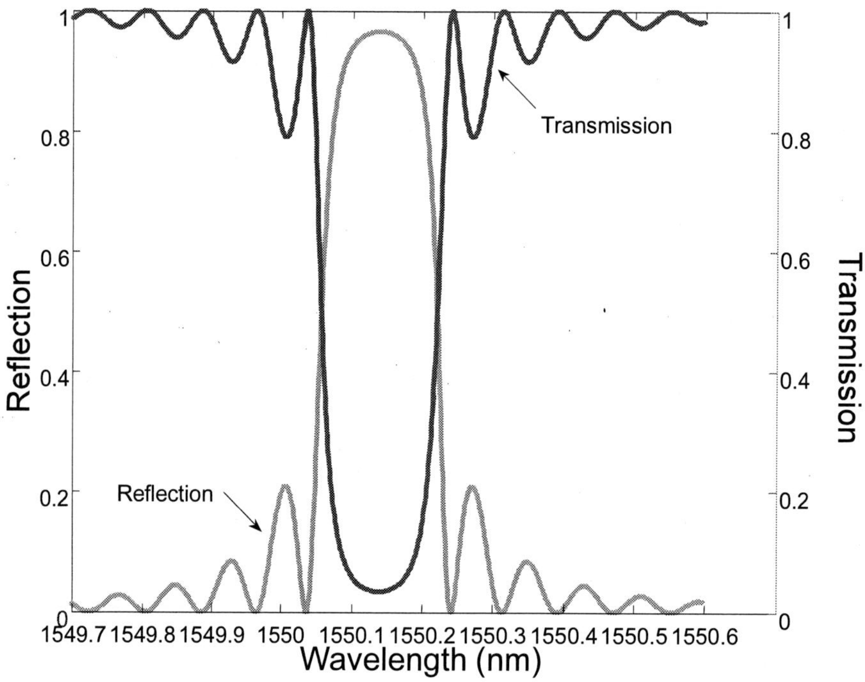

Figure 4. Simulated spectral response of a typical Bragg reflector with a uniform period.

According to the different grating pitch or tilt, fiber Bragg gratings can be classified into the common Bragg reflector, the blazed Bragg grating, and the chirped Bragg grating. The simplest and most frequently used FBG is the Bragg reflector, which has a constant pitch. A typical spectral response of a uniform period is shown in Fig. 4. The Bragg reflector can function as a narrowband transmission or reflection filter and sensor applications [48-51]. The blazed grating has phase fronts tilted with respect to the fiber axis. The applications of blazed gratings are in mode conversion [52] and sensing field [53]. The chirped grating has an periodic pitch, displaying a monotonous increase in the spacing between grating planes. The applications of the chirped Bragg gratings are for dispersion compensation in high-bit-rate transmission systems [54, 55] and sensing elements [56-59].

3. Fiber Bragg Grating Sensors

Sensors based on FBGs have attracted considerable attention since the early stage of the discovery. FBG sensors exceed other conventional electric sensors in many aspects, for instance, immunity to electromagnetic interference, compact size, light-weight, flexibility, stability, high temperature tolerance, and resistance to harsh environment. Additional advantages of FBG sensors include very low insertion loss, narrowband wavelength reflection, linearity in response over many orders of magnitude and compatibility with the existing fiber optics system, especially their absolute wavelength-encoding of measurand information, making FBG sensors interrupt immune [60].

FBG sensors can measure many physical parameters, in which strain [61-64] and temperature [65, 66] measurements are the major fields of interest. Meltz and Morey propounded that the shift in the Bragg grating wavelength was mainly due to strain and temperature changes [67]. The strain response is induced due to both the fractional change in a grating period due to the physical elongation of the optical fiber and the change in fiber index due to photo-elastic effects. The thermal response is induced due to both the inherent thermal expansion of the fiber material and the temperature dependence of the refractive index. It is apparent that any shift in the Bragg wavelength is the sum of the strain and temperature factors. The sensing measurements of other measurands can be realized by transforming to the strain and temperature factors. FBG sensors used to measure humidity [68, 69], vibration [70], pressure [71-73], and refractive index [74-77], and salinity [78] have also been reported recently.

Various kinds of methods for temperature measurement have been developed by using optical fibers, which include blackbody radiation and pyrometry, absorption, intrinsic scattering, various forms of interferometry, and fluorescence based techniques [79], in addition to the newly emerged FBG techniques. Excellent reviews of optical fiber-based temperature sensing techniques can be found in Refs. [80,81]. Current research on optical fiber temperature sensing is focused on improving accuracy and/or the measurement range, and also reducing the cost. One particularly important topic is on identifying approaches to realize simultaneous measurement of multiple parameters, i.e., temperature and other environmental parameters. In many practical applications, it is required to monitor the temperature in order to achieve temperature compensation while measuring the environmental parameters, for example, flow rate, pressure, refractive index, and salinity.

Humidity is another important physical parameter for environmental control, industrial applications, and daily life. Lee *et al.* reviewed different types of optical humidity sensors [82]. One type of fiber-optic humidity sensors responses to the colorimetric interaction of materials immobilized on the surface of the fiber core or its cladding in the humidity-sensing section, in which the transmitted optical power changes as a function of the relative humidity due to the humidity-induced changes in the refractive indices of the materials. Kharaz *et al.* reported an optical fiber humidity sensing system based on the colorimetric interaction of cobalt chloride with water molecules [83]. The use of a hydrophilic gel (agarose) deposited on the thinner zone of a biconically tapered single-mode optical fiber to achieve humidity measurement has been reported [84]. Gupta *et al.* discussed a fiber-optic humidity sensor based on the moisture-dependent absorption of light by using a phenol red doped polymethylmethacrylate (PMMA) film over a small portion of the core of the plastic clad silica fiber with a low refractive index [85]. Yeo *et al.* reported the use of a FBG together with a moisture sensitive polymer coating for humidity sensing [86]. Some polymer materials, such as polyimide, undergo volume expansion as a result of humidity absorption and desorption [87]. Once these polymers are coated on FBGs, the strain effects occur due to the volume changes of these coatings. Mrotek *et al.* discussed the diffusion of moisture through optical fiber coatings [88]. Other optical techniques have also been reported for humidity sensors, for example, changes in the incident light due to absorption [89], cooled mirror principles [90], and photo-acoustic technique [91]. Although optical sensors possess an advantage of high sensitivity, their drawbacks include hysteresis, large size, high power consumption for some optical fiber sensors other than in the forms of optical fibers, and especially missing of temperature compensation. The superiority of optical humidity sensors in tracking low moisture level has been noted [82].

As discussed above, FBGs respond to changes in both strain and temperature. It is necessary to discriminate these effects in order to reveal each physical parameter and various methods have been proposed. A practical approach is to use a reference grating or grating pairs [92, 93]. The reference grating or grating pairs, which is isolated from one parameter, e.g., strain, is placed near the sensor FBG. The reference grating can be on the same fiber as the sensor FBG [94]. Another method is to use two FBGs with a large difference in their Bragg wavelengths, which show different responses to the same measurands [95]. FBGs written on fibers of different diameters have also been proposed, which give different strain responses, while the temperature responses are the same [96,97]. A sensing head for simultaneous measurement of strain and temperature is demonstrated based on two Bragg gratings arranged in a twisted configuration [98]. By writing FBG with close wavelengths in undoped and boron doped fibers, different temperature sensitivities are obtained while the strain sensitivities remain the same [99]. In these reported techniques, it is necessary to use special fibers (specialty fibers with different doping elements, microstructured fibers, and photonic crystal fibers), complicated configurations (external lasers), or special spectroscopic techniques (fluorescence and interferometry) in order to distinguish temperature and strain, which result in bulky sensor systems targeting for the sole purpose of simultaneous measurement of temperature and strain only. Lu *et al.* reported an approach to resolve the cross-sensitivity between temperature and strain of FBGs [100], in which acrylate and polyimide polymers were used as the coating materials for different FBGs to achieve simultaneous measurement of axial strain and temperature. Since the standard FBG fabrication technique needs to strip the protective plastic coating off the fiber before FBG inscription and recoat a polymeric layer afterwards to protect the grating, the coating of FBGs with different polymeric materials proposed does not complicate the procedures or add extra cost in the FBG fabrication, which can be easily realized by fiber recoaters or dip coating technique. Furthermore, without additional optical devices or spectroscopic techniques, this approach provides one-fiber solution for multiple applications of FBGs in addition to resolve the cross-sensitivity of axial strain and temperature.

A large number of FBG sensors may be integrated at different locations along a single optical fiber or several fibers to form a quasi-distributed sensor [101-103]. The advantage of such kind of FBG array is that each FBG has a unique Bragg wavelength and can be individually interrogated. These FBG arrays can be fabricated with an arbitrary linear distance between the gratings, while the wavelength separation between the adjacent Bragg wavelengths is determined only by the wavelength shift of the neighboring higher and lower Bragg wavelengths.

Field applications of FBG sensors have also been reported [104-108]. FBG sensors possess a number of advantages for applications in spacecraft with embedded sensors that monitor the performance of reinforced carbon fiber composites as well as advanced testing of gas turbine engines [104]. There are numerous applications of FBG sensors for structural health monitoring in civil engineering, including monitoring of bridges [105], crack detection [106], and power transmission lines [107]. The FBG sensors can also be used in harsh environments [108].

4. Fiber Bragg Grating Temperature and Humidity Sensors

4.1. Theory

4.1.1. Fiber Bragg Grating Temperature Sensitivity

The Bragg grating resonance wavelength, λ_B, which is the center wavelength of light back reflected from a Bragg grating, depends on the effective index of refraction of the core (n_{eff}) and the periodicity of the grating (Λ) through the relation $\lambda_B = 2n_{eff}\Lambda$. The effective index of refraction, as well as the periodic spacing between the grating planes, will be affected by changes in temperature and strain. The shift in the Bragg grating wavelength $\Delta\lambda_B$ due to temperature and strain changes is given by [21]

$$\Delta\lambda_B = 2\left(\Lambda\frac{\partial n}{\partial T} + n\frac{\partial \Lambda}{\partial T}\right)\Delta T + 2\left(\Lambda\frac{\partial n}{\partial l} + n\frac{\partial \Lambda}{\partial l}\right)\Delta l. \tag{9}$$

The first term in Eqn. (9) represents the temperature effect on an optical fiber. The changes in the grating spacing and the index of refraction caused by thermal expansion result in a shift in the Bragg wavelength. This fractional wavelength shift for a temperature change ΔT may be written as [109]

$$\Delta\lambda_{B,T} = \lambda_B(\alpha + \zeta)\Delta T, \tag{10}$$

where $\alpha = (1/\Lambda)(\partial\Lambda/\partial T)$ is the thermal expansion coefficient of the fiber (~ 0.55×10^{-6} °C^{-1} for silica). The quantity $\zeta = (1/n)(\partial n/\partial T)$ represents the thermo-optic coefficient, which is 8.6×10^{-6} °C^{-1} for a germanium doped silica-core fiber. Clearly, the index change is by far the dominant effect. From Eqn. (9), the expected temperature sensitivity of the FBG at 1550 nm is 0.0142 nm/°C.

4.1.2. Temperature and Humidity Sensitivity of Polyimide-coated Bragg Gratings

Using equation $\lambda_B = 2n_{eff}\Lambda$, the shift in the Bragg wavelength due to thermal expansion changes in the grating spacing and changes in the index of refraction can also be expressed as [109]

$$\frac{\Delta\lambda_B}{\lambda_B} = S_{RH}\Delta RH + S_T\Delta T = \left[\beta_{cf} - \hat{P}_e\left(\beta_{cf} - \beta_f\right)\right]\Delta RH + \left[\alpha_{cf} - \hat{P}_e\left(\alpha_{cf} - \alpha_f\right) + \zeta\right]\Delta T \tag{11}$$

where S_{RH} and S_T are the sensitivities to relative humidity and temperature, respectively. ΔRH and ΔT are the changes in the relative humidity and temperature accordingly. β_i is the hygroscopic longitudinal expansion coefficient, which is zero for a bare fiber, and α_i is the thermal longitudinal expansion coefficient. The subscript stands for bare fiber ($i = f$) and coated fiber ($i = cf$). ζ is the thermo-optic coefficient of the fiber core, and $\hat{P}_e$ is the effective photoelastic coefficient of the coated fiber [108]

$$\hat{P}_e = \frac{n^2}{2}\left[P_{12} - \upsilon(P_{11}+P_{12})\right] = 0.213 \quad (12)$$

where P_{11} and P_{12} are the components of the strain-optic tensor, n is the index of refraction of the core, and υ is the Poisson's ratio. For a typical optical fiber $P_{11} = 0.113$, $P_{12} = 0.252$, $n = 1.482$, $\upsilon = 0.16$, and $\upsilon = -\varepsilon_{f,r} / \varepsilon_{f,z}$ where $\varepsilon_{f,r}$ and $\varepsilon_{f,z}$ are the radial and axial elastic fiber strains, respectively.

Table 1. Properties of the fused silica fiber and the polyimide coating.

Parameter	Value
Thermal expansion coefficient α_f (K^{-1})	5×10^{-7}
Thermal expansion coefficient α_c (K^{-1})	4×10^{-5}
Thermo-optic coefficient ξ (K^{-1})	$(55\pm4.8)\times10^{-7}$
Young's modulus, E (Fiber) (GPa)	72
Young's modulus, E (coating) (GPa)	2.45
Poisson's ratio, v (fiber)	0.17
Poisson's ratio, v (coating)	0.41
Hygroscopic expansion coefficient β_f ($\%RH^{-1}$)	0
Hygroscopic expansion coefficient β_c ($\%RH^{-1}$)	7×10^{-5}
Thermal longitudinal expansion coefficient of coated fiber α_{cf} (K^{-1})	1.39×10^{-6}
Hygroscopic longitudinal expansion coefficient of coated fiber β_{cf} ($\%RH^{-1}$)	1.58×10^{-6}

The temperature sensitivity S_T can be expressed as

$$S_T = \alpha_{cf} - \hat{P}_e\left(\alpha_{cf} - \alpha_f\right) + \zeta \qquad (13)$$

Table 1 lists some thermo-optic parameters of the fused silica fiber and the polyimide coating [78]. According to the table, the temperature sensitivity of the grating can be calculated to be $S_T = (6.70\pm0.48)\times10^{-6}$ K^{-1} °C^{-1}. The temperature coefficient K_T, defined as $K_T = \Delta\lambda_B / \Delta T$, is $K_T = 0.0104$ nm/ °C at 1550 nm.

As we know that the hygroscopic longitudinal expansion coefficient of the bare fiber β_f is zero, the relative humidity sensitivity S_{RH} can be expressed as:

$$S_{RH} = \left(1 - \hat{P}_e\right)\beta_{cf}. \qquad (14)$$

Following the parameters in Table 1, the relative humidity sensitivity S_{RH} is 1.676×10^{-6} (%RH)$^{-1}$. The humidity coefficient K_{RH}, defined as $K_{RH} = \Delta\lambda_B/(\Delta RH)$, is 2.6×10^{-3} nm/(%RH) at 1550 nm.

4.2. Experimental Details

A standard telecommunication optical fiber (SMF-28, Corning Inc.) was stored in a high hydrogen pressure environment (1900 psi) at room temperature for two weeks. Afterwards, the fiber was stored in a freezer at -70 °C before use. This process can prevent the hydrogen from diffusing out of the optical fiber. Hydrogen loading of a fiber satisfactorily enhances the photosensitivity of a fiber. A phase mask of 10 mm in length was used to write 1 cm FBG gratings onto the fiber with the laser irradiation from an ArF excimer laser (193 nm). The transmission spectrum of the FBG was monitored *in situ* during the laser exposure with a white light source and an optical spectrum analyzer. Since the photosensitivity of an optical fiber also resulted in further changes of the central Bragg wavelength and the peak loss of the attenuation bands generated after fabrication when the hydrogen diffused out of the fiber, the FBG was baked at 150 °C overnight to eliminate the residual hydrogen and to stabilize the UV-induced index changes. The experimental setup for inscribing gratings is shown in Fig. 5.

The transmission spectrum of a typical FBG is shown in Fig. 6. A broadband light travels through the optical fiber and enters into the FBG, from which one specific wavelength is reflected back by the FBG. Full width at half maximum (FWHM) is a parameter to characterize the bandwidths of laser beams or optical devices. For a strong FBG, it is normally measured at 3 dBm from the reflectivity peak.

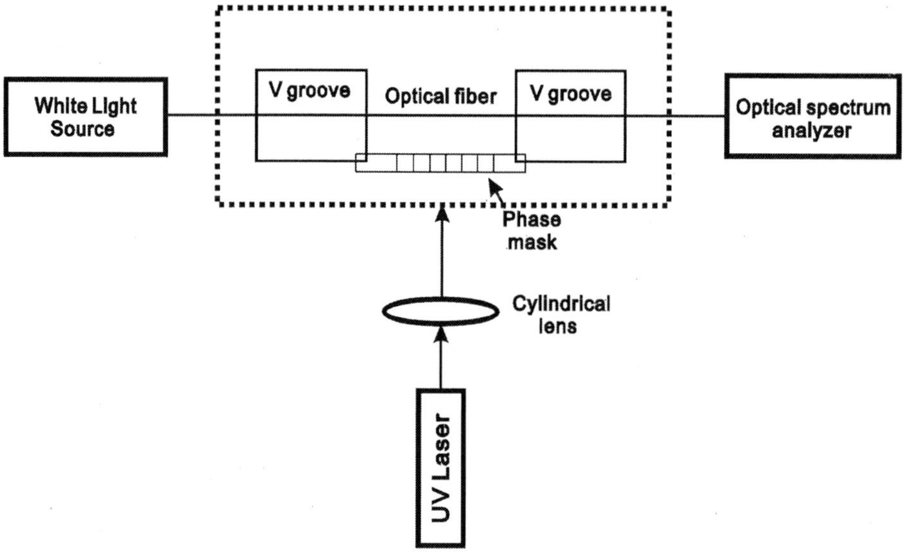

Figure 5. Schematic diagram of the fabrication of a fiber Bragg grating.

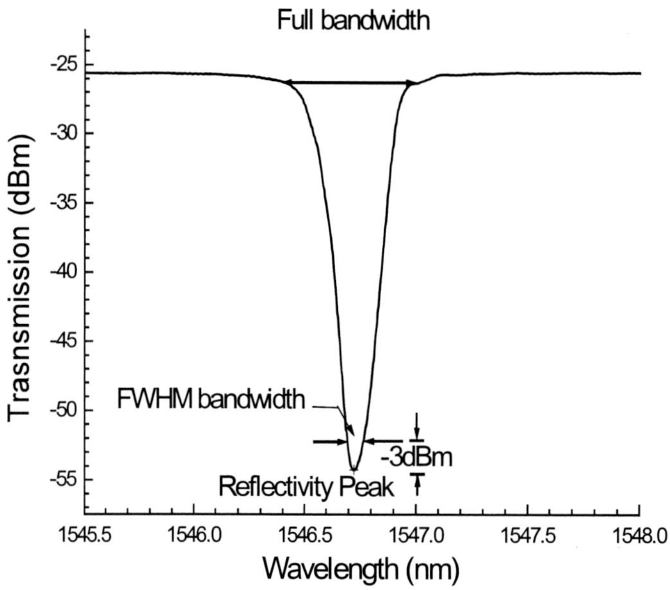

Figure 6. Transmission spectrum of a FBG at room temperature.

Figure 7 illustrates the setup used to determine the temperature-induced shift in the resonance band of the FBG transmission spectrum. Light from a broadband light source (EBS-7210, MPB Communication, Inc.) was launched into one end of the fiber containing the grating and the transmission spectrum was recorded by an optical spectrum analyzer (ANDO AQ-6315E, Yokogawa Co.). The emission spectrum of the Er^{3+} broadband source is shown in Fig. 8, which indicates an output in the wavelength range of 1525-1600 nm. The temperature of the grating was controlled by employing a microprocessor-controlled water bath (Precision 281). The FBG sample was immersed in the water bath for which the temperature has been kept constant for at least half an hour before measurement in order to ensure that the transmission spectrum of the FBG sample was immovable and the system was in a stable condition.

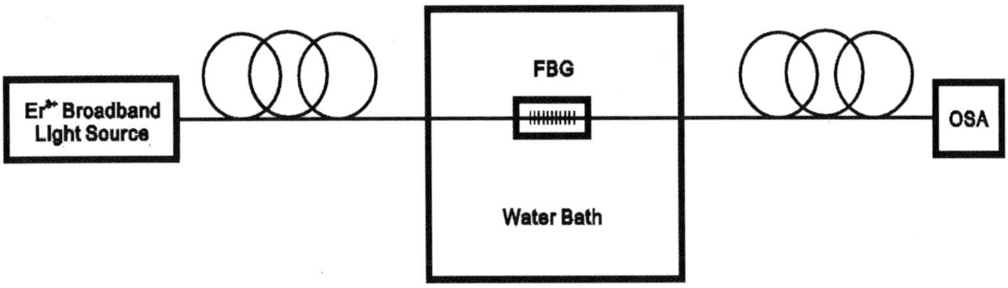

Figure 7. Schematic diagram of a temperature sensing measurement system.

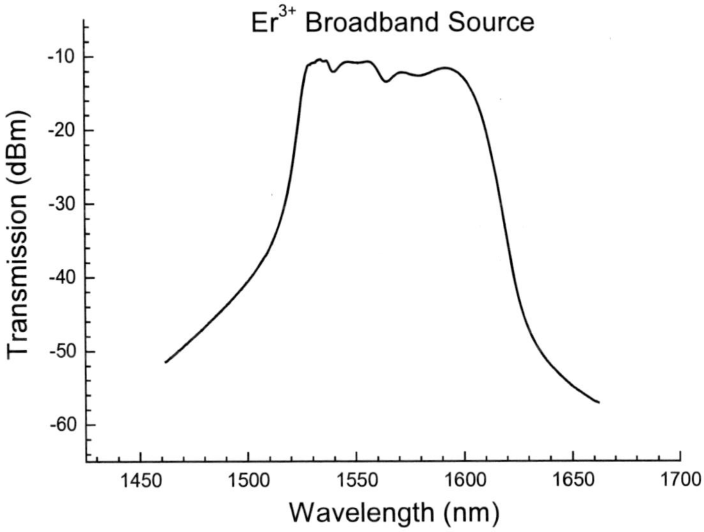

Figure 8. Emission spectrum of the Er^{3+} broadband light source used in this study.

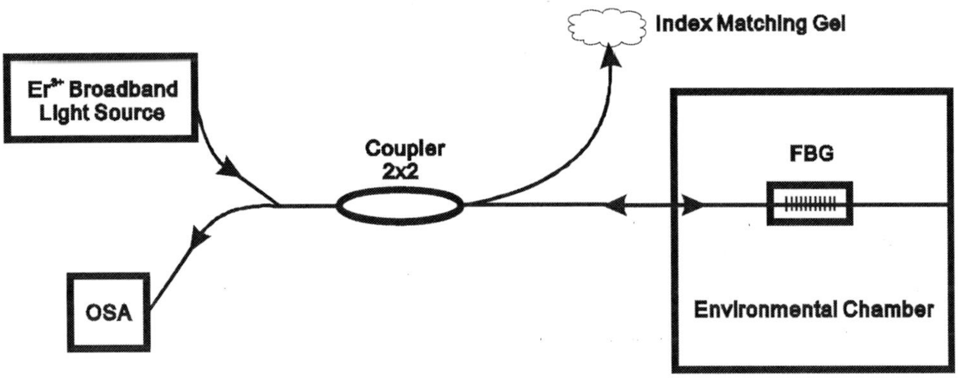

Figure 9. Schematic diagram of the humidity measurement system.

Once the FBG is inscribed onto the SMF-28 fiber, either acrylate or polyimide can be re-coated on the grating section using fiber recoaters or dip-coating method to protect the grating. The experimental setup used to characterize the humidity response of the sensor is shown in Fig. 9. The broadband light source was connected to the FBG sensor using a 2×2 fiber coupler. The reflected signal from the sensor was monitored using the optical spectrum analyzer. The polyimide-coated FBG sensor was stored inside an environmental chamber with controllable relative humidity. All experiments performed in this work were carried out at a room temperature of 20 °C. Sufficient time was given to allow the test environment to stabilize before readings of the wavelength from the optical spectrum analyzer were recorded.

4.3. Temperature Sensing Response of Fiber Bragg Gratings

For testing the strong FBG used here, the water temperature in the water bath was first changed from 5 °C to 90 °C with a step of 5 °C, the temperature was then decreased from 90 °C to 35 °C with a step of 5 °C. At a temperature of 25 °C, the transmission spectrum is shown in Fig. 10.

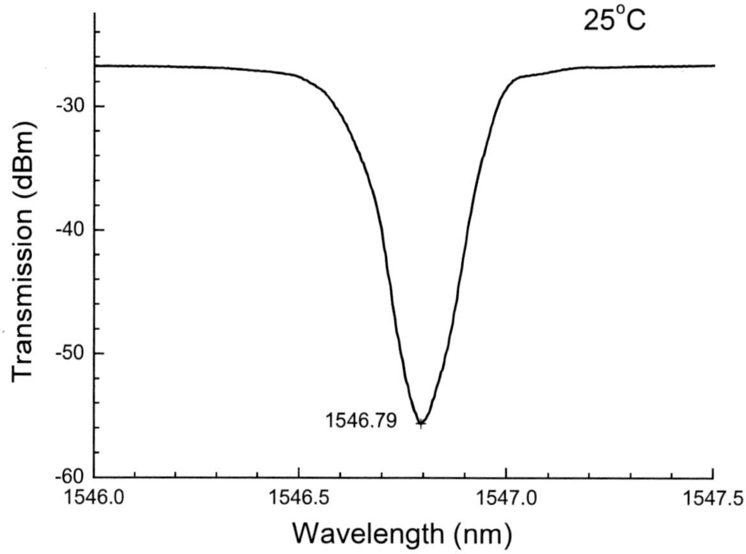

Figure 10. Transmission spectrum of a strong FBG at room temperature.

For the heating process, the observed transmission spectra and the change of the Bragg resonance wavelength as a function of increasing temperature are shown in Figs. 11 and 12, respectively. The dependence of the red-shift of the resonance wavelength of the FBG on the increasing temperature was measured to be 0.0091 nm/ °C.

For the cooling process, the measured change of the Bragg resonance wavelength as a function of decreasing temperature is shown in Fig. 13. The shift in the Bragg wavelength with respect to the decreasing temperature was measured to be 0.0098 nm/ °C. The experimental results of the shift in the Bragg wavelength as a function of temperature are close to the expected value of 0.0142 nm/ °C for a FBG recoated with an acrylate coating. The difference of the temperature sensitivities between the experimental result and theoretical value is mainly due to a discrepancy between different coating materials induced thermal expansion coefficients of the cladding materials and the thermo-optic coefficients of different core materials. Results obtained from the heating and cooling processes demonstrated high repeatability of the temperature sensing measurement.

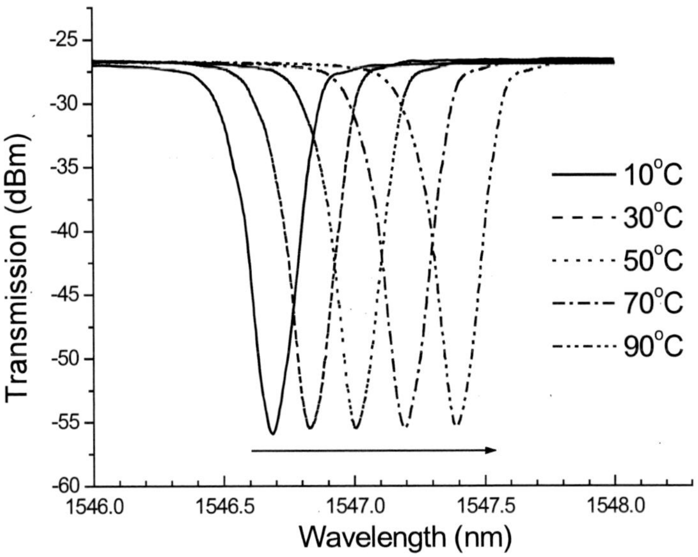

Figure 11. Transmission spectra of the FBG as a function of increasing temperature.

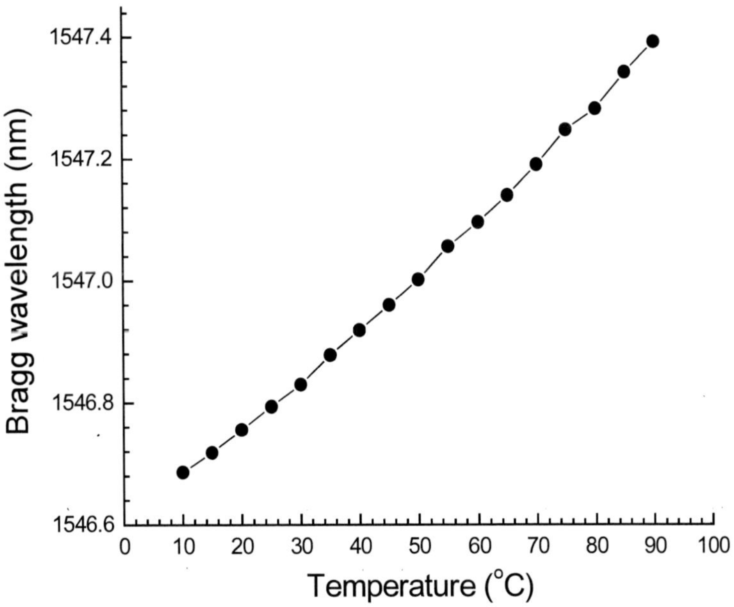

Figure 12. Bragg wavelength of the FBG as a function of increasing temperature.

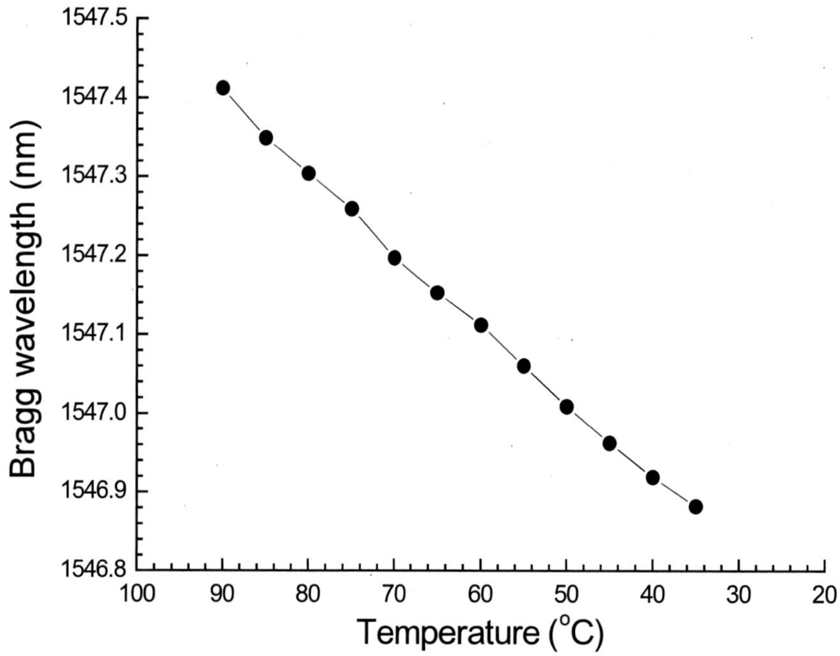

Figure 13. Bragg wavelength of the FBG as a function of decreasing temperature.

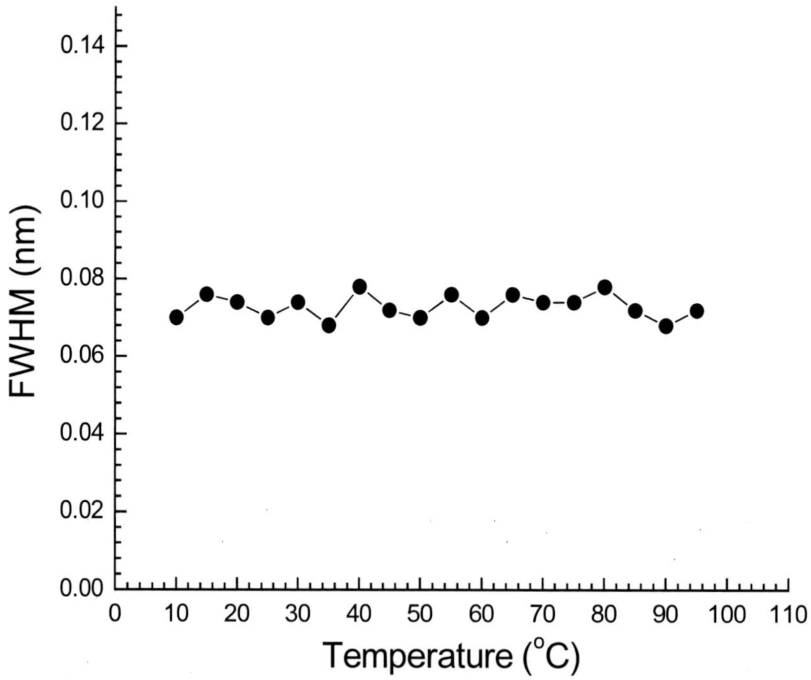

Figure 14. Change in the FWHM bandwidth of the FBG during heating process.

A general expression for the approximate FWHM bandwidth of a grating is given by [21]

$$\Delta\lambda = \lambda_B s \sqrt{\left(\frac{\Delta n}{2n_0}\right)^2 + \left(\frac{1}{N}\right)^2} \tag{15}$$

where N is the number of the grating planes. The parameter s is ~1 for strong gratings (for grating with near 100% reflection) whereas s ~0.5 for weak gratings. The FWHM bandwidth of the FBG loss peak in the heating process is shown in Fig. 14 with a standard deviation of 4.3%, which indicates that the FWHM bandwidth is independent of the changing temperature.

4.4. Humidity Sensing Response of Fiber Bragg Gratings

4.4.1. Humidity Sensing Response of a Polyimide-Coated Fiber Bragg Grating

The polyimide-coated FBG was placed in the environmental chamber where the temperature was kept at 20 °C. At each fixed humidity setting, a reading was taken after the output signal from the optical spectrum analyzer had stabilized for 1 hour. The process was carried out three times to ensure reproducibility. As shown in Fig. 15, the data shows the behavior of the FBG spectral characteristics when the grating was exposed to different humidity conditions. From the figure, the Bragg wavelength shift was observed when the humidity in the test environment was varied from 13 %RH to 91 %RH. As the humidity level increased, the wavelength was found to shift toward the longer wavelength, which is consistent with the elongation of the FBG caused by the expansion of the polymer coating. This humidity change resulted in a wavelength shift of 0.22 nm.

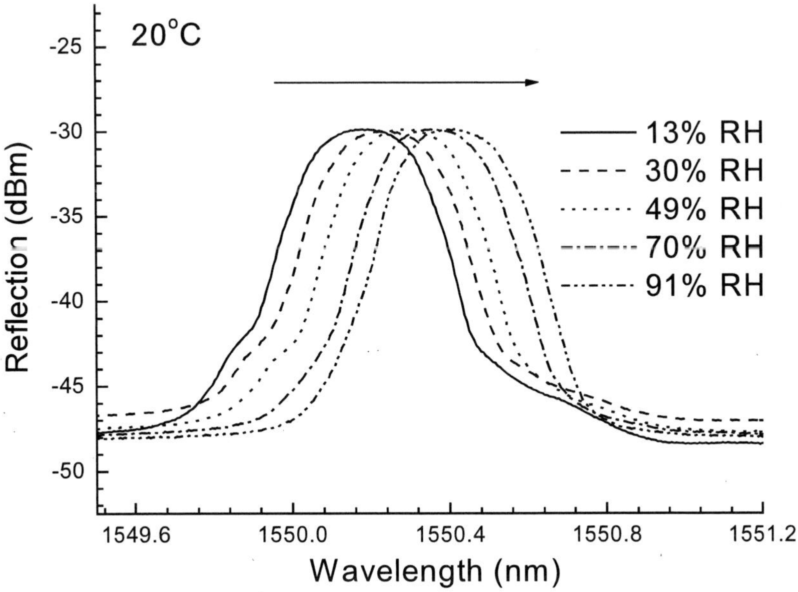

Figure 15. Reflection spectra of the polyimide-coated FBG at different relative humidities at room temperature.

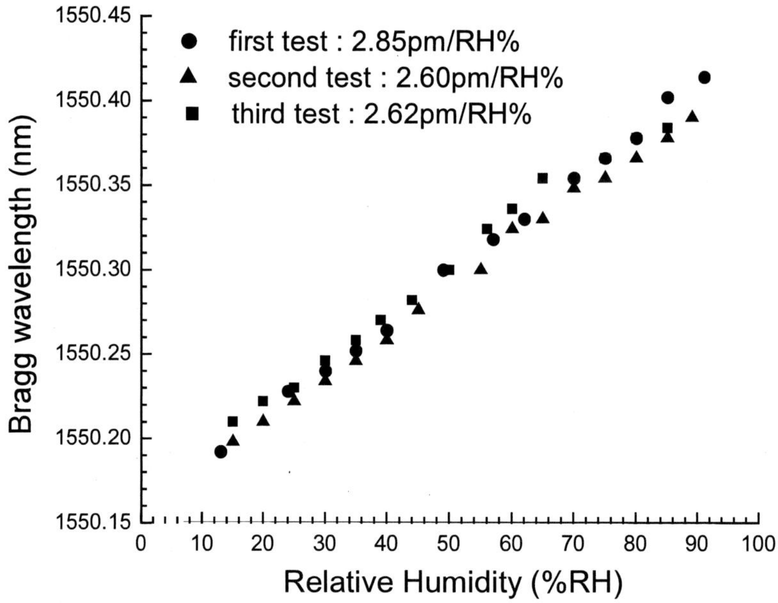

Figure 16. Bragg wavelength of the polyimide-coated FBG wavelength as a function of relative humidity in the environmental chamber.

Figure 16 shows the shifts of the Bragg wavelength versus relative humidity during three repeated tests. A linear regression was performed to establish the relationship between the wavelength shift due to the material expansion and the humidity level in the test chamber. The linear fitting reveals a linear dependence between the two parameters, which agrees well with the observations made by Kronenberg *et al.* [68] and Sager *et al.* [110], where the volume expansion of the polyimide film varies linearly with humidity. The humidity sensitivities of the FBG at 1550 nm were calculated to be 2.85, 2.62, and 2.60 pm/(%RH) respectively. The discrepancy among these test results may be ascribed to the incomplete adhesion between the silica surface and the polyimide layer due to the weak physical bonding at the interface between the fiber and the coating.

Bare silica fibers are not sensitive to humidity. Polyimide polymers, however, are hygroscopic and will swell in aqueous media as the water molecules migrate into them. The swelling of the polyimide coating strains the fiber, which modifies the Bragg condition of the FBG and thus serves as the basis as the humidity sensor.

4.4.2. Humidity Sensing Response of an Acrylate-coated Fiber Bragg Grating

Following the same procedures mentioned above, FBG wavelength of an acrylate-coated FBG as a function of increasing and decreasing relative humidity were measured with the results shown in Figs. 17 and 18, respectively. The results in these two figures clearly indicate that the change of relative humidity will not change the center wavelength of the FBG at room temperature. The acrylate-coated FBG is therefore not sensitive to the change in relative humidity.

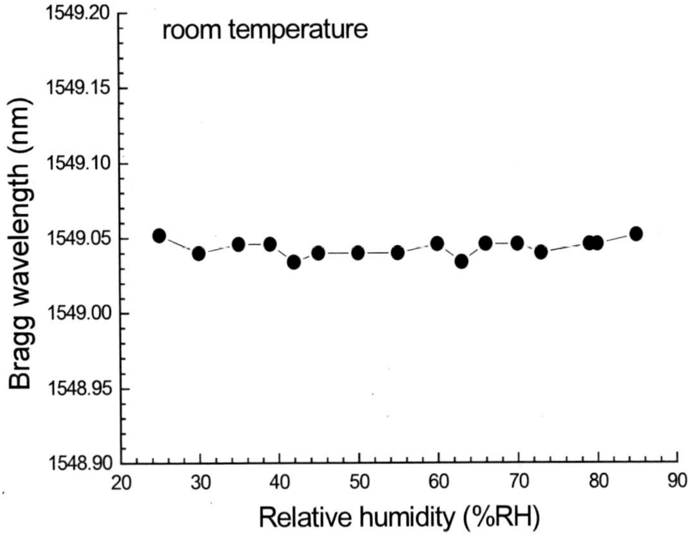

Figure 17. Bragg wavelength of the acrylate-coated FBG wavelength as a function of increasing relative humidity.

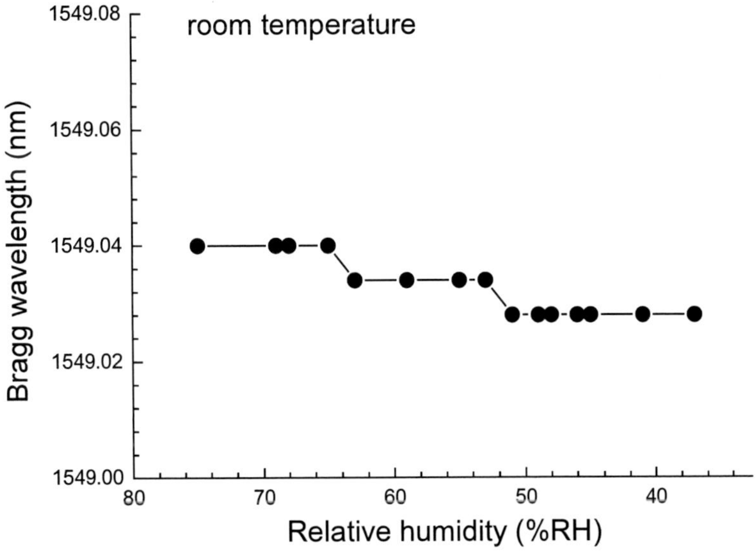

Figure 18. Bragg wavelength of the acrylate-coated FBG wavelength as a function of decreasing relative humidity.

5. Conclusions

Fiber Bragg gratings as temperature and humidity sensors have been reviewed and demonstrated, which possess many unique advantages over conventional techniques. Due to the nature of the fiber grating, it is possible to realize simultaneous measurement of these two parameters or to achieve quasi-distribution monitoring. Investigation on FBGs for applications in high temperature or harsh environment has been one of the research foci nowadays. For FBG humidity sensors, further research is needed for a systematic study of different polymer materials in order to identify materials with better performance as well as the sensing mechanisms including diffusion processes and response time.

Acknowledgments

This work has been supported by the Natural Sciences and Engineering Research Council of Canada (NSERC), Canada Research Chairs Program, Canada Foundation for Innovation, the Province of Newfoundland and Labrador, and the Memorial University of Newfoundland.

References

[1] Snyder, A. W.; Love, J. D. *Optical Waveguide Theory*; Chapman and Hall: London, UK, 1983, 3-5.

[2] Kao, K. C.; Hockham, G. A. *Proc. IEE*, 1966, 113, 1151-1158.

[3] Keck, D. B.; Schultz, P. C.; Zimar, F. W. U. S. Patent 3, 737, 393.

[4] Kapron, F. P.; Keck, D. B.; Maurer, R. D. *Appl. Phys. Lett.* 1970, 17, 423-425.

[5] Iizuka, K. Elements of Photonics, Vol. II *For Fiber and Intergrated Optics*; Wiley-Interscience, New York, NY, 2002, 741.

[6] Hill, O.; Fujii, Y.; Johnson, D. C.; Kawasaki, B. S. *Appl. Phys. Lett.* 1978, 32, 647-649.

[7] Meltz, G.; Morey, W. W.; Glenn, W. H. *Opt. Lett.* 1989, 14, 823-825.

[8] Feng, X.; Tam, H. Y.; Wai, P. K. A. *Photon. Technol. Lett.* 2006, 18, 1088-1090.

[9] Liu, Y.; Chiang, K. S.; Chu, P. L. *Appl. Opt.* 2005, 44, 4822-4829.

[10] Liaw, S. K.; Dou, L.; Xu, A. *Opt. Exp.* 2007, 15, 12356-12361.

[11] Lausten, R.; Rochon, P.; Ivanov, M.; Cheben, P.; Janz, S.; Desjardins, P.; Ripmeester, J.; Siebert, T.; Stolow, A. *Appl. Opt.* 2005, 44, 7039-7042.

[12] Littler, I.; Rochette, M.; Eggleton, B. *Opt. Exp.* 2005, 13, 3397-3407.

[13] Bilodeau, F.; Johnson, D. C.; Thenault, S.; Malo, B.; Albert, J.; Hill, K. O. *IEEE Photon. Technol. Lett.* 1995, 7, 388-390.

[14] Dong, L.; Hua, P.; Birks, T. A.; Reekie, L.; Russell, P. St. J. *IEEE Photon Technol. Lett.* 1996, 8, 1656-1658.

[15] Sheu, L. G.; Chuang, K. P.; Lai, Y. *IEEE Photon. Technol. Lett.* 2003, 15, 939-941.

[16] López-Higuera, J. M. ed. *Handbook of Optical Fibre Sensing Technology*; John Wiley & Sons, Ltd: West Susses, England, 2002.

[17] Buck, J. A. *Fundamentals of Optical Fibers*; Wiley-Interscience: Hoboken, NJ, 2004.

[18] Bass, M. ed. Handbook of Optics Vol. IV. *Fiber Optics & Nonlinear Optics*, Second Edition; McGraw-Hill: New York, NY, 2001.

[19] Truesdale, C. M. Optical Fibers, in Waynant, R. W. and Ediger, M. N. ed. *Electro-Optics Handbook*, Second Edition, McGraw-Hill: New York, NY, 2000.

[20] Kashyap, R. *Fiber Bragg Gratings*; Academic Press: San Diego, CA, 1999.

[21] Othonos, A.; Kalli, K. *Fiber Bragg Gratings*; Artech House: Norwood, MA, 1999.

[22] Lam, D. K. W.; Garside, B. K. *Appl. Opt.* 1981, 20, 440-445.

[23] Mizrahi, V.; Sipe, J. E. *IEEE J. light. Technol.* 1993, 11, 1513-1517.

[24] Erdogan, T.; Sipe, J. E. *J. Opt. Soc. Am. A* 1996, 13, 296-313.

[25] Lemaire, P. J.; Atkins, R. M.; Mizrahi, V.; Reed, W. A. *Electron. Lett.* 1993, 29, 1191-1193.

[26] Atkins, R. M.; Lemaire, P. J.; Erdogan, T.; Mizrahi, V. *Electron. Lett.* 1993, 29, 1234-1235

[27] Awazu, K.; Kawazoe, H.; Yamane, M. *J. Appl. Phys.* 1990, 68, 2713-2718.

[28] Bilodeau, F.; Malo, B.; Albert, J.; Johnson, D. C.; Hill, K. O. *Opt. Lett.* 1993, 18, 953-955.

[29] Williams, D. L.; Ainslie, B. J.; Armitage, R.; Kashyap, R.; Campbell, R. *Electron. Lett.* 1993, 29, 45-47.

[30] Williams, D. L.; Ainslie, B. J.; Kashyap, R.; Maxwell, G. D.; Armitage, J. R.; Campbell, R. J.; Wyatt, R. *Proc. SPIE*, 1993, 2044, 55-68.

[31] Albert, J.; Malo, B.; Bilodeau, F.; Johnson, D. C.; Hill, K. O.; Hibino, Y.; Kawachi, M., *Opt. Lett.* 1994, 19, 387-389.

[32] Dyer, P. E.; Farley, R. J.; Gied, R.; Byron, K. C.; Reid, D. *Electron. Lett.* 1994, 30, 860-862.

[33] Atkins, R. M.; Mizrahi, V. *Electron. Lett.* 1992, 28, 1743-1744.

[34] Hand, D. P.; Russell, P. St. *J. Opt. Lett.* 1990, 15, 102-104.

[35] Bernandin, J. P.; Lawandy, N. M. *Opt. Commun.* 1990, 79, 194-199.

[36] Douay, M.; Xie, W. X.; Taunay, T.; Bernage, P.; Niay, P.; Cordier, P.; Poumellec, B.; Dong, L.; Bayon, J. F.; Poignant, H.; Delevaque, E. *IEEE J. light. Technol.* 1997. 15, 1329-1342.

[37] Fonjallaz, P. Y.; Limberge, H. G.; Salathe, R. P.; Cochet, F.; Leuenberger, B. *Opt. Lett.* 1995, 20, 1346-1348.

[38] Ky, N. H.; Limberge, H. G.; Salathe, R. P.; Cochet, F.; Dong, L. *Phys. Lett.* 1999, 74, 516-518.

[39] Limberge, H. G.; Fonjallaz, P. Y.; Salathe, R. P.; Cochet, F. *Appl. Phys. Lett.* 1996, 68, 3069-3071.

[40] Malo, B.; Hill, K. O.; Bilodeau, F.; Johnson, D. C.; Albert, J., *Electron. Lett.* 1993, 29, 1668-1669.

[41] Hill, K. O.; Malo, B.; Bilodeau, F.; Johnson, D. C.; Albert, *J. Appl. Phys. Lett.* 1993, 62, 1035-1037.

[42] Anderson, D. Z.; Mizrahi, V.; Erdogan, T.; White, A. E. *Proc. of the Conf. on Optical Fibre Communication, OFC'93,* Tech. Dig., 1993, 68.

[43] Dragomir, A.; Nikogosyan, D. N.; Zagorulko, K. A.; Kryukov, P. G.; Dianov, E. M.; *Opt. Lett.* 2003, 28, 2171-2173.

[44] Mihailov, S. J.; Smelser, C. W.; Grobnic, D.; Walker, R. B.; Lu, P.; Ding, H.; Unruh, J. *J. Light. Technol.* 2004, 22, 94-100.

[45] Martinez, A.; Dubov, M.; Khrushchev, I.; Bennion, I., *Electron. Lett.* 2004, 40, 1170-1172.

[46] Violakis, G.; Konstantaki, M.; Pissadakis, S. *IEEE Photon. Technol. Lett.* 2006, 18, 1182-1184.

[47] Chen, Q.; Chen, K. P.; Xu, W.; Nikumb, S. *Opt. Commun.* 2006, 259, 123-126.

[48] Kersey, A. D.; Davis, M. A.; Patrick, H. J.; LeBlanc, M.; Koo, K. P.; Askins, C. G.; Putnam, M. A.; Friebele, E. J. *IEEE J. light. Technol.* 1997, 15, 1442-1463.

[49] Ball, B.; More, W. W. *Opt. Lett.* 1992, 17, 420-422.

[50] Othonnos, A.; Alavie, A. T.; Serge, S. M.; Karr, S. E.; Measures, R. M. *Opt. Eng.* 1993, 32, 2841-2846.

[51] Alavie, A. T.; Karr, S. E.; Othonos, A.; Measures, R. M., *IEEE Photon. Technol. Lett.* 1993, 5, 1112-1114.

[52] Hill, K. O.; Bilodeau, F.; Faucher, S.; Malo, B.; Johnson, D. C. *Electron. Lett.* 1991, 27, 1548-1550.

[53] Laffont, G.; Ferdinand, P. *Meas. Sci. Technol.* 2001, 12, 765–770.

[54] Williams, J. A. R.; Bennion, I.; Sugden, K.; Doran, N. J. *Electron. Lett.* 1994, 30, 985-987.

[55] Kashyap, R.; Chernikov, S. V.; Mckee, P. F.; Taylor, J. R. *Electron. Lett.* 1994, 30, 1078-1080.

[56] Fallon, R. W.; Zhang, L.; Gloag, A.; Bennion, I. *Electron. Lett.* 1997, 33, 705-706.

[57] Putnam, M. A.; Williams, G. M.; Friebele, E. J. *Electron. Lett.* 1995, 31, 309-311.

[58] Leblanc, M.; Huang, S. Y.; Ohn, M.; Measures, R. M.; Guemes, A.; Othonos, A. *Opt. Lett.* 1996, 21, 1405-1407.

[59] Huang, S.; Leblanc, M.; Ohn, M. M.; Measures, R. M. *Appl. Opt.* 1995, 34, 5003-5009.

[60] Othonos, A.; Kalli, K. Fibre Bragg gratings Fundamentals and Applications in Telecommunications and Sensing, 1999, (Boston, MA: Artech House), 98-99

[61] Betz, D. C.; Thursby, G.; Culshaw, B.; Staszewski, W. J. *IEEE J. light. Technol.* 2006, 24, 1019-1026.

[62] Botsis, J.; Humbert, L.; Colpo, F.; Giaccari, P. *Opt. Laser. Eng.* 2005, 43, 491-510.

[63] Kersey, A. D.; Berkoff, T. A.; Morey, W. W. *Opt. Lett.* 1993, 18, 1370-1372.

[64] Fernandez, A.; Berghmans, F.; Brichard, B.; Megret, P.; Decreton, M.; Blondel, M.; Delchambre, A. *Meas. Sci. Technol.* 2001, 12, 1-4.

[65] Jung, J.; Nam, H.; Lee, B.; Byun, J. O.; Kim, N. S. *Appl. Opt.* 1999, 38, 2752-2754.

[66] Flockhart, G. M. H.; Maier, R. R. J.; Barton, J. S.; MacPherson, W. N.. Jones, J. D. C., Chisholm, K. E., Zhang, L., Bennion, I., Read, I. and Foote, P. D. *Appl. Opt.* 2004, 43, 2744-2751.

[67] Meltz, G.; Morey, W. W. *Proc. SPIE* 1991, 1516, 185-199.

[68] Kronenberg, P.; Rastogi, P. K.; Giaccari, P.; Limberger, H. G. *Opt. Lett.* 2002, 27, 1385-1387.

[69] Yeo, T. L.; Sun, T.; Grattan, K. T. V.; Parry, D.; Lade, R.; Powell, B. D. *IEEE Sensors J.* 2005, 5, 1082-1089.

[70] Dong, X.; Huang, Y.; Lang, K.; Zhang, W.; Kai, G.; Dong, X. Microw. *Opt. Tech. Lett.* 2004, 42, 474-476.

[71] Bock, W. J.; Urbańczyk, W. *Appl. Opt.* 1998, 37, 3897-3901.

[72] Hsu, Y. S.; Wang, L.; Liu, W.; Chiang, Y. J. *IEEE Photon. Technol. Lett.* 2006, 18, 874-876.

[73] Peng, B.; Zhao, Y.; Yang, J.; Zhao, M. *Measurement* 2005, 38, 176-180.

[74] Zhao, C.; Yang, X.; Demokan, M. S.; Jin, W. *IEEE J. Light. Technol.* 2006, 24, 879-883.

[75] Iadiciccoa, A.; Campopianoa, S.; Cutoloa, A.; Giordanob, M.; Cusanoa, A. *Sens. Actuators B,* 2006, 120, 231-237.

[76] Shu, X.; Gwandu, B. A. L.; Liu, Y.; Zhang, L.; Bennion, I. *Opt. Lett.* 2001, 26, 774-776.

[77] Iadicicco, A.; Campopiano, S.; Cutolo, A.; Giordano, M.; Cusano, A. *IEEE Photon. Technol. Lett.* 2005, 17, 1250-1252.
[78] Men, L.; Lu, P.; Chen, Q. *J. Appl. Phys.* 2008, 103, 053107.
[79] Wade, S. A.; Collins, S. F.; Baxter, G. W. *Appl. Phys. Lett.* 2003, 94, 4743-4756.
[80] Grattan, K. T. V.; Meggitt, B. T. Optical Fiber Sensor Technology; Chapman & Hall, London & Kluwer Academic, Dordrecht Netherlands, 1994/98/99, Vols. 1-4.
[81] Dakin, J.; Culshaw, B., Optical Fiber Sensors; Artech House, Boston, 1988/89/96, Vols. 1-3.
[82] Lee, C. Y.; Lee, G. B. *Sensors Lett.* 2005, 3, 1-14.
[83] Kharaz, A.; Jones, B. E.; *Sens. Actuators A* 1995, 46, 491-493.
[84] Bariáin, C.; Matías, I. R.; Arregui, F. J.; López-Amo, M. *Sens. Actuators B* 2000, 69, 127-131.
[85] Gupta, B. D.; Ratnanjali *Sens. Actuators B* 2001, 80, 132-135.
[86] Yeo, T. L.; Sun, T.; Grattan, K. T. V.; Parry, D.; Lade, R.; Powell, B. D.; *IEEE Sensors J.* 2005, 5, 1082-1089.
[87] Sager, K.; Schroth, A.; Nakladal, A.; Gerlach, G. Sens. Actuators A 1996, 53, 330-334.
[88] Mrotek, J. L.; Matthewson, M. J.; Kurkjian, C. R. *IEEE J. Light. Technol.* 2001, 19, 988-993.
[89] Schirmer, B.; Venzke, H.; Melling, A.; Edwards, C. S.; Barwood, G. P.; Gill, P.; Stevens, M.; Benyon, R.; Mackrodt, P. *Meas. Sci. Technol.* 2000, 11, 382-391.
[90] Sorli, B.; Pascal-Delannoy, F.; Giani, A.; Foucaran, A.; Boyer, A. *Sens. Actuators A* 2002, 100, 24-31.
[91] Bozóki, A.; Szakáll, M.; Mohácsi, Á.; Szabó, G.; Bor, Z., *Sens. Actuators B* 2003, 91, 219-226.
[92] Tang J. L.; Wang, J. N. *Sens. Transducers J.* 2006, 68, 597-605.
[93] Xu, M. G.; Reekie, L.; Chow Y. T.; Dakin, J. P. *Electron. Lett.* 1993, 29, 389-399.
[94] Song, M.; Lee, S. B.; Choi, S. S.; Lee, B. *Opt. Fiber Technol.* 1997, 3, 194-196.
[95] Xu, M. G., Archambault, J. L.; Reekie, L.; Dakin, J. P. *Electron. Lett.* 1994, 30, 1085-1087.
[96] James, S. W.; Dockney, M. L.; Tatam, R. P. *Electron. Lett.* 1996, 32, 1133-1134.
[97] Song, M.; Lee, B.; Lee, S. B.; Choi, S. S. *Opt. Lett.* 1997, 22, 790-792.
[98] Frazao, O.; Ferreira, L. J. Opt. A: *Pure Appl. Opt.* 2005, 7, 427-430.
[99] Cavaleiro, P. M.; Araujo, F. M.; Ferreira, L. A.; Santos, J. L.; Farahi, F. *IEEE Photon. Technol. Lett.* 1999, 11, 1635-1637.
[100] Lu, P.; Men, L.; and Chen, Q. 2008, *Appl. Phys. Lett.* 92, 171112.
[101] Kersey, A. D.; Dandridge, A. *Proc. SPIE*, 1988, 985, 113-116.
[102] Berkoff, T. A.; Davis, M. A.; Bellemore, D. G.; Kersey, A. D. *Proc. SPIE*, 1995, 2444, 288-294.
[103] Askins, C. G.; Putnam, M. A.; Friebele, E. J. *Proc. SPIE*, 1995, 2444, 257-266.
[104] Friebele E. J.; Askins C. G.; Bosse A. B.; Kersey A. D.; Patrick H. J.; Pogue W. R.; Putnam M. A.; Simon W. R.; Tasker F. A.; Vincent W. S.; Vohra S. T. *Smart Mater. Struct.* 1999, 8, 813-838.
[105] Zhou, Z.; Graver, T. W.; Hsu, L.; Ou, J. *Pacific Science Review* 2003, 5, 116-121
[106] Okabe, Y.; Tanaka, N.; Takeda, N. *Smart Mater. Struct.* 2002, 11, 892–898.
[107] Bjerkan, L. *Appl. Opt.* 2000, 39, 554-560
[108] Wnuk, V. P.; Mendez, A.; Ferguson, S.; Graver, T. *Proc. SPIE*, 2005, 5758, 46-53.

[109] Meltz G. and Morey, W. W., *Proc. SPIE*, 1991, 1516, 185-199.
[110] Sager, K., Schroth, A., Nakladal, A., and Gerlach, G., *Sens. Actuators* 1996, 53, 330-334.

Reviewer:

Yanfei Ding, Department of Physics and Physical Oceanography, Memorial University of Newfoundland, St. John's, NL A1B 3X7, Canada.

In: Nuclear Chemistry: New Research
Editor: Axel N. Koskinen, pp. 181-211

ISBN: 978-1-60456-957-5

Chapter 6

OXA-PENTANEDIAMIDE AND OXA-GLUTARAMIC ACID: NEW AND PROMISING LIGANDS IN NUCLEAR INDUSTRY

Ping Zhang and Hairong Wang
Institute of Nuclear and New Energy Technology, Tsinghua University
P.O.Box 1021, Beijing 102201, China

Abstract

N,N-Dialkyl-3-Oxa-Glutaramic Acid ($RR'N(COCH_2)_2OHO$) (DAOGA) and N,N,N'N'-tetra-alkyl-3-oxa-pentanediamide ($(RR'NCOCH_2)_2O$) (TAOPDA) are two serials of powerful ligands to lanthanides (III) and actinides(III and IV) . Their solubility in organic phase or aqueous phase and other properties vary with their different substituted alkyl groups, R and R'. Due to these merits, they are able to be used as extractants or stripping agents in nuclear chemistry, specifically treatment of nuclear spent fuel. In this chapter, the progress of lately researches on organic-soluble oxa-pentanediamide as extractants, which is world widely studied including in our lab, is reviewed. Secondly, the synthesis, properties of aqueous-soluble DAOGA and TAOPDA, and their applications are summarized, of which the following work were initiated in our lab: synthesis and identification of six ligands of these two series; structure identification with IR and EXAFS of the extracted complexes of Ln(III) with N, N'-dimethyl-3-oxa-glutaramic Acid (DMOGA); determination of stability constants of Np(IV) and Pu(IV) complexes with DMOGA; and measurements of the distribution ratios of actinides and some fission products between 30% trialkyl phosphine oxide (TRPO) in kerosene and HNO_3-DMOGA solutions. Based on these results, DMOGA is proposed to be used as an effective stripping agent for An(III, IV) and therefore the TRPO process, well-known as one of the promising processes for treatment of high level liquid waste(HLLW) , is simplified.

Keywords: TODGA, DMOGA, lanthanides, actinides, extraction, HLLW, TRPO

1. Synthesis, Extraction Behaviors, Process Development and Radiolysis of Dialkylamides

1.1. Introduction

Treatment and disposal of high level liquid waste (HLLW), which is generated from the reprocessing of nuclear spent fuel, are crucial to the sustainable development of nuclear energy[1]. Partitioning-transmutation is considered as one of the most important strategy for HLLW treatment[2]. For the reprocessing of spent nuclear fuel and the partitioning of HLLW, several processes based on various extractants have been developed during the past decades. These processes utilize different extractants, which mainly include octyl(phenyl)-N,N-diisobutylcarbamoylmethylphosphine oxide (OΦD(iB)CMPO, or CMPO), dimethyldibutyltetradecylmalonamide(DMDBTDMA), trialkylphosphine oxide (TRPO), and diisodecylphosphoric acid (DIDPA). Based on these phosphorus contained extractants, different processes were developed, they are TRPO process, TRUEX process, DIMAX process, and DIDPA process[3-8].

Recently, diamides derivatives named oxapentanediamide(OPDA) caused high interests due to their desirable coordination properties to metal ions[9-12]. Oxapentanediamide, which can be generally expressed as (RR'-NCO-CH_2)$_2$-O, may have strong extraction properties to actinides and lanthanides with its three oxygen donors and two nitrogen donors. The extraction efficiency of the diglycolamides is thought to be related to both the electro-inductive effects as well as the steric hindrances around the binding site. Therefore, the nature of the two groups, R and R', has important effect on the extraction ability of diamides. By modifying the alkyl groups attached to carbonyl carbon or amidic nitrogen atoms, the coordination properties of OPDA may be easily adjusted. Furthermore, the solubility of diamides in organic solvents or aqueous solutions could be changed by varying the R or R's group. Another attractive advantage is that this kind of compound is composed of CHON, which means it can totally thermal decomposed without generating secondary waste. With these merits, OPDA is considered as a kind of good extractant.

OPDA were intensively studied across the world. Several groups, including Sasaki Y from Japan atomic energy agency (JAEA), Ye G from Chinese atomic energy academy (CEAE), a group from the institute of nuclear energy technology(INET), China, Modolo G from Julich, Germany, Mowafy from Egypt, Ansari M from India, et al., carried out researches in various aspects on OPDA, especially on octyl- substituted OPDA, i.e., N,N,N',N'-Tetraoctyl-3-oxapentanediamide (TODGA) . Their works include the syntheses of OPDA and its analogues, extraction of metal ions with OPDA, radiolysis of TODGA, and process development based on OPDA.

In this part, we review the research works on OPDA. The synthesis method is introduced in detail, the identification data are collected. The extraction of metal ions with OPDA and other analogues are reviewed and the process development based on TODGA or other analogues are summarized and commented. In addition, the radiolysis of TODGA is reviewed.

For many compounds were involved in different works, and scientists usually use their own abbreviations, we remain the original abbreviations of those extractants used in reference papers. For the convenience of reading, we present an abbreviation list shown in table 1.

Table 1. Abbreviation and structure formula of extractants

Abbreviation	Compounds
CMPO	(phenyl))-N,N-diisobutylcarbamoylmethyl phosphine oxide
DMDBTDMA	*N,N,N',N'*-dimethyl dibutyl tetradecyl malonamide
DIDPA	Diisodecylphosphoric acid
TRPO	trialkyl phosphine oxide
TOOPDA	*N,N,N',N'*-tetraoctyl-3-oxapentanediamide
TDOPDA	*N,N,N',N'-tetradecyl-3-oxapentanediamide*
TBOPDA	*N,N,N',N'-tetra-n-butyl-3-oxapentanediamide*
TiBOPDA	*N,N,N',N'-tetra-iso-butyl-3-oxapentanediamide*
THOPDA	*N,N,N',N'-tetra-n-hexyl-3-oxa-pentanediamide*
TEHOPDA	*N,N,N',N'-tetra-(2-ethylhexyl)-3-oxa-pentanediamide*
TBDOODA	*N,N,N',N'-tetra-n-butyl-3,6-dioxaoctanediamide*
TiBOGA	*N,N,N',N'*-tetra-iso-butyl-3-oxa-glutaramide.
THDGA	*N,N,N',N'*-tetrahexyl-3-oxapentanediamide
TODGA	*N,N,N',N'*-Tetraoctyl-3-oxapentanediamide
DMDHGA	*N,N'-d*imethl-*N,N'*-dihexyl-3-oxa-glutaramide
DMDOGA	*N,N'-d*imethl-*N,N'*-dioctyl-3-oxa-glutaramide
DMDL GA	*N,N'-d*imethl-*N,N'*-dilauryl-3-oxa-glutaramide
DMDHOA	*N,N'-d*imethl-*N,N'*-dihexyl-3,6-dioxa-octanediamide
TEHDGA	*N,N,N',N'*-tetra-2-ethylhexyl-3-oxapentanediamide
TDDGA	*N,N,N',N'*-Tetradecyl-3-oxapentanediamide
DGA	diglycolamide
DMDOHEMA	2-(2-hexyloxy-ethyl)-N,N.-dimethyl-N,N.-dioctyl-malonamide
DMDBTDMA	N,N.-dimethyl-N,N.-dibutyl-tetradecylmalonamide
(DHOA)	N,N,-dihexyloctanamide
DBDA	dibutyl decanamide
D2EHAA	di(2-ethylhexyl) acetamide
D2EHPRA	di(2-ethylhexyl) propionamide
D2EHIBA	di(2-ethylhexyl) isobutyramide
DHDA	Dihexyl decanamide
TPOGA	*N,N,N',N'*-Tetrapropyl-3-oxapentanediamide
TiPOGA	*N,N,N',N'*-Tetra-iso-propyl-3-oxapentanediamide
DHDMOGA	*N,N'-d*iheptyl-*N,N'*-dimethyll-3-oxa-glutaramide

1.2. Synthesis and Characterization of TODGA and Its Analogues

Diglycolamides are generally synthesized by the method initially proposed by Pretsch E et al [13], In his work, eleven diglycolamides with various substituted groups, including alkyl, cycloalkyl, and benzene groups were prepared. In recent years, Sasaki Y, Zhang P, and Mowafy et al. respectively reported the improved synthesis method of diglycolamides with different substituted alkyl group[14-16]. The process consists of the following three steps: firstly, the initial material, diglycolic acid, is dehydrated to form diglycolic anhydride; secondly, it reacts with respective amines to form corresponding mono-substituted amide, and thirdly, the mono-substituted amide reacts with amines to process diglycolamides. The reactions could be expressed as follows:

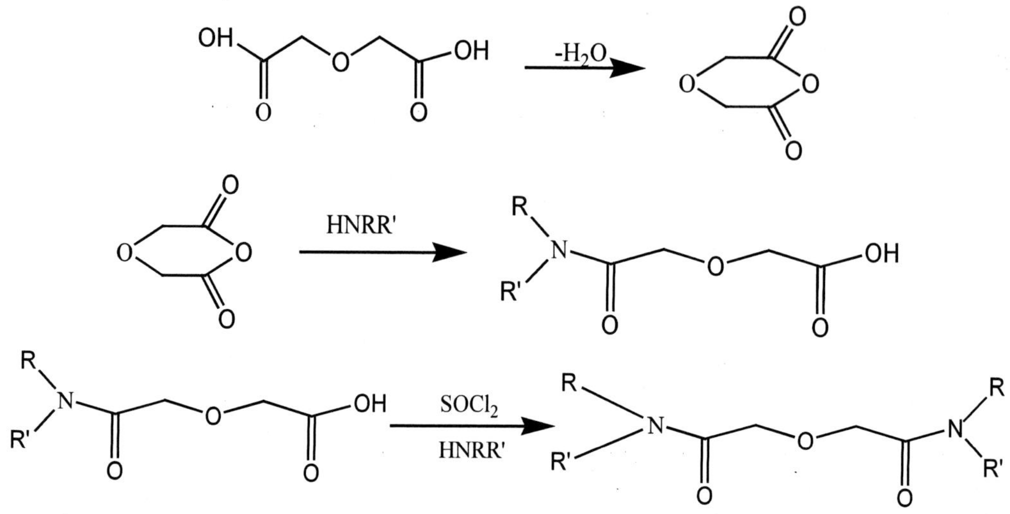

The synthesis procedures are described in detail with TBOPDA as an example.

(1) Preparation of diglycolic anhydride

Dissolving the diglycolic acid (about 50 g) with an excess of redistilled acetic anhydride (140 ml), drops of phosphorus acid are added in as an initiating agent. The temperature rises to boiling point of acetic anhydride and the reaction mixture is stirred for 1 hour to ensure the completeness of the reaction. Acetic anhydride is evaporated when the temperature rises to 125°C. The residue is recrystallized from toluene after most solvent is evaporated out. The yield is about 90%. The melting point of product diglycolic anhydride is 92-94°C.

(2) Synthesis of mono-substituted oxa-pentaneamide

The required quantity of diglycolic anhydride (58.5 g) is dissolved in an excess amount of dioxane (450 mL). The mixture is stirred until the solution becomes transparent. The mixture of 87.5 ml dibutyl amine and 40.6 mL pyridine is added in the solution dropwisely in ice water bath. After complete addition, the mixture is stirred and refluxed for about 3 hours. Then 1,4-dioxane is distilled and the crystal appears when concentrated hydrochloric acid is dropped into the residual solution. The crystal is separated by filtration and dried. The yield of the mono-substituted oxa-pentaneamide is 89% and the m.p is 82—85°C.

(3) Synthesis of TBOPDA

The mono-substituted oxa-pentaneamide (67 g) is added to excess of $SOCl_2$ (34 mL) and refluxes for about one hour. When the mixture is cooled down to room temperature, 600 mL aether is added under the intensive stirring. The reaction vessel is cooled down to below 0°C with ice bath. Subsequently, the mixture of 100 mL dibutyl amine, 50 mL pyridine and 40 mL aether is slowly added. The reaction mixture is refluxed for about 3 hours and after cooling, the precipitate is removed by filtration. The crude product is washed alternately with 1 mol/L hydrochloric acid and deionized water twice. Solvent is removed by distillation, the remaining thick brown liquid obtained is dried overnight over anhydrous sodium sulphate. The final purification involves distillation under reduced pressure.

The synthesis processes reported by different researches are quite similar, only the solvents and electrophilic reagents employed in the third step are different. In addition, the purification steps are different, too. Sasaki used dicyclohexylcarbodiimide as the electrophilic reagent and the reaction proceeded in CH_2Cl_2, the final product was purified by distillation and silica gel column separation. Zhang and Mowafy used thionyl chloride to react with mono-substituted oxa-pentaneamide, diglycolamides produced were purified by washing alternately with hydrochloric acid and deionized water and following vacuum distillation. These synthesis processes have no essential differences; however, various procedures and purification methods lead to differences in yield or time consumption.

Table 2. Physical data and elemental analysis results

Acronym	Structure formula	M.W (g/mol)	Yield (%)	Elemental analysis found(calcd.)/%	Ref.
TPOGA	$O[CH_2CON(C_3H_7)_2]_2$	300	68	C 63.74(63.98), H10.79 (10.74), N9. 15 (9. 33)	[13]
TiPOGA	$O[CH_2CON(CH(CH_3)_2]_2$	300	14	C 63.88(63.98), H10.51 (10.74), N8.97 (9. 33)	[13]
DHDMOGA	$O[CH_2CON(CH_3(C_7H_{15})_2]_2$	356	90	C 67.10(67.37), H11.37 (11.31), N7.68 (7.86)	[13]
TPDGA	$O[CH_2CON(C_3H_7)_2]_2$	300	48	C64.89 (64), H 10.11 (10.7), N 8.95 (9.3)	[16]
T2MPDGA	$O[CH_2CON(CH_2CH(CH_3)_2]_2$	356	47	C66.53 (67.41), H10.59 (11.23) N8.16 (7.86)	[16]
TBDGA	$O[CH_2CON(C_4H_9)_2]_2$	356	51	C66.73 (67.4), H 11.1 (11.3) N7.34 (7.86)	[16]
THDGA	$O[CH_2CON(C_6H_{13})_2]_2$	468	50	C70.27 (71.79), H 11.12 (11.96), N 6.15 (5.98)	[16]
T2EHDGA	$O[CH_2CON(C_2H_5C_6H_{13})_2]_2$	580	50	C73.42 (74.4), H11.65 (12.4) N4.59 (4.81)	[16]
TODGA	$O[CH_2CON(C_8H_{17})_2]_2$	580	49	C73.71 (74.41), H 12.11 (12.4) N4.94 (4.82)	[16]
TODGA	$O[CH_2CON(C_8H_{17})_2]_2$		65–70	C73.1 (74.5), H 12.8 (12.4),N 5.2 (4.8)a	[33]
TBOPDA	$O[CH_2CON(C_4H_9)_2]_2$	356	40	C 67.2(67.4); H 11.5(11.3).. N 7.57 (7.86)	[15]
TMPOPDA	$O[CH_2CON((CH_3)_2CHCH_2)]_2$	356	42	C 67.2(67.4), H 11.4(11.3), N 7.97(7.86).	[15]
THOPDA	$O[CH_2CON(C_6H_{13})_2]_2$	468	45	C 70.4(71.7), H 12.2(12.0), N 6.18(5.98)	[15]
TEHOPDA	$O[CH_2CON(C_4H_9CH(C_2H_5)CH_2)_2]_2$	580	48	C 73.7 (74.4), H 12.5 (12.5), N 4.98(4.82)	[15]
DMDHOPDA	$O[CH_2CON(CH_3C_6H_{13})]_2$		70	C 63.9(65.8)	[11]
DHOPDA	$O[CH_2CONH(C_6H_{13})]_2$		20		[11]

Many diamides were prepared and characterized with elemental analysis, IR and H-NMR, or C-NMR spectra. Here we summarize the physical data, elemental analysis results (table 2), NMR and IR data (table 3) for reference.

Table 3. NMR and IR data

Acronym	C-NMR /H NMR	IR(cm^{-1})	Ref.
TPOGA	H-NMR($CHCl_3$): 0.9(t,12H), 1.5(m, 8H), 3.2(m, 8H), 4.3(s, 4H) C-NMR($CHCl_3$): 11.1and 11.4(2q), 20.8 and 22.1(2t), 47.3 and 48.5(2t), 69.2(t), 168.6(s)	1650	[13]
TiPOGA	H-NMR($CHCl_3$): 1.3(8,24H), 3.7(b, 4H), 4.2(s, 4H),	1640	[13]
DHDMOGA	H-NMR($CHCl_3$): 0.9(t,6H), 1.3-1.5(b, 20H), 2.9, 2.94(2s, 6H), 3,2 and 3.4(2t, 4H), 4.27 and 4.28(2s, 4H) C-NMR($CHCl_3$): 14.1(q), 22.6(t), 26.7, 26.8, 27.1, 28.4, 29.0 and 31.8 (6t), 33.0, 34.1(2q) 47.8 and 49.0(2t), 69.2 and 69.4(2t), 168.5 and 168.6(2s)	1670	[13]
TPDGA		1650, 1255, 1293	[16]
T2MPDGA		1646, 1301, 1256	[16]
TBDGA		1650, 1293, 1255	[16]
THDGA		1646, 1299, 1258	[16]
T2EHDGA		1652, 1262, 1288	[16]
TODGA		1649, 1308, 1249	[16]
TODGA	H-NMR:δ 0.92(12H,t), 1.26(40H, m), 1.56(8H,m), 3.18(8H, t), 4.39(4H, s)	1640	[33]
TBOPDA	C-NMR :δ12.7(1C), 18.8-19.0(2C), 28.5-29.8(3C), 44.3-45.5(4C), 167.3(5C), 67.8(6C)	1650 (νC=O), 1293, 1255 (νC−N)	[15]
TMPOPDA	C-NMR: δ20(1C), 26-27(2C), 53-55(3C), 170(4C), 68(5C)	1646 (νC=O),	[15]
THOPDA	C-NMR :δ12.8-12.9(1C), 21.4(2C), 30.4(3C), 25.3-25.5 (4C), 26.4-27.7(5C), 44.6-45.7(6C), 167.3(7C), 67.9 (8C).	1646 (C=O), 1299, 1258 (C−N)	[15]
TEHOPDA	C-NMR :13.8-13.9(1C), 22.7-22.9(2C), 28.2-28.6(3C), 30.1-30.3(4C), 36.4-37.6(5C), 23.4-23.5(6C), 10.3-10.7(7C), 44.6-45.7(8C), 169.3(9C), 68.9(10C)	1652 (νC=O), 1262 (νC−N),	[15]
DMDHOPDA	H-NMR($CDCl_3$): δ 0.85(t,6H),1.2-1.6(16H), 2.9(s,6H),3.2-3.3(t,4H), 4.3(s,4H)	1600 (C=O), 1200 (C−N),	Sasaki [11]
DHOPDA	H-NMR($CDCl_3$) δ0.88(t,6H), 1.2-1.8(16H), 3.3(t,4H), 4.0(s,4H), 6.5(s, 2H)	1600 (C=O),1200 (C−N),	Sasaki [11]

Although the synthesis of diglycolamides was developed and improved, the process is usually time-consuming; also the yield is not satisfactory, which may lead to high cost. Considering that TODGA is expected to be used industrially in the future, simpler synthesis procedures, especially effective purification methods, should be further explored.

1.3. Extraction of Metal Ions with Dialkylamides

Generally, the researches on extraction chemistry include following: the determination of distribution coefficient or ratio (D_M), extraction equilibrium constant, extraction reaction formula, calculation of thermodynamics parameters (ΔG_o, ΔH_o, and ΔS_o), studies on structure of extracted complex, et al.

The extraction reaction from nitric acid solution could usually be expressed as follows:

$$aM^{n+}{}_{(a)} + bS_{(o)} + iNO^-_{3\,(a)} \Leftrightarrow M_a(NO_3)_i \bullet bS_{(o)}$$

Where M^{n+} is metal ions, S is extractant, (a) refers to aqueous phase, (o) refers to organic phase.

Distribution ratio is expressed as:

$$D_M = \frac{C_{(M)(o)}}{C_{(M)(a)}}$$

Where $C_{(M)(o)}$ and $C_{(M)(o)}$ refer to the concentration of metal ions in organic phase and aqueous phase, respectively.

Extraction equilibrium constant (K_{ex}) is expressed as:

$$K_{ex} = \frac{C_{(M_a(NO_3)_i \bullet bS)(o)}}{C^a(M^{n+})C^i(NO_3^-)C^b(S_{(o)})}$$

$$D_M = \frac{C_{(M)(o)}}{C_{(M)(a)}} = \frac{aC_{(M_a(NO_3)_i \bullet bS_{(o)})}}{C_{(M^{n+})}} = \frac{aK_{ex}C^a{}_{(M^{n+})}C^i{}_{(NO_3^-)}C^b{}_{(S_{(o)})}}{C_{(M^{n+})}}$$
$$= aK_{ex}C^{a-1}{}_{(M^{n+})}C^i{}_{(NO_3^-)}C^b{}_{(S_{(o)})}$$

$$\log D_M = b\lg C_{(S_{(o)})} + \lg\left[K_{ex}C^{a-1}{}_{(M^{n+})}C^i{}_{(NO_3^-)}\right] + \lg a$$

By determining the dependency of D against various parameters, such as concentration of nitric acid, concentration of extractants, the coefficient of the reaction formula, i.e., a, b, and c, could be calculated.

The enthalpy and entropy of extraction reaction could be calculated by Van't Hoff relation.

$$\ln K = -\frac{\Delta H^0}{R}\left(\frac{1}{T}\right) + \frac{\Delta S^0}{R}$$

From the dependence of the extraction constant (K) on temperature, the thermodynamics parameters could be calculated.

Because the dialkylamides extractants will be used for partitioning of actinides and fission products from high level liquid waste, it is the most important to their application to possess strong ability of extraction to the main metal ions in HLLW, especially to actinides and radioactive fission products.

In this section, the extraction behaviors of dialkylamides to metal ions with various valences contained in HLLW are presented and discussed.

1.3.1. Extraction of Sr(II) and Other Divalent Ions

Sr(II) is an important fission product (FP) in HLLW, for the radioactive ^{90}Sr is a β emitter and main heat releasing nuclide among FPs.

Using 40% octanol/kerosene as diluents, Ye et al[17] studied the extraction of Sr(II) with T-n-BOGA, T-i-BOGA and TBDOODA from nitric acid solution of Sr(II). They also examined the composition of the extracted complexes and measured the apparent equilibrium constant of the extraction reaction . Furthermore, they calculated the values of thermodynamic functions, i.e., ΔH, ΔS, and ΔG., which are shown in table 1. By studying the dependency of distribution ratio of Sr(II) on concentration of Sr(II) and the extractant, the coefficient of extraction reaction is obtained. The reaction formula is expressed as following:

$$Sr^{2+} + 2NO_3^- + iS_o \Leftrightarrow Sr(NO_3)_2 \bullet iS_o$$

Where S_o refers to the extractant. For TBOGA, i =3, for TBDOODA, i = 2.5.

Tian et al[18] investigated the extract property of TiBOGA to Sr(II), attention was paid on the effect of different diluents on the extraction of Sr(II). The effects of several diluents, including kerosene, 1-octanol, toluene, and chloroform, and some mixtures of them, on the distribution ratio were tested. Results show that the mixture of 40/60 (V/V)% 1-octanol/kerosene is a feasible diluent. In addition, based on the results of the dependence of D_{Sr} on the concentrations of the TiBODA and temperature, the composition of species of Sr(II) formed in the extraction is assumed, and the proposed extraction reactions is as follows:

$$Sr^{2+}{}_{(aq)} + 2NO^{3-}{}_{(aq)} + 3TiBOGA_{org} \xleftrightarrow{k} Sr(TiBOGA)_3 \bullet (NO_3)_{2(org)}$$

The extraction equilibrium constant K is 1.04×10^3 by calculation.

Tian investigated the possibility of recovering Sr from the raffinate of Chinese TRPO process after TRU extraction. The test results show that 0.2 mol/L TiBOGA in 40/60% 1-octanol/KO has good extraction ability and selectivity for Sr(II) from HLLW, and the system may be used to separate ^{90}Sr from the raffinate.

By determine the K value at different temperatures, the enthalpy and entropy of extraction are calculated from van't Hoff relation and shown in table 4. It can be observed that Tian's results are in agreement with Ye group's, and the data suggested that the extraction process is exothermic.

Table 4. Thermodynamics data of Sr(II) extraction reaction

Extractant	ΔH/ kJ.mol^{-1}	ΔS/J/mol.K	ΔG/kJ / mol	Ref.
TBOGA	- 50.48	117.6	- 15.44	17
TiBOGA	- 55.79	- 122.9	- 19.16	17
TiBOGA	-48.9	-107	-17.1	18
TBDOODA	- 53.11	- 141.6	- 10.91	17

Solvent extraction of alkaline earth metal ions was studied by Sasaki Y et al[19]. TODGA and its homologues, TDDGA and TEHDGA, were synthesized and applied to the extraction of divalent alkaline earth metal ions, including Sr, Ca and Ba, from nitric acid solutions. The stoichiometry of the extracted Sr(II)-DGA species is found to depend mainly on the concentration of nitric acid and molecular structure of DGA. The degree of extraction of M(II) by TODGA is in the order of Ca(II)>Sr(II)>Ba(II) (ionic radius: 112, 125, and 142 pm for Ca, Sr, and Ba), the smaller the ion radius is, the easier the metal ion is extracted. This tendency implies that the interaction between TODGA and M(II) is dominantly electrostatic.Based on the results of logD vs. log[HNO_3] and log[TODGA], the supposed extraction mechanism is as follows:

$$M^{2+} + 2NO_3^- + 2TODGA + HNO_3 \Leftrightarrow M(NO_3^-)_2(TODGA)_2(HNO_3)$$

Which is different from those proposed by Ye and Tian[17,18].

Besides extraction of Sr(II), Sasaki Y extensively studied the extraction of other divalent ions, including Ca(II), Sr(II), Ba(II),Cd(II), Pb(II), Pd(II), and Hg(II)[20]. By slope analysis, the useful information on the composition of the extracted chemical species is obtained. The slopes obtained for these ions are almost the same, except for Hg (II), 0.85. These data indicate that the molar ratios of the extracted species are mainly 1:2 for metal: TODGA; the divalent metal ions with coordination number four, e.g., Be(II), Cu(II), Zn(II) and Cd(II), are unextractable. Moreover, Sasaki Y elucidated the relationship of distribution ratio and ionic radius.Sasaki also studied the mutual separation of Ca, Sr, and Ba, and it was shown that the average separation factors of Ca/Sr and Ca/Ba are 15 for Ca/Sr and 78 for Sr/Ba. These results indicate that Sr could be easily recovered from HLLW without cross-decontamination of other ions.

With its strong complexing ability to Sr2+, dialkylamides, especially TODGA and its analogues, become an effective extractant for recovering Sr(II) from HLLW as well as separation Sr(II) from other divalent ions.

1.3.2. Extraction of Trivalent Metal Ions

Trivalent metal ions in HLW mainly include lanthanides(III) and Am(III). Many works involved in the extraction of Ln(III) and Am(III) by dialkylamides, particularly TODGA and TOOPDA.

Sasaki group conducted comprehensive researches on the extraction of trivalent lanthanides and actinides with diglycolamides[21-26]. Six diglycolamides expressed as R,R–N-CO-CH_2-O-CH_2-CO-N–R,R ($R=CnH_{2n+1}$, $3\leqq n\leqq 10$) were synthesisized, which were abbreviated to TPOPDA, TBOPDA, TAOPDA, THOPDA, TOOPDA, and TDOPDA. The dependence of the distribution coefficients of Eu(III) and Am(III) on the length of alkyl chain in the extractants was examined. Among these diglycolamides, TOOPDA and TDOPDA present sufficient solubility in *n*-dodecane, due to their appropriate lipophilicity modified by increasing the length of the alkyl chains attached to amidic N atoms. The distribution ratios of actinides by TOOPDA increase with the increasing of HNO_3 concentration. The number of diglycolamide molecules coordinated to the actinide ions was believed to be three for Th(IV), U(VI), Pu(IV), and to be four for Am(III) and Cm(III) by slope analysis. The order of *DM* for the actinides at high nitric acid concentrations is An(III), An(IV) > An(VI) > An(V) and that for lanthanides, heavier>lighter. The conclusion is TOOPDA diluted in *n*-dodecane can extract both actinides and lanthanides completely from aqueous HNO_3 solution.

In addition to TDOPDA and its homogeneous compounds, Sasaki et al synthesized some diamide extractants with the different backbones and studied their extractability for lanthanides(III) and actinides(III), (IV), (V) and (VI). The diglycolamide introduced an ether oxygen into the main frame, $O\text{-}(CH_2\text{-}CONR_1,R_2)_2$, exhibits the highest extractability for An(III), (IV) and (VI), compared to the other bidentate diamides. Also, the present results show that the thiadiglycolamide, TDGA, which introduced ether sulfur instead of oxygen in DGA molecule, enhances the extraction of actinides.

After having tested many kinds of compounds, Sasaki et al. considered TODGA as the most promising extractant and conducted more researches on TODGA . For extraction of Nd(III) by TODGA-n-dodecane solvent, they chose N,N,-dihexyloctanamide (DHOA) as the phase modifier to eliminate the third phase formed. Their works show that the loading capacity of the solvent depends not only on the acidity and the temperature but also on molecular size of alkane solvent and kinds of aqueous anions. In addition, the extraction of Eu(III) and Am(III) with TODGA was investigated further, and different diluents, including 1,2-dichloroethane, octanol, chloroform, toluene and dodecane, were employed to investigate the influence of polarity of diluent on extraction mechanism and species formed in extracted complexes. In polar diluents, the chemical species is $M(TODGA)_2$, while in non-polar diluents, the number of coordination for the metal ions is three or four.

In China, several groups, such as institute of nuclear and new energy of technology(INET) of Tsinghua University, Chinese academy of atomic energy(CIAE) and Sichuan University, carried out the study on extraction of lanthanides and actinides.

Several diamide derivates were synthesized at INET[15,27] and the extraction of actinides and some fission products were studied. N,N',N'-tetra-(2-ethylhexyl)-3-oxa-pentanediamide (TEHOPDA) was proved to be a suitable extractant for removing actinides from nitric acid solution and the actinides can be stripped from the loaded solvent by the dilute nitric acid. TEHOPDA shows high loading capacity to actinides and lanthanides when a mixture of n-octanol and kerosene is taken as the diluent.

Ye et al[28,29] studied the extraction of Eu(III) and Am(III) with N ,N ,N',N'-tet rabuty1-3-oxa-pentanediamide (TBOPDA), N,N,N,N'-tet raisobuty1-3 -oxa pentanediamide (TiBOPDA) and N,N,N',N'-tet rabuty1-3 ,6-dioxa-octanediamide (TBDOODA). The distribution ratio dependences of Am and Eu with the nitric acid concentration were measured. They determined the compositions of the extracted complexes and supposed the extraction mechanism. The extraction reaction supposed is as follows:

$$M^{3+} + 3NO_3^- + iS_{(o)} \Leftrightarrow M(NO_3)_3 \bullet iS_{(o)}$$

Where S is the extracant. For TBOPDA and TiBOPDA, i =3; for TBDOODA, i = 2.5.

By measuring the electro-conductivity of organic phase loaded with metal ion, they assumed that the extraction is neutral complexation. In addition, the extraction thermodynamics and absorption spectra of the extracted complexes were studied. The measured extraction enthapy of Am(III) is - 80.54, - 81.99 and -75.88 kJ / mol for TBOPDA ,TiBOPDA and TBDOODA , respectively. Therefore, the extraction is an exothermic reaction. In addition, they investigated the infrared spectra changes of the extractant after extraction.

Another group at Sichuan University of China, lead by Chen W studied the influence of structure of diamides on their extractive properties to Eu(III) and Am(III) [30-32],. They synthesized diamides, $(R_1R_2NCO)_2R_3$, which can be expressed with following formulae:

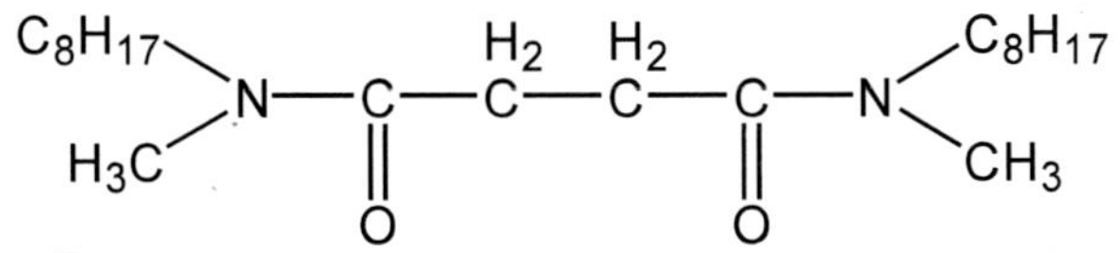

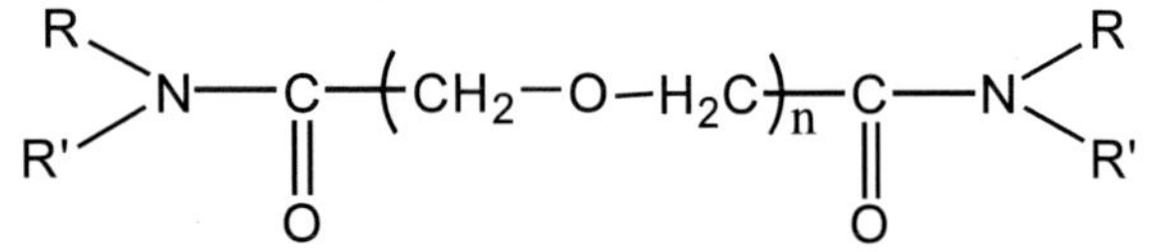

(Amido podant)

R= CH3, R′= C8H17, *n*= 1: *N,N*'-dimethyl-*N,N*-3-oxa-dioctylglutaramide (2) ;

R= R′= C4H9, *n*= 1: *N,N,N',N*- tetrabutyl-3-oxy-glutaramide (3) ;

R= R′= C4H9, *n*= 2: *N,N,N',N*- tetrabutyl-3,6-dioxy-octanediamide (4).

It is shown that the extractive ability of the diamides to trivalent actinides and lanthanides ions is in the order of R_3= - CH_2OCH_2 - > -(CH_2OCH_2)$_2$->>(CH_2)n. Using diamides with R_3= $(CH_2OCH_2)_n$ as extractants, the distribution ratios of Eu (III) and Am (III) increase with the increasing of concentrations of nitric acid in aqueous phases. For amido podands with R_3= _ CH_2OCH_2, D_{Eu} is obviously higher than D_{Am}; With the extractive system developed, the amido podands could be utilized in separation or recovery of trivalent actinides and lanthanides in nuclear industry or of rare earth metals in hydrometallurgical industry.

Bisides, Chen's group also carried out researchs on extraction of Ln(III) and Am(III) with similar extractants. They prepared amido podand with following structure:

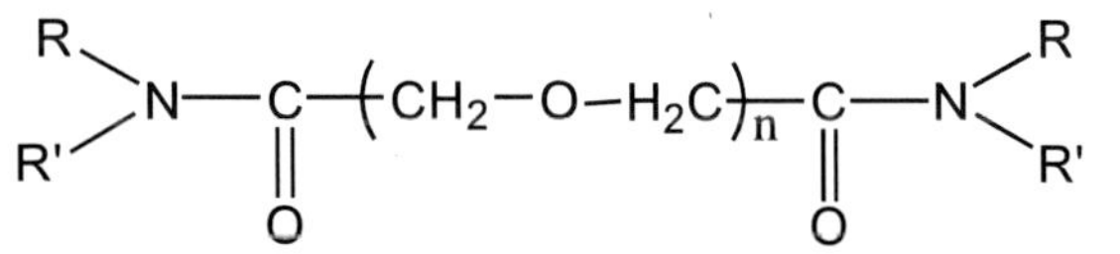

R,R' = alkyl, cycloalkyl, substituted alkyl, aryl and substituted aryl.

The effects of the concentrations of the nitric acid in aqueous phase and the structure of amido podands on distribution ratios were examined. Under the same nitric acid concentration, the extractive capability of amido podands of different structures was compared. The extractive capability of asymmetric amine is higher than that of symmetric amine. In addition, they investigated the extraction of trivalent lanthanides, such as Y^{3+}, Eu^{3+}, and La^{3+} with some asymmetric extractants, such as N,N'-dimethyl-N,N'-dioctyl-3-oxy-glutaramide (DMDOGA), and it shows that DMDOGA presents good extractability and selectivity.

In India, Ansari et al. [33,34] studied effect of diluent on the distribution behavior of Am(III) by employing diluents of different dielectric constants, including n-dodecane, toluene, 1-octanol, 1,2-dichloroethane, and nitrobenzene. Also, extraction behavior of various metal ions was investigated from nitric acid medium as well as from simulated high level waste (SHLW) solution, employing 0.1M TODGA and 0.5M DHOA in n-dodecane as the extractant. To solve the problem of the limiting organic concentration (LOC) for Nd(III), or appearance of the third phase, which was also studied by Sasaki, they chose N,N-dialkyl amides with varying alkyl groups and chain lengths, including dibutyl decanamide (DBDA), di(2-ethylhexyl) acetamide (D2EHAA), di(2-ethylhexyl) propionamide (D2EHPRA), di(2-ethylhexyl) isobutyramide (D2EHIBA), dihexyl octanamide (DHOA) and dihexyl decanamide (DHDA), to evaluate their possibilities as the phase modifier, and DHOA is shown to be a promising one.

Mowafy et al[16] synthesized six structurally related diglycolamides compounds, including TPDGA, T2MPDGA, TBDGA, THDGA, T2EHDGA and TODGA , and applied them to the extrcaction of selected actinides and lanthanides from nitric acid solution, using benzene as diluent. The dependence of the extraction distribution ratio on the length of alkyl chain of the extractants, along with the type of diluents was investigated. Distribution ratios of the investigated metal ions have been studied as a function of aqueous HNO_3 concentrations, diglycolamides concentration, and temperature. Except for TPDGA ($R=C_3H_7$), the extraction of Nd(III), La(III), Am(III), Eu(III) were found to significantly decrease with the increase of the alkyl chain length, R, from C_4H_9 to C_8H_{17}. The extraction of the investigated metal ions is also affected by the branching of the alkyl group (2-methylpropyl and 2-ethylhexy), compared with the corresponding diglycolamides. La(III), Nd(III), Eu(III) and Am(III) extraction by TODGA and THDGA show a sharp increase as the nitric concentration increase till 3 M HNO_3, and a slight increase with increase nitric acid concentration up to 5M HNO_3. The order of extractability of these trivalent ions is Eu(III) > Am(III)> Nd(III)> La(III) in all the nitric acid concentration. The general extraction ability of TODGA and THDGA is in order: Eu(III) >Am(III) >Nd(III) > La(III) >Sr(II) > Ba(II) >Cd(II)≥Co(II).

Although many works have been done on the extraction of trivalent ions, few of them involves in the structures of the extracted complexes. Sasaki[25] employed infrared spectra and luminescence lifetime determination to study the coordination of Eu(III) with TODGA, the results suggested that the carbonyl oxygen coordinates with Eu(III), and luminescence lifetimes indicate that there is no water molecules in the inner coordination sphere of the extracted Eu-complex. The extracted species of Eu-TODGA complexes are different in polar and non-polar solvent, chemical species: 1:2 for metal:TODGA complex is present in polar diluents. The metal complexes need three or more TODGA molecules to remain stable in non-polar diluents. Earlier studies suggested that the Nd(III)-TODGA complex in n-dodecane is composed of two extractant molecules bonded in tridentate fashion with some water molecules present in the inner coordination sphere, and depending on the solvent used, more TODGA molecules could be organized in the outer sphere. Similar observations have been reported on the basis of EXAFS study. It is therefore suggested that in Am(III)–TODGA complex, the two TODGA molecules present in the inner coordination sphere, whereas other TODGA molecule could be present in the outer sphere in case of dodecane and toluene. It appears that polar diluents, e.g. 1-octanol, 1,2-dichloroethane and nitrobenzene, do not facilitate the interaction of TODGA molecules in the outer sphere.

To sum up, TODGA is an effective extractant of trivalent ions; there are many researches involved in the extraction of trivalent ions, especially lanthanides(III) and Am(III), with various dialkylamides extractants. The main factors of influence on the distribution ratios of trivalent ions, including concentration of nitric acid, concentration of extractants, were tested; the extraction reaction formulae were given, and thermodynamics data were measured or calculated. But because these works were carried out under various conditions, it is difficult to compare them and reach conclusions. In addition, the study on the structure of extracted complexes is not much. The further studies should focus on the structure and coordination chemistry of extracted complex, which is of importance to elucidate the extraction mechanism and optimize the extraction process.

1.3.3. Extractions of Tetravalent Ions

The main tetravalent ions studied include Th(IV), Pu(IV), and Zr(IV).

Sasaki Y et al[E1] measured the distribution ratio of Th(IV) and Pu(IV) in TOOPDA/dodecane-HNO_3 extraction system, and presented a D_M order for the actinides at high nitric acid concentration, that is An(III), An(IV) > An(VI) > An(V) and that for lanthanides, heavier > lighter. By slope analysis, the number of diglycolamide molecules coordinated to the actinide ions was estimated to be three for Th(IV), U(VI), Pu(IV), and four for Am(III) and Cm(III). The principal species were estimated to be $Th(DA)_3(NO_3)_4$, $UO_2(DA)_3(NO_3)_2$, $Pu(DA)_3(NO_3)_4$, $Am(DA)_4(NO_3)_3$, and $Cm(DA)_4(NO_3)_3$.

The structurally tailored diamides were employed to study their extraction behavior to tetravalent ion, Th^{4+}. The distribution ratios of Th(IV), D(Th), by various diamides, were plotted as a function of diamide concentration. The order of the ion-pair extractability of diamides for Th(IV) is: DGA>>TDGA>MA, ODPA>MLA>SCA0 GLA>TDPA>OXA. DGA also exhibits the highest potential for Th(IV) extraction. By slope analysis, the number of diamide molecules associated with the main extraction reactions is estimated to be one or two.

Sasaki summarized the extraction behavior of some tetravalent ions, such as Zr^{4+} and Hf^{4+}.When the concentration of nitric acid is greater than 1M, Zr(IV), Hf(IV) and An(IV) can be extracted quantitatively by 0.1M TODGA/n-dodecane, but other tetravalent cations are weakly extracted . The slopes, 3.4, 3.1, 3.1 and 3.0 for Zr(IV), Hf(IV), Th(IV) and Pu(IV), respectively, are approximately 3, and slopes of the other metals: Pt(IV) and Sn(IV), were 2.0 and 0.64. It is well known that Zr and Hf have very similar chemical behaviors, under the same experimental conditions, the distribution ratio of Zr(IV) is higher than that of Hf(IV), showing high separation factors (SF) of about 7.5.

Mowafy [16] also studied the extraction of Th(VI) with structurally related diglycolamides compounds which were synthesized in his lab. He measured the distribution ratio of metal ions with various valences and gave the sequence for the extraction of those ions as follows:

TPDGA: Th(IV)> Am(III) > Eu(III) > U(VI) > Nd(III) > La(III) > Sr(II) > Ba(II)

T_2MPDGA: Th(IV) >Am(III) > Eu(III) > Nd(III) > U (VI) > La(III) > Sr(II) > Ba(II)

TBDGA: Th(IV) > Eu(III) > Am(III) > Nd(III) > La(III) > U(VI) > Sr(II) > Ba(II)

THDGA: Th(IV) > Eu(III) > Nd(III) > Am > La(III) > U(VI) > Sr(II) > Ba(II)

T_2EHDGA: Eu(III) > Th(IV) > Nd(III) > Am(III)> U(VI) > La(III) > Ba(II) > Sr(II)

TODGA: Th(IV) > Eu(III) > Nd(III) > Am(III) > U(VI) > La(III) > Sr(II) > Ba(II)

These sequences indicate that, except for T2EHDGA, Th(IV) is better extracted by diglycolamides than Eu(III), Nd(III) and Am(III). The latter trivalent ions are better extracted over La(III), U(VI), Sr(II) and Ba(II). This extraction sequence for diglycolamide is different from that reported for malonamides that prefer the extraction of UO_2^{2+} to trivalent lanthanides and actinides. The weak extraction of UO_2^{2+} can be related to steric hindrance for the tridentate coordination of diglycolamides in the equilateral position around the central uranium atom of the linear UO_2^{2+} cation.

1.3.4. Extraction of Metal Ions from Simulated High Active Waste and Process Development

The composition of high level liquid waste is very complicated. Most of actinides, almost all fission products and packing materials of spent fuels are included in HLLW. In addition, HLLW from various sources has different composition.

Considering many factorssuch as the extraction kinetics, extraction capacity, separation factors, etc., would affect the extraction of metal ions, it is not enough to only know the extraction behavior of metal ion from HNO_3 solutions without influences of other ions. Therefore, it is necessary to investigate the extraction of metal ions from SAW.

Modolo studied the separation of actinides(III) and lanthanides(III), using TODGA from a high active raffinate (HAR)[35,36]. The effect of oxalic acid and HEDTA complexant on the extraction of actinides(III), lanthanides(III), and important fission products (e.g. Mo, Pd, Sr, Zr, Ru etc.) from synthetic HAR has been studied with 0.2 mol/L TODGA in TPH. With an extractant mixture of TODGA and tributyl phosphate (TBP), the amount of oxalic acid can be reduced to less than 0.3 mol/L for the effective complexation of zirconium, whereas the distribution ratios of actinides(III) and lanthanides(III) are still high for the separation from HAR. Furthermore, the maximum loading of lanthanides (e.g. Nd) can be significantly increased by adding TBP to the extractant. However, the extraction of oxalic acid and nitric acid also increased by the addition of TBP, which can lead to problems during back extraction of the loaded extractant. Extraction studies after radiolysis and hydrolysis reveal that the TODGA+TBP mixture is a sufficient stable extraction system suited for further process development studies.

In addition, Modolo et al[37] used extraction chromatography with DMDOHEMA and TODGA impregnated resins to separate actinides from Low Level Liquid Wastes (LLLW). The uptake of several actinides [U(VI), Th(IV), Am(III), Cm(III)] and fission products was investigated from nitric acid solutions by two novel extraction chromatographic sorbents containing 2-(2-hexyloxy-ethyl)-N,N.-dimethyl-N,N.-dioctyl-malonamide (DMDOHEMA) and N,N,N',N'-tetraoctyl-3-oxapentane-1,5-diamide (TODGA), respectively. The kinetics of the uptake of actinides was studied. The sorption of metal ions from simulated low level liquid waste (LLLW) solutions was evaluated. The results of these experiments revealed that the actinides and lanthanides can be separated from the bulk of other fission products in simulated LLLW solutions on the both sorbents

Distribution behaviors of various metal ions, i.e., Am(III), U(VI), Pu(IV), Fe(III), Eu(III), Sr(II) and Cs(I) as a function of acidity in the absence as well as in the presence of SHLW using 0.1 M TODGA + 0.5 M DHOA solution in *n*-dodecane as the extractant, has been evaluated in BARC[33] .

In INET, Chen J et al [27] developed a process for separation of transuranium. Considering the effective-extraction and easy-stripping of actinides, 0.25 mol/l TEHOPDA – 30% n-octanol + 70% kerosene was selected as the solvent. A cascade extraction experiment was carried out with the simulated dissolver solution of spend fuel as feed. 99.99% U and 99.999% Am, Pu, and Np were extracted in a 4-stage test. Based on the experimental results, a conceptual reprocessing process was proposed, which includes extracting actinides and Tc from dissolver solution of spent fuel with 0.25 mol/l TEHOPDA – 30% n-octanol + 70% kerosene, stripping U from loaded organic phase with 0.3-0.7mol/L HNO_3, stripping transuranium with 0.001 mol/L HNO_3, and recycling of spent solvent.

Tian G et al [18] studied the Separation of Sr(II) from Simulated HLLW. A simulated raffinate of the TRPO process was prepared and the distribution ratios of major cations in the simulated HLLW were determined. Except Ba^{2+}, the distribution ratios of all cations in the simulated HLLW are smaller than 10^{-3}, and the separation factors of Sr^{2+} over these cations, $SF_{Sr/M}$, are greater than 4×10^3. The separation factor for Sr/Ba is lower than other cations, due to the chemical similarities between Ba^{2+} and Sr^{2+}, but still significant. These test results show that 0.2 mol/L TiBOGA in 40/60% 1-octanol/KO has good extraction ability and selectivity for Sr^{2+} from HLLW and the system may be used to separate ^{90}Sr from the HLLW after the TRPO process.

Through comprehensive researches on extraction chemistry of TODGA, JAEA develop a new chemical process, a ARTIST(abbreviation of "Amide-based Radio-resources Treatment with Interim Storage of Transuranics")[38] process, for the treatment of nuclear spent fuel. The ARTIST process is composed of two main steps. Firstly, uranium contained in the spent fuel solution is recovered by a BAMA(branched alkyl monoamides) process. Secondly, TRUs is separated by TODGA-DHOA extraction. Both actinide fractions produced by the two main processes are respectively solidified by calcination and are allowed to the interim storage for future utilization. The options are i) extractive recovery of Pu, ii) group separation of TRUs and lanthanides, iii) isolation of Cs and Sr, respectively. The first optional process is to be operated for producing MOX fuel. The separation of TRUs from lanthanides is operated to obtain target material for transmutation of TRUs in ADS (Accelerator-driven System) or FBR (Fast Breeder Reactor). The separations of Cs and Sr are effective for reducing the cost of disposal of fission products (FPs).

Although some works have been done on the process development, there are three aspects needed to be further studied. Firstly, cascade test is necessary for the development of process. Secondly, hot test in which real nuclides used is crucial for validation of the reliability of process. Thirdly, probable problems relating to application of the process, such as emulsification, formation of third phase, et al., should be considered and further studied.

1.4. Radiolysis of TODGA

As an extractant to be used for recovery of actinides, such as Am, Pu, Np, etc, from HLLW produced by snuclear fuel reprocessing, TODGA may suffer from strong irradiation. Therefore, radiation stability is essential for its utilization.

Sasaki et al[39,40] studied the hydrolysis and radiolysis of TODGA and its effects on extraction of actinides. Hydrolysis by nitric acid was not observed, whereas radiolysis by gamma irradiation was notably observed. The radiolysis study shows that an amide-bond, an ether-bond, and a bond adjacent to the ether-bond tends to be broken by gamma irradiation, and the main radiolytic degradation products were identified by GC/MS and NMR analyses as N,Ndioctylamine (2) and ,N-dioctyl-3-oxapentan-1,5-amic acid(3). The process is shown as follows.

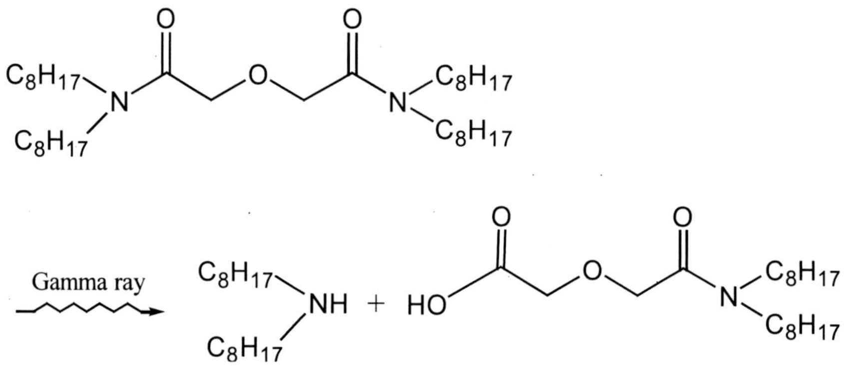

The G-value for degradation of neat TODGA was 8.5 ± 0.9, which was higher than those for N,N′-dioctyl-N,N′-dimethyl-2-(3′-oxapentadecyl)-propane-1,3-diamide and N,N-dioctylhexanamide. N-dodecane had a sensitization effect on the radiolysis of these amidic extractants. In a TODGA–n-dodecane–HNO3 system, the coexisting nitric acid shows an insignificant effect on the radiolysis of TODGA.

Sugo et al. also performed quantitative and qualitative studies on the radiolytic degradation of TODGA. However, they did not perform any extraction experiments, but rather determined the stability of TODGA with the aid of GC (quantitatively) and qualitatively with GC-MS, NMR and FT-IR. In their radiolysis studies, they found that the degradation of TODGA was strongly dependent on both the initial concentration and also on the solvent. The higher the concentration of TODGA in the solutions, the higher the resistance to radiolysis is. This explains the high stability of the irradiated pure TODGA. Sasaki et al observed a significant degradation in n-dodecane as the solvent, whereas higher stability was found in benzene. The addition of an amide (DOHA) also suppressed degradation. Long-time contact of 60 days with 3 molar HNO_3 (each day 30 minutes vigorous mixing and settling) confirmed the high hydrolysis resistance of TODGA, extraction experiments showed that no any significant changes of distribution ratios in comparison with an unirradiated extractants.

Radiation effect on the extraction of Am(III), Pu(IV), and U(VI) was examined using the solution of 0.1 M TODGA in n-dodecane pre-irradiated with g-rays. There is no remarkable decrease in distribution ratios of the actinide ions at high acidity even after irradiation with 422 kGy. On the other hand, those at low acidity increase by irradiation. From the results of the extraction of the actinide ions by N,N-dioctyl-3-oxapentan-1,5-amic acid, which was one of the main degradation products, it is proved that the amic acid causes increase in distribution ratios at low acidity

Giuseppe Modolo[36] investigated the hydrolytic and radiolytic stability of the extractant. The maximum cumulative gamma dose during the whole irradiation time amounted to about 1 MGy at a dose of 1.9 kGy/h. Distribution ratios of Am(III) and Eu(III) were measured and shown as a function of the total dose. With an increasing dose, a slight decline in D_{Am} is registered above 0.6 MGy. It can also be observed that the irradiated samples are apparently more strongly degraded in contact with HNO_3 during irradiation. Pure TODGA samples (without solvent) were also irradiated. The ^{1}H-NMR and the ^{13}C-NMR measurements indicate that no appreciable degradation products were formed even at a total dose of 1 MGy.

As TODGA is expected to be utilized in treatment of HLLW in the future, its radiation stability is crucial to its application. Although these works studied the changes of its extraction ability, deduced some radiolysis products and determined their G values, the study is still not enough. At least the following issues should be considered: firstly, the absorbed dose should be higher and the radiolytic products at higher absorbed dose should be identified. Secondly, the system used for radiation stability should simulate the real solution, for example, TODGA/solvent/HNO_3, which is similar to the real system. Thirdly, retention of actinides caused by radiolytic products should be considered, for the retention usually lead to incomplete stripping of loaded metal ions, just as those occurred in CMPO process and TRPO process[41,42]. Finally, application features, such as the emulsification resulted from irradiation, may be deteriorated.

2. Synthesis, Properties of Aqueous-Soluble DAOGA and TAOPDA, and Their Applications

2.1. Introduction

The treatment and final disposal of high-level liquid waste (HLLW) generated from nuclear fuel reprocessing is one of the key problems that affect the further development of nuclear energy industry. During the past decades, several techniques, including vitrification, partitioning-transmutation (P-T) and partitioning- conditioning (P-C)[43], were proposed and studied worldwide. Vitrification is to vitrify HLLW into a borosilicate glass matrix and encapsulate the glass into a metal container. P-T concept involves chemical separation of transuranium (TRU) elements along with long-lived nuclides (for example, ^{99}Tc, ^{129}I, etc.) from HLLW, and transmutation of them to either stable or short-lived nuclides. The main idea of P-C concept is separating TRU elements and long-lived nuclides from HLLW but temperately conditioning them for future disposal. Studies on the separation of TRU elements from HLLW in the P-T and P-C strategy are the hotspots in the field of nuclear fuel recycling. Several processes have been developed and tested in laboratory scale with real HLLW. In our lab, trialkyl phosphine oxide (TRPO) process based on P-T strategy and simplied trialkyl phosphine oxide (TRPO) process based on P-C strategy have been developed for separating TRU elements along with Tc from HLLW.

In TRPO process, TRU elements are extracted from HLLW with 30% TRPO-Kerosene. To meet the requirements of P-T strategy, the loaded metal ions are respectively stripped, i.e., An(III) and Ln(III) are stripped with 5.5 M HNO_3 solution, Pu and Np with 0.6 M $H_2C_2O_4$ solution, and U with 5% Na_2CO_3 solution, and there is few cross-contaminations present among these nuclides in the process stream, an important advantage of TRPO process.

However, because transmutation of TRU seems much far away from putting into action, P-C strategy becomes more and more attractive. That is one of reasons we improved the TRPO process based on P-T strategy to simplify TRPO process based on P-C strategy. In addition, another disadvantage of TRPO process is the use of concentrated nitric acid (5.5 M) to strip An (III) and Ln(III), causing some problem for the following An(III)/Ln(III) separation. To achieve STRPO process, the nuclides extracted by TRPO except U, must be stripped together and stored. Therefore, it is crucial to develop new water-soluble stripping agents having complexing capacity. To find the proper stripping agents, various compounds were tested, such as oxalic acid, salvolatile, acetic acid, hydrazine carbonate, acetohydroxamic acid, etc[44,45]. However, none of them gave satisfactory results. Considering the strong complexing capacity of the oil-soluble oxa-glutaramic acid and oxa-pentanediamide to metal ions, it is reasonable to assume that the water-soluble oxa-glutaramic acid and oxa-pentanediamide may be effective stripping reagents. Besides, they are completely incinerable and thus do not produce secondary solid wastes, because they are composed of C, H, O and N elements,. These advantages imply that the diamide derivatives might be very promising as effective ligands and therefore stripping reagents.

A series of water-soluble oxa-glutaramic acid and oxa-pentanediamide were synthesized and screened in our lab; the structures of their complexes with Ln were studied; stability constants of their complexes were determined. Based on these investigations, its application in the simplification of TRPO process is proposed [46-51].

Synthesis of water-soluble N,N-dialkyl-3-oxa-glutaramic acid (DAOGA) and N,N,N'N'-tetra-alkyl-3-oxa-pentanediamide (TAOPDA)

DAOGA is a multi-dentate ligand. Its synthesis consists of the following three steps: (1) diethylene glycol is oxidized to diglycolic acid; (2) diglycolic acid is converted into diglycolic anhydride by dehydration; (3) diglycolic anhydride reacts with diamine to yield mono-substituted amide. The reaction scheme is illustrated as follows:

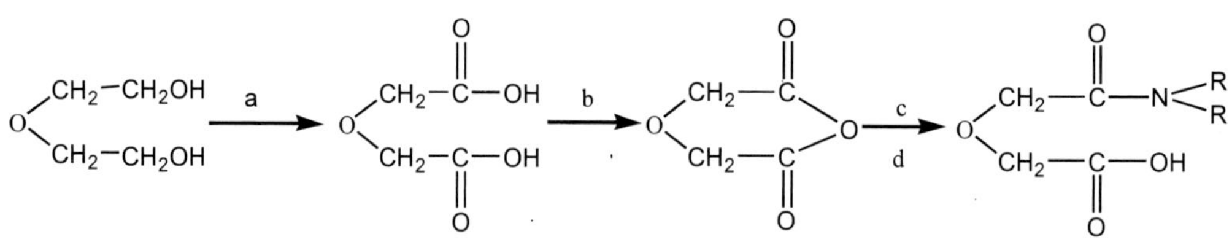

(1a-1c)

a). HNO_3, b) Ac_2O, c). HNR_2: $HN(CH_3)_2$, $HN(CH_2CH_3)_2$,$HN(CH_2CH_2CH_3)_2$ d) C_6H_5N

The synthesis of TAOPDA consists of following three steps: (1) diethylene glycol is oxidized to produce diglycolic acid; (2) diglycolic acid is converted into diglycolic acyl choride; (3) diglycolic acyl choride reacts with diamine in the presence of pyridine to form TAOPDA. The reaction scheme is shown as follows:

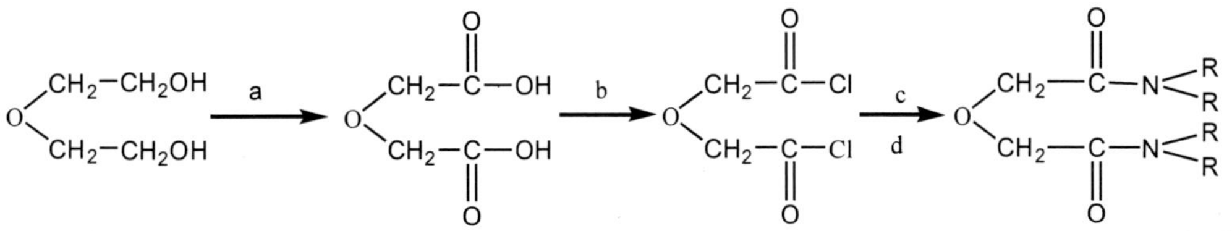

(II)

a). HNO_3, b) $SOCl_2$, c).HNR_2: $HN(CH_3)_2$, $HN(CH_2CH_3)_2$, $HN(CH_2CH_2CH_3)_2$ d) C_6H_5N

Six ligands of these two series were synthesized in our lab with the methods. Their physical data, elemental analysis results, and NMR and IR spectral data are shown in Table 5 and 6.

Table 5. Physical data and elemental analysis results

Compd.	Formula	m.p/℃	Yield/ %	Elemental analysis found(calcd.)/%
1a	$C_6H_{11}O_4N$	71~72	75	calcd C 44.71, H 6.88, N 8.69; found C 44.58, H 6.82, N 8.72.
1b	$C_8H_{15}O_4N$	69~70	69	calcd: C 50.78, H 7.99, N 7.41; found C 50.64, H 8.03, N 7.45.
1c	$C_{10}H_{19}O_4N$:	52~54	69	calcd:C 55.28, H 8.82, N 6.44; found C 55.16, H 8.87, N 6.47.
2a	$C_8H_{16}O_3N_2$		90	calcd: C 51.04, H 8.57, N14.89; found C 50.91, H 8.58, N 14.92.
2b	$C_{12}H_{24}O_3N_2$		75	calcd: C 58.98, H 9.90, N 11.47; found C 58.91, H 9.82, N 11.52.
2c	$C_{16}H_{32}O_3N_2$		80	calcd: C 63.96, H 10.74, N 9.32; found C 63.87, H 10.71, N 9.40.

Table 6. ^{1}HNMR and IR spectral data of compounds 1a~2c

Compd.	^{1}HNMR	IR
1a	δ: 4.40 (s, 2H), 4.21(s, 2H), 3.03(s, 3H), 2.95(s, 3H)	ν: 3600~2500, 1723, 1641, 1119 cm^{-1}.
1b	δ: 4.39(s, 2H), 4.20(s, 2H), 3.44~3.42(q, 2H, J=6.9Hz), 3.19~3.17(q, 2H, J=6.9Hz), 1.23~1.15(m, 6H).	ν: 3600~2500, 1746, 1613, 1132 cm^{-1}.
1c	δ: 4.41(s, 2H), 4.21(s, 2H), 3.35~3.30(t, 2H, J=15.1Hz), 3.12~3.06(t,2H, J=15.1Hz), 1.62~1.55(m, 4H),0.96~0.88(m,6H),	ν: 3600~2500, 1751, 1618, 1132 cm^{-1}.
2a	δ: 4.30(s, 4H), 3.05(s, 6H), 2.96(s, 6H).	ν: 1644, 1122 cm^{-1}
2b	δ: 4.30(s, 4H), 3.40~3.30(m, 8H), 1.21~1.10(m, 12H).	ν: 1649, 1111 cm^{-1}.
2c	δ: 4.30(s, 4H), 3.28~3.15(m, 8H), 1.58~1.53(m, 8H), 0.90~0.87(m, 12H)	ν: 1649, 1122 cm^{-1}.

2.3. Study on Structures of the Complexes of Ln(III) with DMOGA

Among the six ligands of the two series, methyl-substituted ligand, i.e, N,N'-dialkyl-3-oxa-glutaramic acid (DMOGA), has the biggest solubility in aqueous solution. Furthermore, preliminary experiments showed that it has strongest complexing ability to An(III) and to fission products, such as Ln, Fe, Zr, etc. Therefore, DMOGA was chosen as a sample to study the structure of its complexes with Ln(III), to determine the stability constant of the complexes, and to investigate the decomposition pathway.

The structures of crystal of the complex of DMOGA with Nd(III) are studied by infrared spectrometry (IR); nitric acid solution of complex of DMOGA with La , Nd , and Eu are studied by mass spectrometry (MS), and extended X-ray absorption fine spectrometry (EXAFS).

IR measurements

Figure 1 shows the IR spectrum of the complex of Nd (III) with DMOGA, comparing to the one of pure DMOGA. The difference between the two spectra is clear.

The absorption peak of carboxylic acid (C =O) at 1723 cm^{-1} of pure DMOGA disappears when DMOGA coordinates with Nd (III) . In the spectrum of the complex, three new absorption peaks rise at 1625, 1607 and 1383 cm^{-1}, while the two peaks at 1625 and 1607 cm^{-1} merge. These changes are attributed to the coordination of Nd(III) with DMOGA. When the coordinated complex forms, H^+ in the COOH is replaced by Nd (III), and the CO_2^- changes to two equal C-O bonds. As a result , two new absorption peaks at 1383 cm^{-1} (vs ,CO_2^-) and 1607 cm^{-1} (vas , CO_2^-) appear in the spectrum of the complex, and the peak at 1625 cm^{-1} is the stretching vibration absorption peak of C = O of amido. Because Nd(III) has empty electronic orbit , the lone pair electrons in the R - C =O may migrate to the Nd(III) when the O atom coordinates with Nd(III), which weakens the strength of the C = O band and reduces the extent of red transference of C = O stretching vibration. These IR data indicate that DOMGA is a bridge-quadri-dentate ligand.

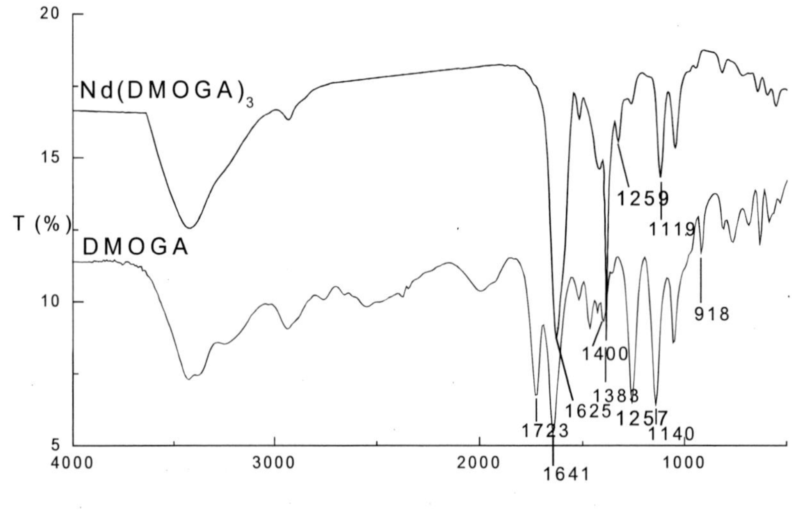

Figure 1. IR spectra of DMOGA and $Nd(DMOGA)_3$

MS measurements

The MS of the complexes DMOGA with Ln(III, Ln= Eu, Nd, La) was acquired with electrospray mass spectrometry in positive mode, and the results are shown in Table 7. From the MS data, it can be deduced that the complexes of Ln(III) with DMOGA are in the molecular formula of Ln $(DMOGA)_3$ with a small amount of Ln(III) in the form of [Ln $(DMOGA)_2]^+$; However for La (III), HLa $(DMOGA)_4$ complex forms..

Table 7. Results of MS measurements of Ln(III) complexes with DMOGA

Ln(III)	m/e	Corresponding species
La	459	$La(DMOGA)_2^+$
	620	$La(DMOGA)_3$
	781	HLa $(DMOGA)_4$
Eu	473	$Eu(DMOGA)_2^+$
	634	$Eu(DMOGA)_3$
Nd	462	$Nd(DMOGA)_2^+$
	625	$Eu(DMOGA)_3$

EXAFS spectra

Figure 2 shows the L_{III} edge EXAFS spectra (after Fourier filter) of La in the complexe and the corresponding Fourier transforms (FT) (before Fourier filter). The results of the curve fitting are also shown. The FT reveals the relative positions and the numbers of the nearest neighbors around the central ion in the complexes.

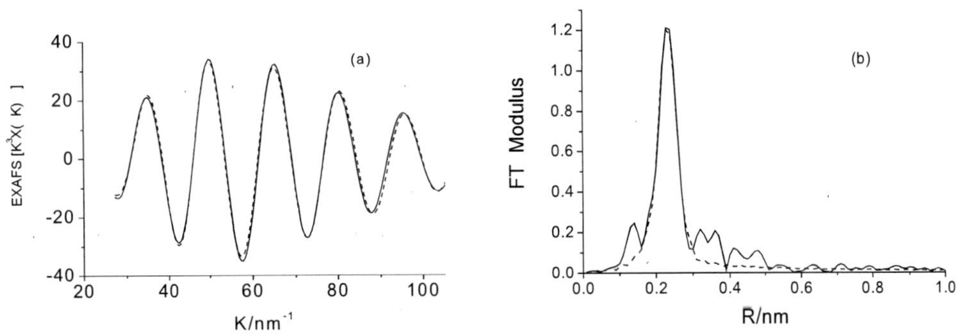

Eu $(DMOGA)_n$

(— Experimental data, ·· theoretical signal)

(a) k^3 weighted EXAFS curve fitting of Ln(DMOGA), b) Fourier transform

Figure 2. Comparison of Fourier transforms (FT) of experimental data with theoretical signal and Comparison of k^3 weighted EXAFS spectra Fourier filtered.

EXAFS spectra of La and Nd in the extracted complexes were measured too, and, and the summary of the curve fitting data are showed in Table 8. The EXAFS data was fitted by using a suite of programs EXCURVE92 reviewed elsewhere [10].

Table 8. Summary of EXAFS curve fitting results

samples	shell	CN	λ/nm	$\Delta\sigma^2$/nm^2	ΔE_0/ev
La(III)	La-O	6.1	0.258	0.0001	6. 96
	La-O	4.2	0.280	0.0001	
Eu(III)	Eu-O	6.0	0.240	0.00008	9.73
	Eu-O	2.4	0.260	0.00001	
Nd(III)	Nd-O	6.0	0.245	0.0001	10.61
	Nd-O	2.4	0.262	0.0001	

From the MS data and the IR spectrum, it can be deduced that the coordination atoms in the complexes of Ln should be oxygen atoms from DMOGA rather than from NO_3^- or H_2O. The EXAFS curve fitting data indicate that there are six oxygen atoms in the first coordination shell and the average length of the O - La bond is about 0.258 nm , while there are about 4.2 oxygen atoms in the second shell of coordination and the average length of the O - La bond is 0.280 nm.. From the results of EXAFS and MS, it can be deduced that about four DMOGA molecules presented in the complex, and three of the four DMOGA molecules provide six oxygen atoms coordinating with La (III) in the first coordination shell in mode of chelate structure of carboxylic acid, and the about 4 oxygen atoms in the second coordination shell are possibly provided by the amide and the ether link.

For Eu (III) or Nd (III) , the results indicate that the coordination number of oxygen atoms in the first coordination shell is 6 , and the average lengths of the O - Ln bond is 0.240 and 0.244 nm, respectively. There are about 2.4 oxygen atoms in the second shell of coordination, and the average lengths of the O – Ln bond is 0.260 and 0.262 nm, respectively. The distance between the Ln (III) and coordination shells decreases in the order of La, Nd and Eu. These results are consistent with the order of the radius of Ln (III): R_{La} (III)> R_{Nd} (III) > R_{Eu} (III) .

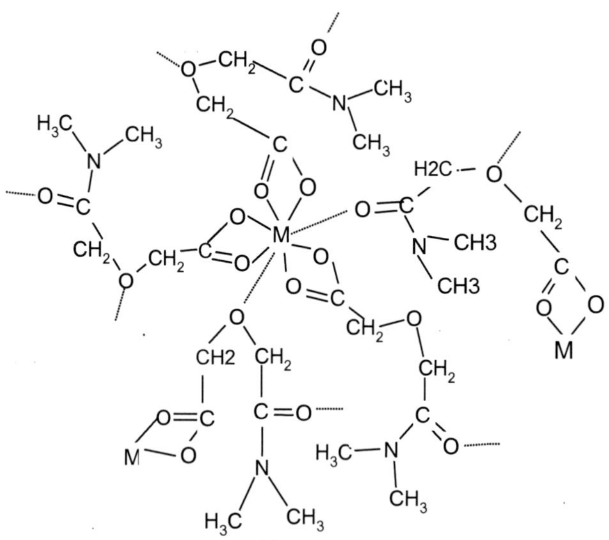

Figure 3. Possible structure of extracted complex of lanthanides with TODGA.

From the discussion above, it is postulated that the molecular formula of the complexes of Ln (III) with DMOGA, is Ln $(DMOGA)_3$ and some of Ln(III) can form the HLn $(DMOGA)_4$ under the condition of the high consistency of the DMOGA , except Eu (III) and Nd (III).

The IR, MS, and EXAFS measurements illustrate that the structures of the complexes of Ln(III) with DMOGA are similar. Because of the coordination property of DMOGA, they can form complexes in state of polymer. For the complexes $Ln_n(DMOGA)_{3n}$, the probable structure is shown in Figure 3.

2.4. Determination of Stability Constant of Complexes of DMOGA with Np(IV) and Pu(IV)

DMOGA is a weak acid, and its Ka is 6. 532 $\times 10^{-4}$, measured at 25°C. Because of its strong polarity, solubility of DMOGA in nonpolar organic solvent is quite small (<3%). The coordination characteristics of DMOGA with Np(IV) and Pu(IV) were studied with solvent extraction, and their stability constants(β) were determined. β is defined as follows:

$$\beta_n = \frac{c\left(M(DMOGA)_n^{4-n}\right)_{(a)}}{c\left(M^{4+}\right)_{(a)} c^n\left(DMOGA^-\right)_{(a)}}$$

Where c ($DMOGA^-$) $_{(a)}$, $c(M^{4+})_{(a)}$, $c(M(DMOGA))_{(a)}$ refers to the equilibrium concentrations of the corresponding ions in aqueous phase.

The determined stability constants are shown in Table 9.

Table 9. stability constant of reactions

Reaction	Stability constant
$Np^{4+} + 3DMOGA^- = Np\ (DMOGA)^+$	$lg\beta_1 = 6.95 \pm 0.17$ $lg\beta_2 = 10.44 \pm 0.78$ $lg\beta_3 = 14.93 \pm 0.33$
$Pu^{4+} + 3DMOGA^- = Pu\ (DMOGA)^+$	$lg\beta_1 = 6.52 \pm 0.27$ $lg\beta_2 - 9.14 \pm 1.66$ $lg\beta_3 = 15.77 \pm 0.55$

25°C, stability constants of Pu and Np were determined in $HClO_4$ solution of 0.5 and 0.8 mol/ kg respectively.

The stability constant of the complexes indicates that strong chelates forms between Pu(IV) and Np(IV) with DMOGA.

2.5. Thermal Decomposition of DMOGA

After having stripped the metal ions, the stripping reagent needs to be disposed or decomposed. The thermal decomposition of DMOGA was preliminarily studied with TG-DTA. Results show that one crystal water molecule is contained in one DMOGA molecule.

The thermal decomposition process is composed of two steps: as the temperature increases from 40℃ to 90℃, the weight loss is about 10.87%, corresponding to loss of the crystal water(the calculated weigh loss is 10.06%).

$$DMOGA \bullet H_2O \xrightarrow{40-90^o C} DMOGA + H_2O$$

When the temperature continues raising to 150℃, DMOGA starts to decomposing and completely decomposes at 260℃.

Based on the research results stated above, DMOGA has strong complexing ability to Ln(III), Np (IV), and Pu(IV), and hence it is applied as the stripping agent of these ions.

2.6. Measurements of Distribution Ratio of Am(III)/ Ln(III), Pu(IV)/Np(IV), U(VI) with Varying HNO_3 Concentration

As it comes from Purex process, HLLW contains some transuranium elements (TRU), including Am, Np, and Pu, and fission products, such as Ln, Fe, Zr, etc. With high extraction capacity of TRPO to U(VI), Np(IV and VI), Pu(III and IV), and fission products, TRPO process was developed to separate TRU elements from HLLW. However, in order to strip TRU out of loaded TRPO organic phase, the stripping reagent should have stronger complex ability under certain conditions. To assess the stripping performance of DMOGA, the distribution ratios (D_M) of these metal ions between TRPO-kerosene and DMOGA solution were measured.

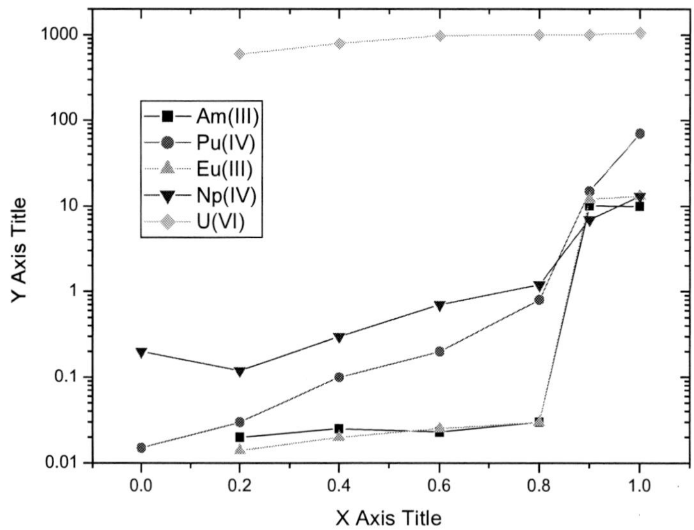

Figure 4. Plot of D vs concentration of HNO_3 containing DMOGA.

Because the extraction of some metal ions by TRPO is strongly dependent on the acidity of the extraction system and the bonding ability of DMOGA is probably affected by the acidity, the D_M was measured at different nitric acid concentrations in aqueous solution. Figure 4 shows the distribution ratio data of Am(III), Eu(III), Pu(IV), Np(IV) and U(VI) between TRPO-kerosene and 0.1M DMOGA at different $[HNO_3]$.

Comparing the D_{Am} between 30% TRPO and aqueous solutions with/without DMOGA, the D_{Am} values with DMOGA are much lower than those without DMOGA when $[HNO_3]$ is lower than 0.8M; while the effect of DMOGA concentration on D_{Am} is almost negligible with $[HNO_3]$ above 0.8M,. The data of D_{Eu} are similar to those of D_{Am} under the same conditions, and present a similar tendency with the increasing acidity or concentration of DMOGA.

The distribution ratio of Am and Eu between 30% TRPO and aqueous solution with DMOGA is a complexing competition of TRPO and DMOGA to the metal ions under various conditions. The extraction reaction of Am from nitric acid solution with TRPO-kerosene can be expressed as (1):

$$Am^{3+}_{(aq)} + 3NO_3^-{}_{(aq)} + 3\ TRPO_{(org)} \leftrightarrow Am(NO_3)_3{\bullet}3TRPO_{(org)} \qquad (1)$$

The complex reaction of Am and DMOGA may be supposed as (2):

$$Am^{3+}{}_{(aq)} + m\ NO_3^-{}_{(aq)} + n\ DMOGA_{(aq)}^- \leftrightarrow Am(NO_3)_m(DMOGA)_{n\ (aq)} \qquad (2)$$

Where (aq) and (org) represent the aqueous phase and organic phase, respectively.

In DMOGA molecule, there are three function groups that may exhibit coordinate ability, i.e., carbonyl group (-C=O), ester group (-O-), and carboxyl group (-COOH); these groups may complex metal ions. However, based on the fact that the bonding ability of DMOGA is dominantly affected by the acidity, it can be deduced that coordinate abilities of the carbonyl and ether group are negligible. That means DMOGA works as an acidic ligand rather than a chelating one. When the acidity is low, the equilibrium constant of equation (1) (K_1) is quite small. When DMOGA can deprotonate and works as an acidic ionic ligand, Am^{3+} and Eu^{3+} are strongly complexed by DMOGA, and the equilibrium constant of equation (2) (K_2) increased. With the increasing of $[HNO_3]$, K_1 increases, while the dissociation of DMOGA is inhibited and K_2 decreases, which means that the metal ions are hard to be complexed by DMOGA, so the D increases

$D_{Pu(IV)}$ and $D_{Np(IV)}$ present the similar tendency with increasing of the $[HNO_3]$. The causes of this tendency may be similar to those of Am^{3+} mentioned above, because the extraction reactions of Pu(IV) and Np(IV) are similar to those of Am(III), where nitrate ions plays important roles in the extraction of them. On the basis of these data, we supposed that both Pu(IV) and Np(IV) could be stripped out of loaded TRPO by the DMOGA in diluent nitric acid.

The distribution ratio of U with TRPO extraction from DMOGA-containing HNO_3 solution is also shown in the figure. It is shown that UO_2^{2+} can be effectively extracted by TRPO with the $[HNO_3]$ range from 0 to 1M, with 0.1 M DMOGA presents in the aqueous solutions, suggesting that DMOGA does not form stable complex with UO_2^{2+}. Therefore, uranium is not stripped by DMOGA and needs to be recovered with additional stripping step. In that case, the uranium is recovered and separated from other metal ions and thus can be easily recycled.

2.7. Potential Application of DMOGA in the Simplification of TRPO Process

2.7.1. Stripping of TRU Elements from Loaded TRPO with DMOGA

With DMOGA in diluent nitric acid solution ($[HNO_3] < 0.8$mol/L), the distribution ratios of Np, Pu, Am and Ln are quite small. Therefore, it is possible to use the diluent HNO_3 solution of DMOGA to strip these elements from loaded TRPO. Simulated active liquid waste (SAW) was prepared with trace amount of Np, Pu and Am added. The main components of SAW are listed in Table 10. SAW was extracted with 30% TRPO three times to load metal ions into TRPO, and then the loaded organic phase was stripped with DMOGA solutions three times in a cross-current way. Table 11 lists the stripping efficiencies of total alpha activities of nuclides.

Table 10. Composition of simulated active liquid waste

Metal ions	Fe	Al	Na	Ni	Cr	Sr	Ba
ent (g/L)	17.4	15.9	51.2	8.2	2.0	0.62	0.1
Metal ions	**K**	**Cs**	**Mo**	**Pr**	**Nd**	**La**	**Zr**
Content (g/L)	0.45	2.0	0.82	0.6	3.3	0.6	36
Total α activity(Bq)				105			

The results show that DMOGA HNO_3 solutions present excellent stripping performance to alpha nuclides loaded in TRPO: the overall stripping efficiencies of 0.2M DMOGA are greater than 99.9%. after 3 times of cross-current back extraction.

Table 11. The stripping efficiency of total α activity

Stipping solution	Stipping percentage of total α activity(%)			
	1st contact	2nd contact	3rd contact	Overall
0.1M DMOGA in 0.1M HNO_3	92.3	86.5	84.2	99.8
0.2M DMOGA in 0.1M HNO_3	97.3	85.3	87.6	99.9
0.1M DMOGA in 0.5M HNO_3	90.0	83.5	79.6	99.6
0.2M DMOGA in 0.5M HNO_3	95.2	87.1	84.9	99.9

2.7.2. The Conceptual Flowsheet of the Simplified TRPO Process

Based on the experimental results, we propose a simplified TRPO process, in which DMOGA is employed as the stripping agent of Np, Pu and Am. The conceptual flowsheet is shown in Figure 5.

The new process uses DMOGA in diluent HNO_3 solution to strip Am(III), Np(IV) and Pu(IV) rather than the concentrated HNO_3 and $H_2C_2O_4$ as used in TRPO process; stripping of U by Na_2CO_3 solution as in TRPO process is adopted. Consequently, in simplified TRPO process, all alpha nuclides but U, along with Ln, are stripped out altogether, while U is separated and recovered for recycling. Comparing to the TRPO process (Figure 1), the simplified TRPO process is more feasible and simpler; moreover the objective of dealphalization and volume reduction of waste are achieved which meets the standard of P-C concept.

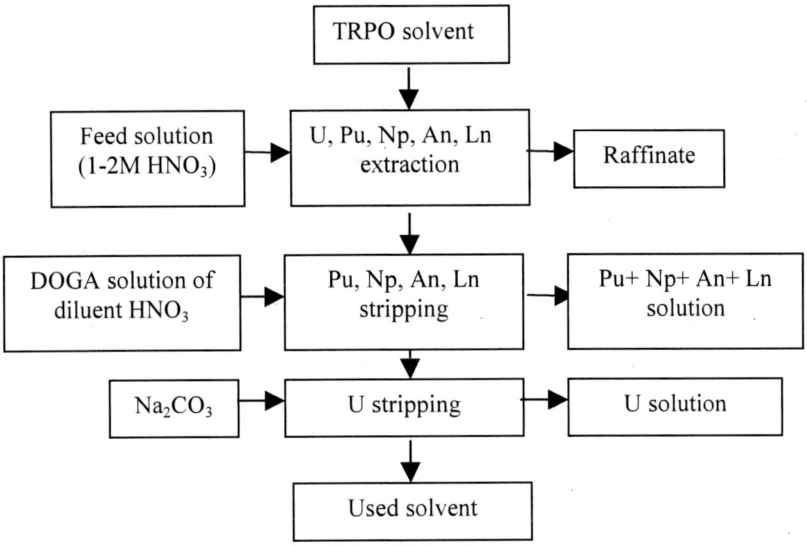

Figure 5. Flowsheet of the simplified TRPO process with DMOGA.

Conclusion

As a promising extractant for recovery of actinides and lanthanides from spent fuel solution or high level liquid waste, diglycoamides, especially TODGA, was intensively studied across the world. In this charpter, we review the research work on diglycoamides. The syntheses method are introduced in detail, the characterization data are collected and presented. The extraction of metal ions with OPDA and other analogues are reviewed and the process development based on TODGA or other analogues are summarized and commented. In addition, the radiolysis of TODGA is reviewed.

The synthesis, properties of aqueous-soluble DAOGA and TAOPDA, especially DMOGA and its applications are summarized, of which the following work were initiated in our lab: synthesis and characterization of six ligands of these two series; structure identification with IR and EXAFS of the complexes of Ln(III) with DMOGA; determination of stability constants of Np(IV) and Pu(IV) complexed with DMOGA; and measurements of the distribution ratios of actinides and some fission products between 30% trialkyl phosphine oxide (TRPO) in kerosene and HNO3-DMOGA solutions of different concentrations. Based on these results, DMOGA is proposed to be used as an effective stripping agent for An(III, IV) and therefore the TRPO process, well-known as one of the promising processes for treatment of high level liquid waste, is simplified.

References

[1] Actinide and Fission Product Partitioning and Transmutation—Status and Assessment Report, OECD-NEA, Paris, France, 1999.

[2] Implications of partitioning and transmutation in radioactive waste management, International Atomic Energy Agency, Technical Reports Series, ISSN 0074– 1914 no. 435, Vienna, Austria, 2004.

[3] Zhu, Y; Song, C. In *Transuranium Element: A Half-Century*; Morss, L.R., Fuger, J., Eds.; American Chemical Society: Washington, DC, 1992, 318.

[4] Schulz, WW; Horwitz, EP. The TRUEX process and the management of liquid TRU waste. *Sep. Sci. and Tech.* 1988, 23, 1191–1210.

[5] Mathur, JN; Murali, MS; Nash, KL. Actinide partitioning—a review. *Solv. Extr. and Ion Exch.* 2001, 19 (3), 357–390.

[6] Mahajan, GR; Prabhu, DR; Manchanda, VK; Badheka, LP. Substituted malonamides as extractants for partitioning of actinides from nuclear waste solutions. *Waste Management.* 1998, 18, 125–133.

[7] Murali, M.S.; Mathur, J.N. Use of mixture of TRPO and TBP for the partitioning of actinides from high-level waste solutions of PUREX origin and its comparison with CMPO and other phosphorus-based extractants. *Solv. Extr. Ion Exch.* 2001, 19, 61–77.

[8] Morita, Y; Glatz, JP; Kubota, M; Koch, L; Pagliosa, G.; Roemer K; Nicholl, A. Actinide partitioning from HLW in a continuous DIDPA extraction process by means of centrifugal extractors. *Solv. Extr. Ion Exch.* 1996, 14, 385.

[9] Sasaki, Y; Choppin, G.R. Extraction behaviors of Eu, Th, U and Am with Diamides+Theonyltrifluoroacetone. *J. Radioanal. Nucl. Chem.* 1996, 207, 383-394

[10] Narita, H; Yaita, T; Tamura, K; Tachimori, S. Solvent extraction of trivalent lanthanoid ions with N,N'-dimethyl-N,N'-diphenyl-3-oxapentanediamide.*Radiochim. Acta.* 1998, 81, 223–226.

[11] Sasaki, Y; Choppin, G.R. Solvent extraction of Eu, Th, U, Np and Am with N,N'-dimethyl-N,N'-dihexyl-3-oxapentanediamide and its analogous compounds. *Anal.Sci.* 1996, 12, 225-230

[12] Narita, H; Yaita, T; Tamura, K; Tachimori, S. Study on the extraction of trivalent lanthanide ions with N, N'-dimethyl-N, N'-diphenyl-malonamide and diglycolamide. *J. Radioanal. Nucl. Chem.* 1999, 239, 381-384.

[13] Pretsch VE, Ammann D, Osswald HF; Guggi M; Simon W. Ionophore vom typ der 3-oxapentandiamide. *Helvetica chimica acta*, 1980, v63, 191-196

[14] Sasaki Y; Choppin GR. Solvent extraction of Eu, Th, U, Np and Am with N,N'-dimethyl-N,N'-dihexyl-3-oxapentanediamide and its analogous compounds. *Anal. Sci.* 1996, v12, 225-230.

[15] Zhang P; Chen J; Li CY; Tian GX. Synthesis of N,N,N',N'-tetra-butyl-3-oxa-pentanediamide and its analogous compounds. *Chemical journal on Internet.* 2003 , V5,52-56

[16] Mowafy EA; Aly H F. Synthesis of some N,N,N',N'-Tetraalkyl-3-Oxa-Pentane-1,5-Diamide and their Applications in Solvent Extraction. *Solv Extra Ion Exch*, 2007, V25, 205 - 224

[17] Ye GA; Luo FX; Jiang YQ; Ding SD; Chen WJ; Study on the Extraction of Strontium With Amido Podand, *Atomic energy science and technology(In Chinese)*, 2001, V35(4). 344-350

[18] Tian GX; Wang JC; Shen YF; Rao LF. Extraction of Strontium from HLLW using N,N,N',N'-Tetraisobutyl 3-Oxa-Glutaramide. *Solv. Extr. Ion Exch.* 2005, V 23, 519 - 528

[19] Tachimori S; Suzuki S; Sasaki Y; Apichaibukol A. Solvent extraction of alkaline earth metal ions by diglycolic amides from nitric acid solutions. *Solv. Extr. Ion Exch,* 2003,V21.707 - 715

[20] Zhu ZX; Sasaki Y; Suzuki H; Suzuki S; Kimura T. Cumulative study on solvent extraction of elements by N,N,N′,N′-tetraoctyl-3-oxapentanediamide (TODGA) from nitric acid into n-dodecane. *Analytica Chimica Acta*, 2004,V 527(2), 163-168

[21] Sasaki Y; Tachimori S. Extraction of actinides (III), (IV), (V), (VI), and lanthanides (III) by structurally tailored diamides. *Solv. Extr. Ion Exch*, 2002,V.(1), 21 – 34

[22] Sasaki Yuji; Sugo Yumi; Suzuki Shinichi; Tachimori Shoichi . The novel extractants, diblycolamides, for the extraction of lanthanides and actinides in HNO3-n-dodecane system. *Solv. Extr. Ion Exch*, 2001,V 19(1) . 91 - 103

[23] Suzuki H; Sasaki Y; Sugo Y; Apichaibukol A; Kimura T. Extraction and separation of Am(III) and Sr(II) by N,N,N′,N′-tetraoctyl-3-oxapentanediamide (TODGA). *Radiochimica Acta* 2004, 92 (8): 463-466

[24] Tachimori S; Sasaki Y; Suzuki S. Modification of TODGA-n-dodecane solvent with a monoamide for high loading of lanthanides(III) and actinides(III). *Solv. Extr. Ion Exch,* 2002, 20(6) 687 - 699

[25] Sasaki Y; Rapold P; Arisaka M; Hirata M; Kimura T; Hill C; Cote G. An additional insight into the correlation between the distribution ratios and the aqueous acidity of the TODGA system. *Solv. Extr. Ion Exch*, 2007,V 25(2). 187 - 204

[26] Sasaki Y; Zhu ZX; Sugo Y; Suzuki H; Kimura T. Extraction capacity of diglycolamide derivatives for Ca(II), Nd(III) and Zr(IV) from nitric acid to n-dodecane containing a solvent modifier. *Analytical Sciences*. 2007, Vol. 23(6). 727

[27] Chen J; Wang SW. New conceptual process for the extraction of actinides based on oxapentane-diamides. *Journal of Radioanalytical and Nuclear Chemistry*, 2007, V 273 (1). 59-63

[28] Ye GA; He JY; Jiang YQ. Study on the extraction of Am(III) and Eu(III) with amido podand. I: Study on extraction mechanism. *Journal of Nuclear and Radiochemistry(In Chinese)*, 2000, V22(2). 65-72

[29] Ye GA; He JY; Luo FX. Study on the extraction of Am(III) and Eu(III) with amido podand. II: Extraction thermodynamics and absorption spectra. *Journal of Nuclear and Radiochemistry(In Chinese)*, 2000, V22(3). 136-143

[30] Jin YD; Feng L; Ding SD; Xia CQ; Chen WJ. Studies of amido podands on extractive behavior of lanthanide and actinides. *Chemical Research and Application(In Chinese)*, 2002. V14(2). 174-177

[31] Ding SD; Chen WJ; Xiao CJ. Investigation of the extraction of uranium(VI) and rare earth(III) with asymmetry amide podands. *Journal of Nuclear and Radiochemistry(In Chinese)*, 2003,V25(3). 188-192

[32] Chen WJ; Zhu L; Ding SD; Liu ZQ. Influence of Structure of Diamides on the Extractive Properties of Eu(III) and Am(III). *Chemical research in Chinese Universities(In Chinese)*. 1998, 19(11):1724-1726

[33] Ansari SA; Pathak PN; Manchanda V K; Husain M; Prasad AK; Parmar VS. N,N,N',N'-Tetraoctyl Diglycolamide (TODGA): A promising extractant for actinide partitioning from high-level waste (HLW). *Solv Extra Ion Exch*, 2005,V23(4). 463 - 479

[34] Ansari SA; Pathak PN; Husain M; Prasad AK; Parmar VS; Manchanda V K. Extraction of actinides using N, N, N′, N′-tetraoctyl diglycolamide (TODGA): a thermodynamic study. *Radiochimica Acta*. 2006, V 94(6-7). 307-312

[35] Modolo G; Asp H; Schreinemachers C; Vijgen H. Development of a TODGA based Process for Partitioning of Actinides from a PUREX Raffinate Part I: Batch Extraction Optimization Studies and Stability Tests. *Solv Extra Ion Exch*, Volume 25, Issue 6 November 2007 , pages 703 - 721

[36] Modolo G; Asp H; Christian Schreinemachers a; Hendrik Vijgen Development of a TODGA based process for partitioning of actinides from a PUREX raffinate Part I: Batch extraction optimization studies and stability tests. *Solv Extra Ion Exch*, 2007,v25(6).703-721

[37] Modolo G; Asp, H; Schreinemachers, H Vijgen. Recovery of actinides and lanthanides from high-level liquid waste by extraction chromatography using TODGA+TBP impregnated resins. *Radiochimica Acta*, 2007,v95(7).391-397

[38] Tachimori S; Sasaki Y; Morita Y; Suzuki S. Recent progress of partitioning process in JAERI: development of amide-based ARTIST process. In:Global'05.

[39] Sugo Y; Sasaki Y; Tachimori S. Studies on hydrolysis and radiolysis of N,N,N',N'-tetraoctyl-3-oxapentane-1,5- diamide. *Radiochim. Acta*, 2002,90 ,161-165

[40] Sugo Y; Sasaki Y; Kimura T; Sekine T; Kudo H.In:Proceedings of GLOBAL 2005, *Radiolysis of TODGA and Its Effect on Extraction of Actinide Ions* 2005 Tsukuba, Japan, Oct 9-13,

[41] Zhang P; Song C L; Liang JF; Xin RX. Identification of radiolytic degradation products of γ-irradiated 30% trialkylphosphine oxide–kerosene Solution. Solv Extra Ion Exch, 2003. V 21(1), 91–108

[42] Nash KL, Gatrone RC, Clark GA. Hydrolytic and radiolytic degradation of OΦD(iB)CMPO, continuous studies. *Separation science and technology*. 1988, V23(12&13) 1355-72

[43] Madic C; Hudson MJ; Liljenzin JO; Glatz JP; Nannicini R. Recent achievements in the development of partitioning progresses of minor actinides from nuclear wastes obtained in the frame of the NEWPART European programme (1996-1999). *Progress in Nuclear Energy*, 2002,V 40(3-4) 523-526.

[44] Zhang P; Liu XQ; Wang JC; Song CL. Application of acethydroximic acid in the simplification of TRPO process. *Nucl. Radiochem (In Chinese)*. 2004, V26 (3) .158-161.

[45] Jiao RZ; Song CL; Zhu YJ. The selection stripping agent in solvent extraction process for treatment of high-level liquid waste. *Atomic energy science and technology (In Chinese)*, 2002, v36(1). 58-61

[46] Zhang P; Wang HR; Wang JC; Chen J. Synthesis and Characterization of Water-soluble Oxa-amide Compounds(in Chinese). *Organic chemistr (In Chinese)y*. 2008, v28(2): 335-338

[47] Huang N; Wang JC; Zhang J. Structures of the complexes of N,N'-dimethyl-3-oxa-glutaramic acid. *J. Tsinghua Univ (Sci & Tech) (In Chinese)* , 2005, V45(9) 1290-93

[48] Huang N; Wang JC ; Zhang P; Sun DZ. Structure of complexes of lanthanides with N, N'- dimethyl-3-oxa-glutaramic acid. *J. rare earths*, 2005, V 23(1). 12-16

[49] Huang N; Wang JC; Wang HR. Zhang P. Determination of stability constants for Np(IV) and Pu(IV) with N,N-dimethyl-3-oxa-glutaramic acid. *J.of Nucl. and Radiochem (In Chinese).* 2005, V27 (2).1 20-124

[50] Sun DZ; Chen J. Thermal decomposition and non-isothermal kinetics of N,N-dimethyl-3-oxaglutaramic acid. *J. Tsinghua U niv (Sci & Tech) (In Chinese)*, 2005, V45(12).1650-1653

[51] Zhang P ; Wang HR; Wang JC, Chen J. Synthesis of N,N'-dimethyl-3-oxa-glutamic acid and application in the simplification of TRPO process. *Journal of Radioanalytical and nuclear Chemistry*, 2007, V273 (1), 65-71

In: Nuclear Chemistry: New Research
Editor: Axel N. Koskinen, pp. 213-240

ISBN: 978-1-60456-957-5

Chapter 7

APPLICATION OF CENTRIFUGAL CONTACTORS IN PARTITIONING OF HIGH LEVEL WASTE IN CHINA

Wuhua Duan, Jianchen Wang and Jing Chen
Institute of Nuclear and New Energy Technology, Tsinghua University, Beijing 102201, China

Abstract

High level waste (HLW) produced from reprocessing of the spent nuclear fuel still contains actinides, ^{90}Sr, ^{137}Cs, ^{99}Tc, ^{129}I, etc., which engender permanent hazard to environment. The safe treatment and disposal of HLW are crucial to sustainable development of nuclear energy. The partitioning and transmutation (P&T) strategy for converting long-life nuclides from HLW into short-life radioactive or stable species has increasingly attracted interest. In recent years, a total partitioning process, including removal of actinides by the TRPO process, removal of Sr by the crown ether strontium extraction (CESE) process, removal of Cs by ion exchange using potassium titanium hexacyanoferrate (II) or by solvent extraction with 25,27-Bis(2-propyloxy) calix[4]-26,28-crown-6 (iPr-C[4]C-6), and separation of lanthanides and actinides with Cyanex 301, has been developed to treat Chinese HLW at Institute of Nuclear and New Energy Technology (INET), Tsinghua University, China. A series of cold and hot tests of both the separate processes and the total partitioning process were carried out with centrifugal contactors, and satisfactory results were obtained from these tests. In particular, a hot laboratory-scale test of the total partitioning process was carried out with the genuine HLW at INET in 1996. In this test, 50 stages ϕ10 mm centrifugal contactors were applied successfully for removal of actinides and Sr. The test results showed that the decontamination factors (DFs) for total α activity, ^{99}Tc, ^{90}Sr and ^{137}Cs were 588, 125, >2500, and >200 respectively. In addition, a cold pilot-scale test of the total partitioning process was carried out with the simulated HLW at INET in 2005. In this test, 24 stages ϕ70 mm centrifugal contactors were applied successfully in the stripping section of the TRPO process for removal of actinides. The test results showed that 99.9% of Nd, Zr and U was stripped respectively. In order to perform successfully these tests, a series of new centrifugal contactors have been developed at INET. In this paper, developments of centrifugal contactors for partitioning of HLW are introduced, and applications of centrifugal contactors in tests for partitioning of HLW in China are reviewed in detail.

1. Introduction

Waste management and waste minimization are of primary importance regarding the future use of nuclear energy. High level waste (HLW) produced from reprocessing of the spent nuclear fuel still contains moderate amounts of transuranium (TRU) actinides and fission products, in addition to the slightly enriched uranium, which engender permanent hazard to environment. [1,2] The safe treatment and disposal of HLW are crucial to sustainable development of nuclear energy. The partitioning and transmutation (P&T) strategy for converting long-life nuclides from HLW into short-life radioactive or stable species constitutes an advanced nuclear fuel cycle, and has increasingly attracted interest. The partitioning of HLW is one of the critical technical issues of the P&T strategy, and can be implemented with transmutation in the future to reduce the long-term risk of nuclear waste therefore reduce the uncertainties associated with geologic disposal. It can also be used as a method to reduce the volume of α waste and highly active waste (HAW). [3] Various partitioning processes have been developed for partitioning of HLW, such as the DIAMEX process in France, [4] the DIDPA process in Japan, [5] the TRUEX process in the USA, [6] the UNEX process in Russia and USA, [7] and the TRPO process in China. [8]

The objective of partitioning of HLW in China is to separate TRU elements, ^{99}Tc as well as ^{90}Sr and ^{137}Cs. After partitioning, the original HLW will become a non-α, low and intermediate level radioactive waste that would be suitable for shallow-land disposal. The required decontamination factors (DFs) are given in Table 1.[3] In Table 1, the α-waste standard for the individual package (4×10^6 Bq/kg) is chosen. For the Chinese highly saline HLW, the required DFs are much lower than those of commercial HLW. The required DFs are given in Table 2. [3] The α-waste standard in Table 2 is also for the individual package (4×10^6 Bq/kg).

Table 1. The required DFs for partitioning of the typical commercial HLW to a waste suitable for shallow land disposal [3]

Nuclides	Specific activity in HLLW	Chinese standard GB-9132-88			American regulation 10CFR61		
		Standard	Required D. F.		Standard	Required D. F.	
		Bq/kg	Vitrification	cementation	Ci/kg	Vitrification	Cementation
TRU	1.26×10^{14}	4×10^6	1.3×10^5	4.0×10^4	1×10^{-4}	1.3×10^5	4.0×10^4
^{99}Tc	4.78×10^{11}	-	-	-	3	33	11
^{90}Sr	2.07×10^{15}	4×10^{10}	215	71	7000	61.4	20
^{137}Cs	3.01×10^{15}	4×10^{10}	313	104	4600	136	45

Table 2. The required DFs for partitioning of the Chinese highly saline HLW [3]

Nuclides	Total α	^{99}Tc	^{90}Sr	^{137}Cs
Required D.F	38	1.1	10	13.5

The TRPO process has been studying in China since 1980s. The hot test of the TRPO process was firstly carried out with ϕ10 mm centrifugal contactors successfully in the hot cell of Institute for Transuranium Element (ITU) of EU in Karlsruhe Germany. [9,10] In recent years, a total partitioning process, including removal of actinides by the TRPO process, removal of Sr by the CESE process, removal of Cs by ion exchange using potassium titanium hexacyanoferrate (II) (KTiFC) or by solvent extraction with iPr-C[4]C-6, and separation of lanthanides and actinides with Cyanex 301, has been developed to treat Chinese HLW on the base of the TRPO process at INET. [3,11] In particular, the total partitioning process including the TRPO process, the CESE process, removal of Cs by ion exchange using KTiFC has been verified by both the hot test for laboratory-scale with ϕ10 mm centrifugal contactors and the cold test for pilot-scale with ϕ70 mm centrifugal contactors at INET. [12,13]

Centrifugal contactors are efficient liquid-liquid extraction equipments. Compared with conventional contactors such as mixer-settlers and pulsed columns, centrifugal contactors offer some advantages including low hold-up volume, short residence time therefore less solvent degradation, excellent phase separation, high mass transfer efficiency, great safety with respect to nuclear criticality, compact therefore low capital costs, no disturbance to the steady state when being shut down, rapid start-up and shut-down, etc. Therefore, centrifugal contactors are favored for reprocessing of the spent nuclear fuel and partitioning of HLW. [14,15]

The primary centrifugal contactor, which was the paddle type, was successfully developed and operated for many years at Savannah River Plant (SRL). In the late 1960s, the paddle type centrifugal contactor was modified to the annular type at Argonne National Laboratory (ANL). The ANL centrifugal contactor was reliable, easy to operate and maintain. [16-18] In particular, it was applied in the hot tests of the TRUEX process at ANL. [19] Recently, USA, Russia, France, Germany, U.K, India and Japan achieved new progress on development and applications of centrifugal contactors in nuclear industry. [20-28]

Since the late 1970s, in order to meet requirements of the TRPO process tests, INET has been developing its own annular centrifugal contactors. Four types of centrifugal contactors with 10 mm, 20 mm, 50 mm and 70 mm of rotor diameter have been developed for partitioning of HLW respectively. [29-32] A series of cold and hot tests for partitioning of HLW were carried out with centrifugal contactors in China. In this paper, developments of centrifugal contactors for partitioning of HLW are introduced, and applications of centrifugal contactors in tests for partitioning of HLW in China are reviewed in detail.

2. Centrifugal Contactors

The centrifugal contactor mainly consists of a motor, a drive part and a rotor part. The main structure of the rotor part is shown in Figure 1.

Two immiscible liquids with different density are fed from the opposite sides into the annular mixing zone between the spinning rotor and the stationary housing. The liquid-liquid dispersion created by turbulent Couette flow in the annular mixing zone flows by gravity to the inlet in the bottom face of the rotor and thus into the centrifugal separating zone inside the rotor. Here the dispersion breaks and separates rapidly under the high centrifugal force. The separated phases flow separately through the heavy phase weir and the light phase weir of the rotor into their collector rings in the housing. Then each liquid leaves its collector ring through a tangential exit, and flows into an adjacent stage respectively. When centrifugal contactors are being operated, all of them must be filled with the aqueous phase before the organic phase is fed to them.

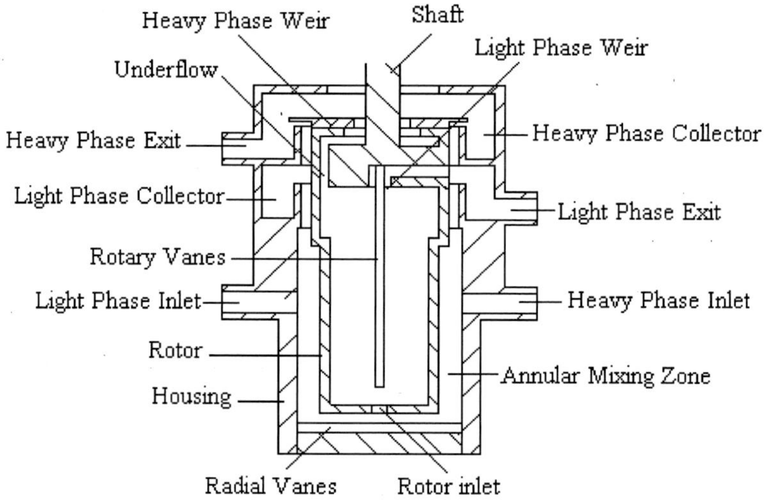

Figure 1. Cutaway view of a centrifugal contactor.

During operation of the centrifugal contactor, the axial thickness of the dispersion between the aqueous phase and the organic phase in the rotor (see Figure 2) decreases with increase of height of the settling section. Both the axial thickness and the radial position of the dispersion change with variation of operational factors such as rotor speed, throughput, flow ratio, and so on. When the dispersion moves to the underflow in the rotor, there will be organic entrainment in the aqueous phase. While the dispersion moves to the light phase weir, there will be aqueous entrainment in the organic phase. When the dispersion spreads the whole rotor, there will be organic entrainment in the aqueous phase and aqueous entrainment in the organic phase. So, the radial position of the dispersion must be controlled in order to avoid entrainment. Generally, the radius of interface is used to represent the radial position of the dispersion in rotor. The radius of interface can be obtained by the following equation: [33-35]

$$r_i = \sqrt{\frac{F_A R_A^2 - (\rho_O / \rho_A) F_O R_O^2}{1 - \rho_O / \rho_A}} \tag{1}$$

where r_i is the radius of interface (cm), $F=(r/R)^2$, R is the radius of phase weir (cm), r is the radius of liquid surface when flowing over the phase weir (cm), footnotes 'A' and "O" represent heavy and light phase respectively.

The nominal throughput of the centrifugal contactor can be obtained by a series of hydraulic tests or by calculation. The dispersion number concept proposed by Leonard et al. can be used to obtain the maximum throughput of a centrifugal contactor. The dispersion number N_{Di} that depends on the system, mixing intensity and the type of dispersion, is defined as: [35,36]

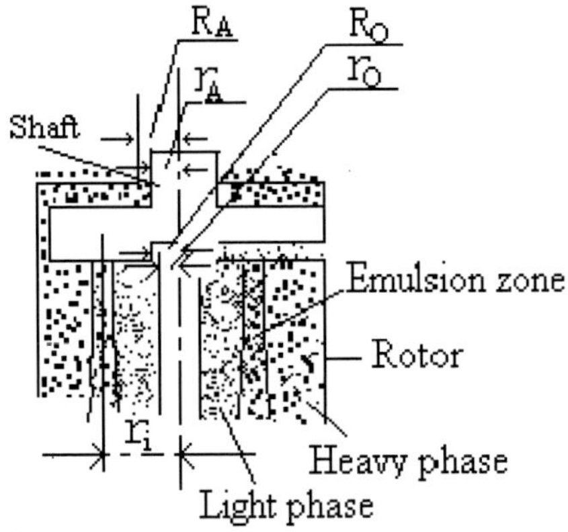

Figure 2. Hydraulic characteristics of the centrifugal contactor.

$$N_{Di} = \frac{Q}{V}\sqrt{\frac{\Delta Z}{\bar{r}\omega^2}} \tag{2}$$

where Q is the maximum throughput of both phases (cm^3/s), V is the volume of the settler (cm^3), ΔZ is the initial thickness of the dispersion band (cm), ω is the rotational velocity (rad/s), and $\bar{r}$ is the average radius (cm) given by: [35,36]

$$\bar{r} = \frac{2(r_{out}^3 - r_{in}^3)}{3(r_{out}^2 - r_{in}^2)} \tag{3}$$

Where r_{out} is the radius to the outer edge of the dispersion band (cm) and r_{in} is the radius to the inner edge of the dispersion band (cm). N_{Di} values for centrifugal units can be obtained for any extraction system, either from batch gravity settling test or by using a small centrifugal settler with interface control provision, so that the dispersion can be positioned to occupy the full separating zone. Leonard et al. present a standard test for measuring N_{Di}. [36] After obtaining the experimental dispersion numbers for both aqueous and organic phase continuous operation, the maximum throughput can be obtained from equation (2).

The extraction stage efficiency of the centrifugal contactor is calculated by Murphree equation as follows: [37,38]

$$E_A = \frac{(C_{A,in} - C_{A,out})}{(C_{A,in} - C_{A,eq})} \times 100\% \text{ (for the aqueous phase)} \tag{4}$$

$$E_O = \frac{(C_{O,out} - C_{O,in})}{(C_{O,eq} - C_{O,in})} \times 100\% \text{ (for the organic phase)} \tag{5}$$

Where E_A and E_O are the extraction stage efficiency of the aqueous phase and the organic phase respectively (%), $C_{A,in}$ and $C_{A,out}$ are the inlet and the outlet concentrations of the metal ion in the aqueous phase respectively (mg/L), $C_{O,in}$ and $C_{O,out}$ are the inlet and the outlet concentrations of the metal ion in the organic phase respectively (mg/L), $C_{A,eq}$ and $C_{O,eq}$ are the equilibrium concentrations of the metal ion in the aqueous phase and in the organic phase respectively (mg/L).

Continuous separation, washing, and extraction processes often require many steps or stages to achieve desired product quality or required extraction efficiencies. Centrifugal contactors can be readily interconnected to allow multistage processes. The extraction rate of the cascade is described as follows: [37,38]

$$\rho = \frac{(C_{O,out} - C_{O,in})V_O}{C_{A,in}V_A} \times 100\% \tag{6}$$

where ρ is the extraction rate (%), V_O and V_A are the volume of the aqueous phase and the organic phase respectively (L), the definition of $C_{O,in}$, $C_{O,out}$ and $C_{A,in}$ is the same as above.

3. Development of Centrifugal Contactors for Partitioning of HLW

3.1. Laboratory-Scale

The laboratory-scale centrifugal contactors include two types of centrifugal contactors, namely, the ϕ10 mm centrifugal contactor and the ϕ20 mm centrifugal contactor. Their main structures are same. The ϕ10 mm centrifugal contactor with rotor diameter of 10 mm was developed for the experimental study of the TRPO process in 1980. Its nominal throughput is about 0.5 L/h and its hold-up volume is 4-5 mL, thereby, the residence time for one stage was less than 1 min. The ϕ20 mm centrifugal contactor with rotor diameter of 20 mm was developed in 2003. Its nominal throughput is about 5 L/h for kerosene-H_2O system. In order to meet the requirements of the hot total partitioning process tests, some advanced designs have been achieved in the laboratory-scale centrifugal contactors to result in lower capital and operating costs, easier manipulation, more reliable and simpler to maintain.[30,39]

3.1.1. Modular Design

In order to make the centrifugal contactor remote maintenance easy and reliable in the hot cell, the mating joint of every part of the contactor need not use screws or pins. The laboratory-scale centrifugal contactor is made up of two modules, namely, the rotor module shown in Figure 3 and the housing module shown in Figure 4. Modules are installed each other by simply inserting on the other one, and they are fitted by the cylindrical joint with the clearance fit without screws and nuts. The centrifugal contactor is assembled and disassembled fast by simply moving modules up and down. Two modules are made within the acceptable deviation and are interchangeable. Surface A is the surface of basic hole and surface B is the surface of the shaft part with suitable lower deviation. Assembly of the centrifugal contactor is completed by mating of surface A and B. In this way, it is much easy to manipulate centrifugal contactors in the hot cell. For instance, if one of the motors failed or any other unexpected things occurred to make one or some of centrifugals contactors stop working when the multi-stage flowsheet is being tested in the hot cell, centrifugal contactors can be fast shut down, then the rotor module with the problem can be replaced with the good rotor module by a manipulator, and centrifugal contactors can be started again soon, then the test is continued. The rotor module can be replaced smoothly and the operation time is less than 5 minutes. This design is similar to that in the 12-mm centrifugal contactor developed by Rivalier, P., et al. [23]. This design is also similar to that in the 2-cm ANL centrifugal contactor developed by Leonard R. A., [19] but two guide posts in the 2-cm ANL centrifugal contactor are not needed in INET centrifugal contactors.

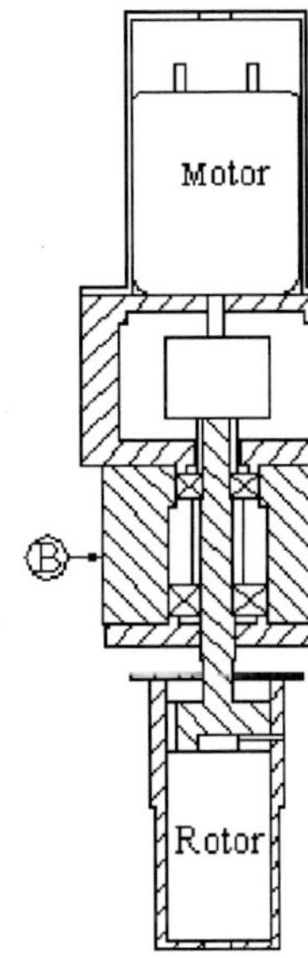

Figure 3. The rotor module.

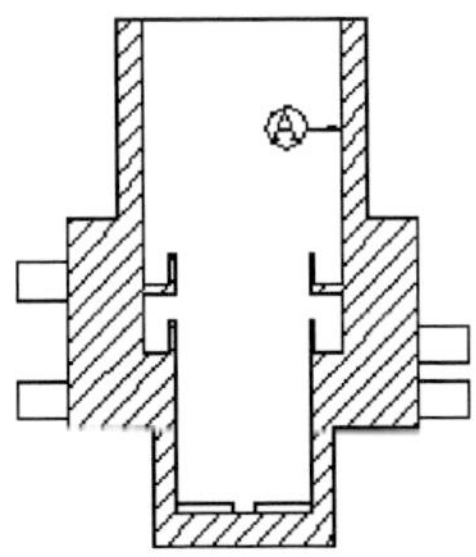

Figure 4. The housing module.

3.1.2. Improved Design of Housing

To further reduce the liquid hold-up of the laboratory-scale centrifugal contactors, the interstage connection should be reformed. The interstage connection change must be preceded by change of housing design. Two types of housing have been developed for the laboratory-scale centrifugal contactors. One is called as the multi-stage group housing shown in Figure 5, the other is called as the single-stage compact housing shown in Figure 6.

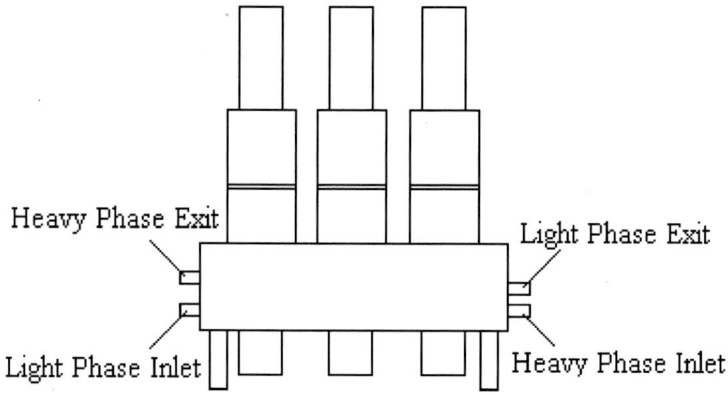

Figure 5. The multi-stage group housing.

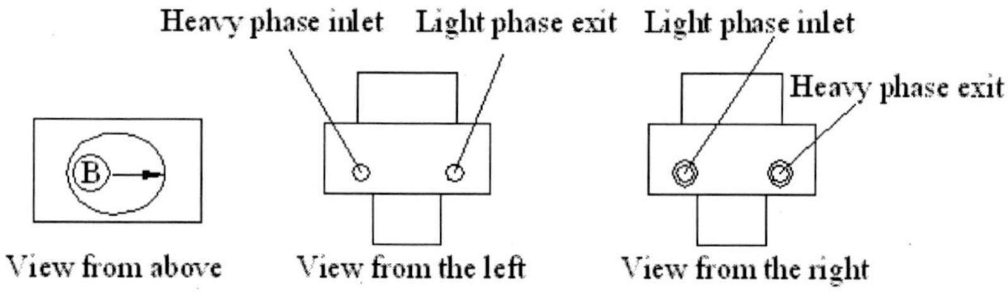

Figure 6. The single-stage compact housing (From Ref. 39, courtesy of He, X. M.).

The multi-stage group housing is a housing system containing multiple stage housings (e.g. 4, 3 or 2 stages) in a set. The principal part of the set is a whole stainless steel block, which consists of heavy phase collecting rings, light phase collecting rings and inside interstage lines. For a set containing 4-stage ϕ10 mm annular centrifugal contactors, its size is 50 mm wide and 210 mm long. For a set containing 3-stage ϕ20 mm annular centrifugal contactors, its size is 70 mm wide and 190 mm long. The advantage of the set is no outside interstage lines between two stages in the set, therefore has very small liquid hold-up and no risk of leakage in interstage lines. The multi-stage group housing also make multi-stage cascade unit more compact for multi-stage processes. The disadvantage of the set is that the number of stage in a set is fixed after the set being made, therefore, it is not flexible to meet requirement of extraction flowsheet with different stages of extraction, scrubbing, stripping and cleaning.

The single-stage compact housing means one set containing one stage. There are four holes on the two opposite sides of housing, which are exits and inlets of both light and heavy phases, respectively. Good parallelism and flatness of the two sides are needed to make interstage sealing well. The grooves are designed on the one side to hold sealing gasket made of polytetrafluoroethylene. The intersage seal is shown in Figure 7. The design was applied in the hot test of the total partitioning process in the hot cell at INET in 1996. [40]

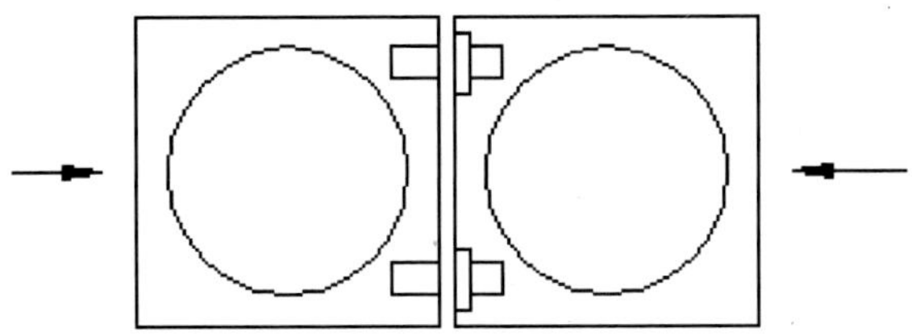

(From Ref. 39, courtesy of He, X. M.).

Figure 7. Interstage joint of the single-stage compact housing.

3.1.3. Sampling System

After the hot test, the liquids left in centrifugal contactors needs to be sampled for analysis. It is not easy to get samples from many centrifugal contactors by the manipulator. A remote sampling system has been developed for the hot tests. Stoppers with sealing rings are used to stop the holes on the bottoms of contactors when contactors are running. The automatic sampling system is shown in Figure 8. The system is attached to the bottom of the housing of the centrifugal contactor. When centrifugal contactors are working, the holes on the bottoms of centrifugal contactors are stopped up by stoppers with sealing rings. When sampling, the stoppers are moved down and liquid samples flow out from centrifugal contactors by gravity. The samples are collected in collecting rings and then flow into sample bottles under the holes. The manipulator was used to drive the stoppers through a sample power transmission mechanism. All bottles with samples are placed on a shuttle plate that can move under centrifugal contactors on the linear motion antifriction bearing guide-way and can be driven by the manipulator. This sampling system was also successfully applied in the hot test of the total partitioning process in the hot cell at INET in 1996. [40]

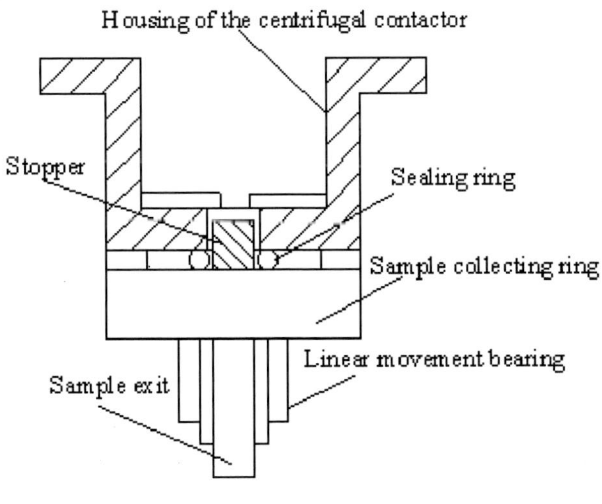

Figure 8. The automatic sampling system (From Ref. 39, courtesy of He, X. M.).

3.1.4. Rotor Speed Acquisition and Monitoring System

The rotor of the centrifugal contactor is directly driven by motor, of which speed depends on the voltage of power supply, so the speed of rotor can be simply controlled by changing voltage. For multi-stage centrifugal contactors, the rotor speed acquisition and monitoring system is a practical and necessary way. With the system, centrifugal contactors can be remotely controlled and problems can be found outside the hot cell in time.

He, et al. designed a rotor speed acquisition and monitoring system for ϕ10 mm centrifugal contactors. [39] A gear disc was fixed on the coupling of the centrifugal contactor and an optic sensor was stalled in the centrifugal contactor. The gear disc was spinning synchronously while the rotor was spinning. The optic sensors gave out pulse signal, of which frequency was proportional to the speed of the rotor. The signal was acquired by computer after amplification, photoelectricity isolation and regulation, and was processed to be speed of the centrifugal contactor. The system shown in Figure 9 was tested in the hot test of the total partitioning process in the hot cell at INET in 1996., [40] but it can be further improved and simplified.

A new rotor speed acquisition and monitoring system for the laboratory-scale centrifugal contactors was designed at INET. The system is shown in Figure 10. The adjustable voltage stabilizer was used to directly adjust voltage of the motor. Accordingly, the rotor speed was changed. The data acquisition system was used to acquire the signal of the output voltage by computer and process to be speed of the centrifugal contactor. Compared with the system developed by He, et al. The new system needs not both the gear disc and the optic sensor, therefore is simpler and more reliable.

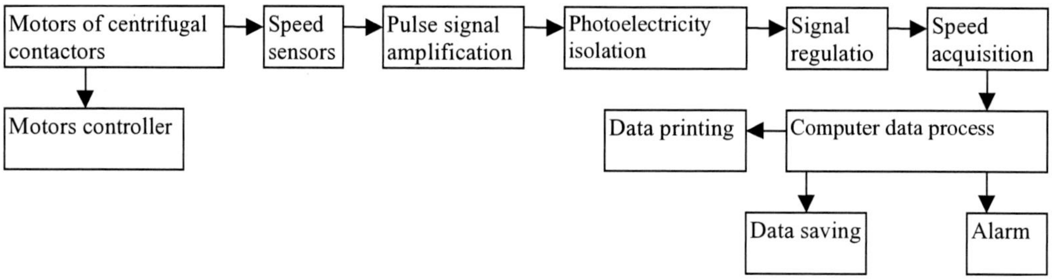

Figure 9. The old rotor speed acquisition and monitoring system (From Ref. 39, courtesy of He, X. M.).

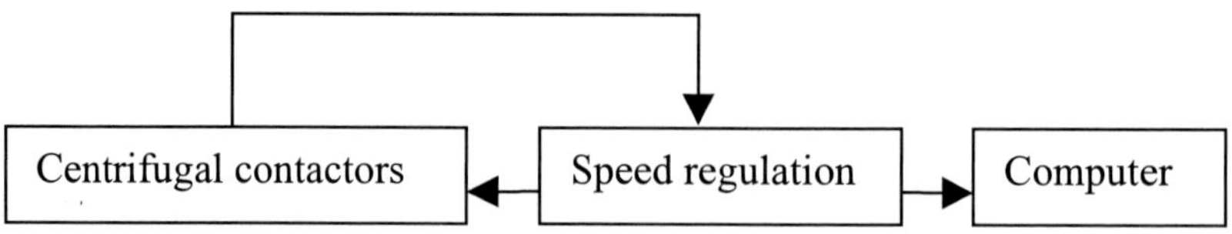

Figure 10. The new speed acquisition and monitoring system.

3.2. Industry-Scale

The ϕ50 mm centrifugal contactor was developed in 2003. The inside diameter of its rotor is 50 mm, and the hold-up volume is about 300 mL. [31] The ϕ70 mm centrifugal contactor was developed in 1996. The inside diameter of its rotor is 70 mm, and the hold-up volume is about 350 mL. [32]

3.2.1. Modular Design

In order to make the centrifugal contactor remote maintenance easy and reliable, the mating joint of every part of the cenntrifugal contactor need not use screws or pins. The centrifugal contactor for industry-scale is made up of three modules (see Figure 11) that are the motor module, the rotor module and the housing module.[47,64] Modules are installed each other by simply inserting in the other one, and they are fitted by mating cone without screws and nuts or pins. The contactor is assembled and disassembled fast by simply moving modules up and down. The magnetic coupling is used in the contactor to provide the torque to rotate the rotor without losing any speed. The application of the magnet coupling to drive the rotor can make the motor replaced independently of the rotor. This feature provids an easier mode of remote maintenance and decreased the space requirement on the upside when replacing the motor. All of these designs make the contactor remote maintenance easy and reliable.

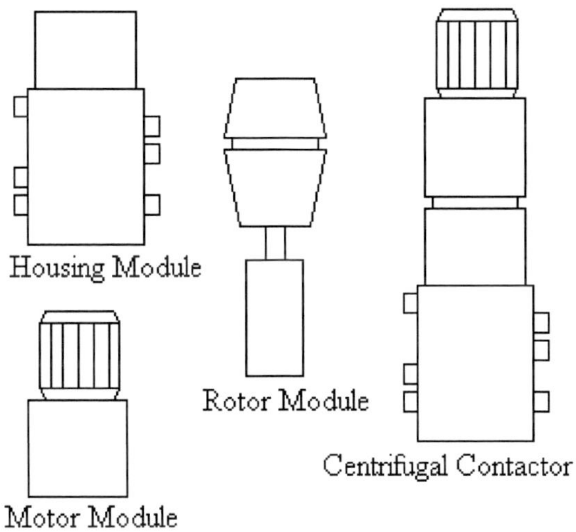

Figure 11. Modular design.

3.2.2. Overflow Structure Design

For uses of the previous centrifugal contactors in a multistage system or cascade, if a stage of the cascade became inoperative such as caused by the failure of a motor, the whole cascade must be shut down. To settle this problem, Jubin, et al. and Jenkins, et al. invented to install horizontally oriented overflow ports below the organic collector rings and above the feed points of the centrifugal contactors in the cascade. [41] The overflow ports provided for the overflow of the mixture of both aqueous and organic phases from the inoperative stage into the mixing zone in the adjacent operating stages. However, the overflow of liquid through the overflow ports consisted of both aqueous and organic phases, so the overflow ports were not useful when the inoperative stage was an end stage in the cascade. Duan, et al. developed another overflow mode in the ϕ70 mm centrifugal contactor. Two kinds of overflow line were designed in the housing module (see Figure 12). One was the light phase overflow, and the other was the heavy phase overflow. The light phase overflow was below the organic collector ring and above the heavy phase overflow. The heavy overflow was below the light phase overflow and above the feed inlets. If a centrifugal contactor was ceased to work due to a motor failure or due to other abnormal accidents, the heavy phase and the light phase could separate by gravity in the housing of the stopped centrifugal contactor and overflow in opposite direction through their own overflow line into adjacent stages respectively. In this way, the overflows were also useful when the stopped stage was the end stage in the cascade. The use of the overflow structure design provides a satisfactory solution to the failure or inoperativeness of one stage or non-adjacent stages of a multistage centrifugal contactor arrangement or cascade, in other words, when a multi-stage cascade is operated, even though one stage or non-adjacent stages are ceased to work, the cascade can continue to operate.

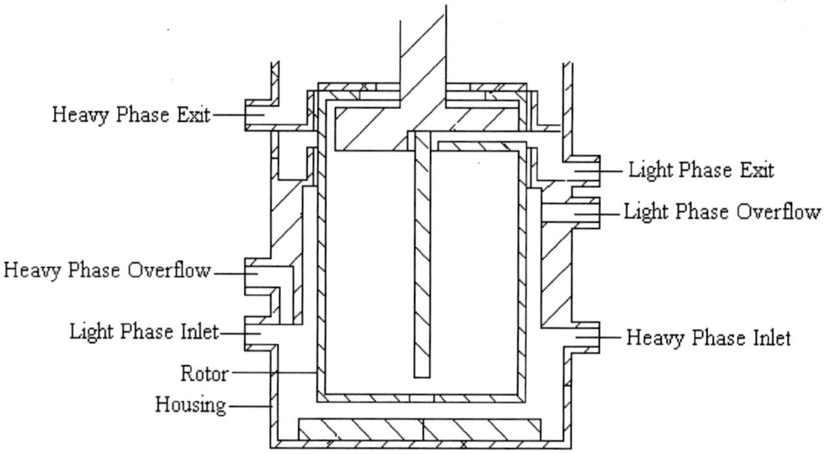

Figure 12. The overflow structure.

4. Application of Centrifugal Contactors in Partitioning of HLW in China

4.1. Tests of the TRPO Process for Removal of TRU Elements

The TRPO process for removal of TRU elements from the commercial HLW has been developing in China since 1980. [3,42,43]

A mixed trialkylphosphine oxide (TRPO, commercial name Cyanex 923) is used as extractant and kerosene is used as diluent in the process. TRPO contains C_6-C_8 mixed alkyl and the average molecular weight is 346. Its density is 0.8796 g/cm^3, melting point is -21°C, and boiling point is from 180 °C to 225 °C at 2-4 mm Hg. Its solubility in water is lower than 0.1 g/L. Until a total dose of 1×10^6 Gy is absorbed, no significant change in extraction properties occurs.

30% TRPO-kerosene is a very good extractant for tetra- and hexa-valent TRU ions in nitric acid. It is also an effective extractant for tri-valent Pu, Am and Cm at low and moderate HNO_3 concentration (<2 mol/L). But the extraction of penta-valent Np is very poor. Among fission product elements, 30% TRPO-kerosene is highly extractable for Zr. The extraction ability of 30% TRPO-kerosene for Tc(VII) is very high at low and medium nitric acid concentration, but it drops sharply with increase of nitric acid concentration. The extraction of Mo is also sufficiently high and is similar to Tc(VII).

When the TRPO process is applied directly for partitioning of the Chinese HLW, the salting-out effect enhances the extraction of Fe(III) and causes the formation of the third phase. The addition of modifiers like TBP or n-octanol can inhibit formation of the third phase. It can also be eliminated by diluting the HLW. The third phase will not appear at room temperature when the HLW is diluted by 3 times and the concentration of nitric acid is above 0.7 mol/L.

There is no any stripping difficulty for Np, Pu, Am and Cm in the TRPO loaded with TRU elements. 5.5 mol/L HNO_3 was used to strip Am (Cm) together with Lns, 0.6 mol/L $H_2C_2O_4$ was then used to strip Np and Pu, and U was finally stripped by 5% Na_2CO_3. The TRPO process was verified firstly with the simulated HLW and separating funnels, and later with 32 stages ϕ10 mm centrifugal contactors at INET. [44,45]

The simulated HLW (see Table 3) was prepared. In the cold test, Nd and Zr were used to simulate Am and Pu respectively. The flowsheet for the cold test is presented in Figure 13. The result of the cold test showed that 99.9% of Nd and 99.9% Zr were extracted. The extracted Nd and Zr were completely stripped out, respectively. The cross contamination between them was small.

Table 3. The main composition of the simulated HLW [45]

Element	Concentration, g/L	Element	Concentration, g/L	Element	Concentration, g/L
Na	51.2	Nd	2.05	Mo	0.82
Cr	2.0	Cs	2.0	Mn	0.13
Fe	17.4	Ba	0.074	K	0.45
Ni	8.2	Y	0.2	Nb	0.03
Sr	0.61	Pr	0.48	Rh	0.43
Zr	0.014	Sm	0.335	Al	15.9
Pd	0.03	Ce	0.78	La	0.59
Ru	0.62				

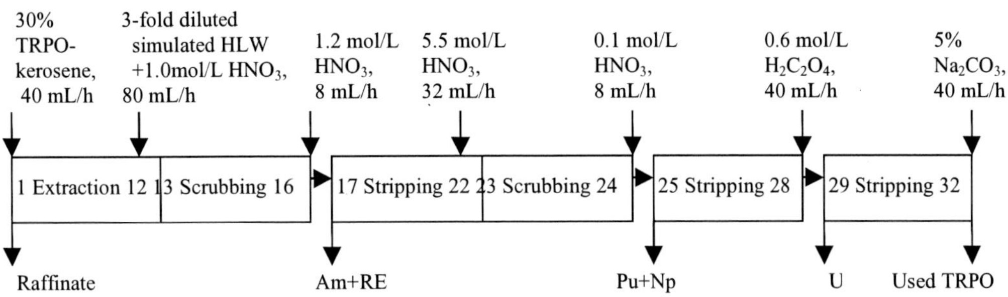

Figure 13. The flowsheet of the TRPO process with the simulated Chinese HLW in the cold test.

Table 4. Composition of the HLW feed solution for hot test 1 [9,10]

Element	Concentration, g/L	Element	Concentration, g/L	Element	Concentration, g/L
TRU Element		Fission Product		Fission Product	
U	0.64	Rb	0.044	Te	0.012
Np	0.043	Sr	0.098	I	0.000
Pu	0.034	Y	0.067	Cs	0.36
Am	0.023	Zr	0.019	Ba	0.39
Cm	0.002	Mo	0.17	La	0.23
Corrosion Product		Tc	0.12	Ce	0.45
Cr	0.104	Ru	0.26	Pr	0.23
Mn	0.04	Rh	0.051	Nd	0.79
Fe	0.95	Pd	0.50	Sm	0.18
Ni	0.077	Ag	0.000	Eu	0.025
Co	0.001	Cd	0.48	Gd	0.043
HNO_3	0.75 mol/L	Sn	0.003		

Table 5. The DF of TRU elements, ^{99}Tc and Nd in the hot tests [9,10]

Element	DFs in hot test 1	DFs in hot test 2
^{144}Nd	>22000	>33000
^{237}Np	12.4	>4100
^{238}U	>5400	>7000
^{239}Pu	>760	>950
^{241}Am/^{241}Pu	>2800	>3200
^{243}Am	>900	>760
^{99}Tc	>1400	>1700

The hot test of the TRPO process were carried out with HLW of WAK and 12 stages ϕ10 mm centrifugal contactors at Institute for Transuranium Elements (ITU) Join Research Center of European Union at Karlsruhe Germany. [9,10] Two different concentrated HLW solutions, originating from WAK, Karksruhe, were used in the hot tests. The HLWs were centrifuged to remove solid suspensions. Before being used as feed solutions, Fe(II) sulfamate along with hydroxylamine were added to reduce Np(V) to Np(IV), and the concentrated HLW solutions were diluted 10 times. In the meantime, the nitric acid concentration was adjusted to desired value. The composition of the feed solution for hot test 1 is given in Table 4. The flowsheet and some of parameters for hot test 1 were given in Figure 14. In hot test 1, the feed acidity was 0.7 mol/L HNO_3 and the extraction section for TRU elements was 6 stages. Hot test 1 was carried out in two separate steps. Extraction and stripping of Am took place in the first step. After the extraction process reached a steady state, 200 mL loaded TRPO were collected. It was then fed into the cleaned centrifugal contactor set for a second step for stripping of Np, Pu as well as U. Very good DF results were obtained for TRU elements except Np. The D.F. value of Np was only 12.4. The cross contamination among the stripping sections is very low. In hot test 2, only extraction section with 10 extraction stages and 2 scrubbing stages was tested. The feed acidity was increased to 1.36 mol/L HNO_3. The DF value of 4100 for Np was obtained. The results for other TRU elements were also very good. The DFs of TRU elements, ^{99}Tc and Nd in the two hot tests are shown in Table 5. The hot tests demonstrate the TRPO process is an effective and prospective process for partitioning of TRU elements as well as ^{99}Tc from HLW.

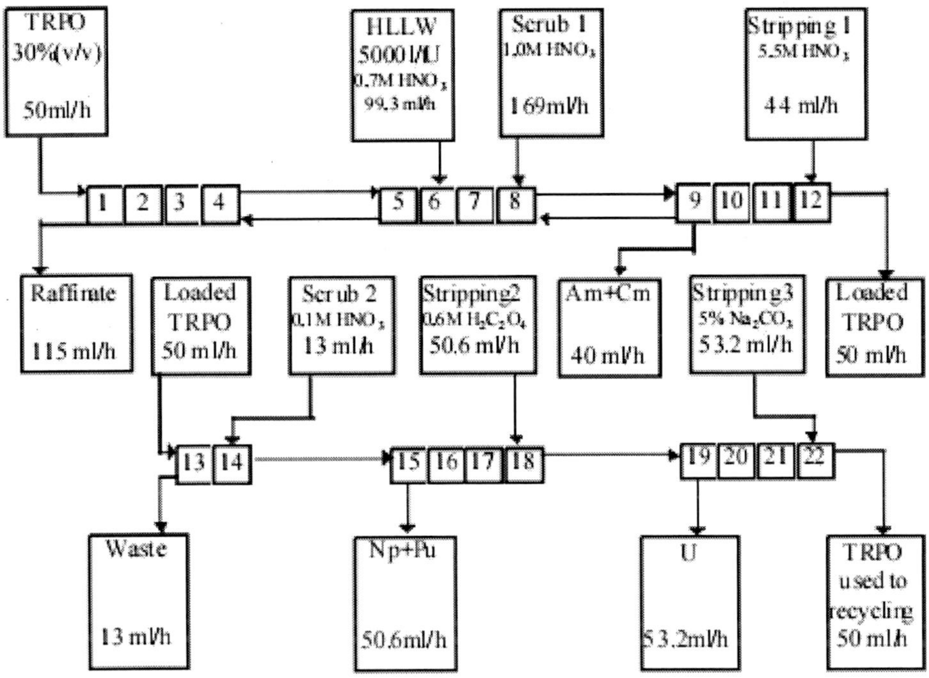

Figure 14. The flowsheet of hot test 1 (From Ref. 10, courtesy of Song, C. L.).

Advantages of the TRPO process are low cost of the trialkylphosphine oxide reagents and high extraction efficiency for several actinide and lanthanide elements. On the other hand, one of the main drawbacks pointed out is the great affinity of phosphine oxide moieties for nitric acid, therefore, there is always a competition of HNO_3 with the metal ions for extraction. Another weakness is the narrow range of nitric acid concentrations that allow an efficient extraction of the trivalent actinides, and for the recycle of actinides from the dissolved spent nuclear fuel, dilution (by a factor of 10) and denitration of the feed solutions are necessary, hence, the accomplishment of these features significantly increase the volume of the liquid wastes to be processed.[2]

4.2. The Cold Test of the CESE Process for Removal of Sr

Dicyclohexyl 18 cown 6 (DCH18C6, see Figure 15) in n-octanol provides efficient and highly selective extraction of Sr from nitric acid solution. Most of cation ions in the nitric acid solution, such as Ni^{2+}, Fe^{3+}, Al^{3+}, Cr^{3+}, etc., are not extracted by DCH18C6/n-octanol, except for K^{+}, Mo^{6+}, Na^{+} and Cs^{+}. The solubility of DCH18C6 in H_2O and diluted HNO_3 is quite high, but it drastically decreases in saline solution and in the presence of n-octanol. 0.1 mol/L

DCH18C6/ n-octanol is used for removal of Sr from Chinese HLW. The extraction of Sr benefits from the salting-out effect of Al^{3+}, Na^{+} and Fe^{3+}, and can be carried out at low nitric acid concentration by DCH18C6/ n-octanol, but DCH18C6/ n-octanol does not extract TRU elements under these conditions. [42,46-48]

A CESE process with DCH18C6/n-octanol for removal of Sr has been developed at INET. The cold test of the CESE process (see Figure 16) was carried out with the simulated HLW (see Table 6) and 16 stages ϕ10 mm centrifugal contactors. [49] The distributions of the main elements in every effluent after extraction, scrubbing and stripping are listed Table 7. It was shown that 98.3% of Sr was extracted from the simulated HLW, in the meantime, 56.6% of K and 3% of Na were extracted from the simulated HLW by DCH18C6/ n-octanol, but other elements were not extracted. After stripping, nearly 100% of Sr was recovered, and 99.1%-99.9% of K was removed, so the used organic phase could be reused. The result of the cold test showed the CESE process was a promising process for partitioning of Sr from the Chinese HLW.

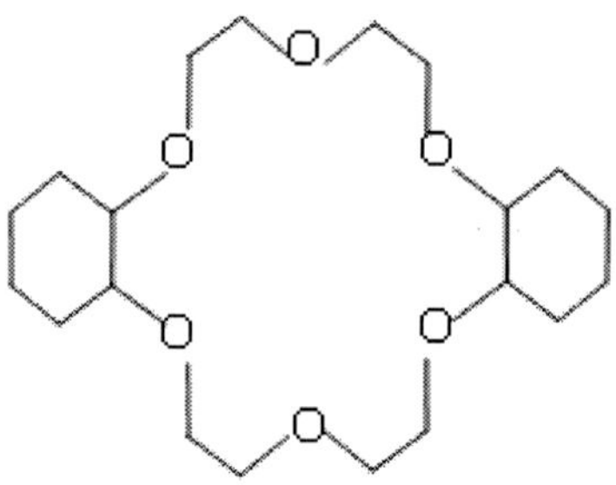

Figure 15. DCH18C6.

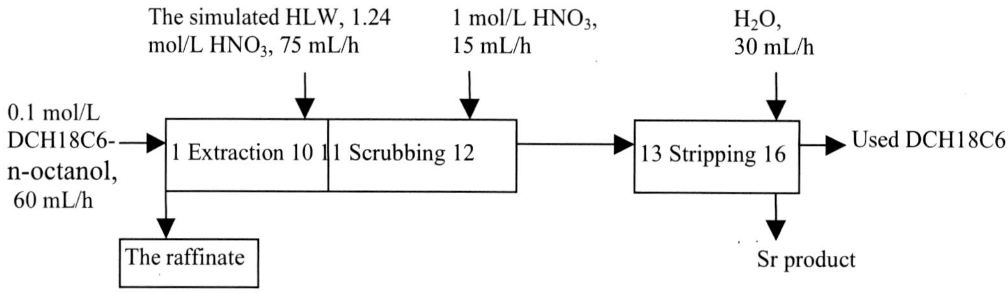

Figure 16. The flowsheet of the cold test for the CESE process.

Table 6. The main composi tion of the simulated HLW [49]

Elements	Sr	K	Cs	Mo	Al	Na	Ni	Fe	Cr	Rh
Concentration, g/L	0.205	0.141	0.565	0.024	5.00	17.1	2.06	4.45	0.496	0.179

Table 7. Distribution of the main elements in every effluent after extraction (%) [49]

Element	The raffinate	The Sr product	The used DCH18C6	Marerial balance
Sr	1.66	95.7	-	97.4
Na	97.0	0.164	0.009	97.3
K	43.3	53.9	0.45	97.8
Fe	114	0.044	0.009	114
Al	99.1	0.059	0.015	99.2
Mo	110	20.0	1.26	131
Cr	106	0.015	-	106
Ni	105	-	-	105
Cs	111	0.3	0.073	111
Rh	105	-	-	105
HNO_3	90.0	12.4	0.2	103

4.3. The Hot Test of the Total Partitioning Process for Laboratory-Scale

In 1990s, a total partitioning process including removal of actinides by the TRPO process, removal of Sr by the CESE process, removal of Cs by ion exchange using KTiFC has been developed to treat Chinese highly saline HLW on the base of the TRPO process at INET. [3,11] KTiFC is dark brown and rather hard spherical particles with 0.40-0.45 mm of diameter, and has excellent selectivity and high Cs absorption capacity from the highly saline nitric acid solution. It has good chemical resistance in 0.5~1.5 mol/L nitric acid solution as well as good radiation stability. Its static capacity and its appearance do not change until a dose of 1×10^6 Gy was adsorbed. The distribution coefficients (K_d) of main components in the simulated HLW are listed in Table 8. [42,50]

Table 8. The K_d of the composition between KTiFC and simulated HLW diluted by 3 times at 0.5 mol/L HNO_3 [42]

Element	Cs	K	Fe	Ni	Sr	Na	Cr	Nd	Mn
K_d/mol·g^{-1}	219.6	35.2	1.6	4.4	<0.01	4×10^{-3}	5×10^{-5}	<0.001	1×10^{-5}

A hot test with the genuine HLW was carried out to verify the total partitioning process with 50 stages ϕ10 mm centrifugal contactors and a small ion exchange column at INET in 1996. [3,11,12,40,51-53] ϕ10 mm centrifugal contactors were specially developed for the hot test (see 3.1 Section). The ion exchange column with 4 mm of an inner diameter and 35 mm of a length was filled with 0.5 g 60-100 meshes spherical KTiFC. The effective bed volume was 0.44 mL. The flowsheet of the total partitioning process is given in Figure 17. The feed solution was 2.7-fold diluted highly saline HLW and its acidity was regulated to 1 mol/L HNO_3. The genuine HLW has a gray-blue color and is distinguished by its high salt content. The total salt content is about 380 g/L and the total oxide content is about 208 g/L. The ionic strength is as high as 8.7 (not including that contributed by HNO_3). The main components of the genuine HLW are non-radioactive elements such as Na, Al, Fe, Cr and Ni. They contribute 88.2% of the total elemental mass and 94.1% of the total nitrates. The weight of the TRU and ^{99}Tc contribute only 0.31% of the total elemental mass and 0.12% of the total nitrates. Am and Pu contribute >99% of the total α activity but Np and Cm were less than 1%. 80% of the total γ radiation comes from ^{137}Cs. The main composition of the genuine HLW is given in Table 9.

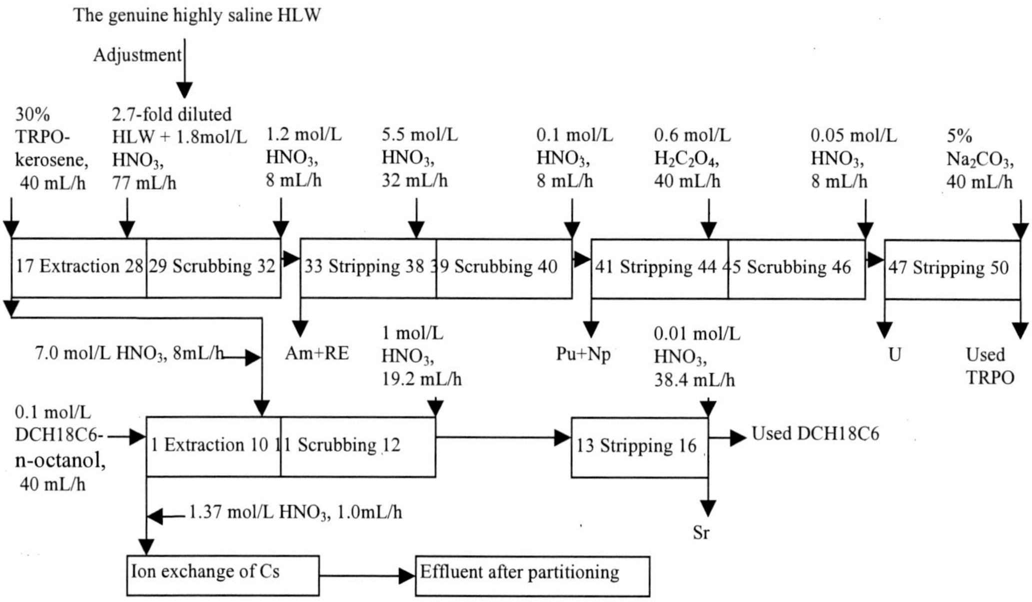

Figure 17. The flowsheet of the total partitioning process in the hot test (From Ref. 11, courtesy of Song, C.L.).

Table 9. The main composition of the genuine HLW [11]

Elements	Na	Al	Fe	Cr	Ni	RE	Sr	Cs	Zr	Ru	Tc
Concentration, g/L	51.2	16.5	17.4	2.0	8.2	4.6	1.62	1.76	0.01	0.62	0.26

The whole hot test had been operated for 6 h. Very good results were obtained from the hot test. 99.85% of Am, 93.4% of Np, 99.86% of Pu, 99.99% of U and 99.2% of ^{99}Tc were removed from the genuine HLW by the TRPO process, namely 99.83% of the total α nuclide was removed. 99.96% of ^{90}Sr was removed from the genuine HLW by the CESE process. Moreover the cross contamination between them was small. The D.F. of TRU was 588 (see Table 10) and that for ^{99}Tc, ^{90}Sr and ^{137}Cs was 125, >2500 and >200 respectively. It means that after partitioning of HLW became non-α intermediate level liquid waste and it could be solidified by cementation and disposed in shallow land with engineering barrier. If the Sr, Cs as well as Am (including RE) waste are solidified by vitrification and the content of waste oxide is 20%, the volume reduction factor of HAW waste will be 39.4 for α waste and 13.5 for the sum of the α waste and HAW.

Table 10. The DFs obtained in the hot test of the total partitioning process [11]

Nuclide	Total α	^{241}Am	^{239}Pu	^{237}Np	U	^{99}Tc	^{90}Sr	^{137}Cs
Material balance,%	93	93	94	124	131	104	119	-
DF	588	666	695	15.2	10000	125	>2500	>200

4.4. The Cold Test of the Total Partitioning Process for Pilot-Scale

The cold test of the total partitioning process for pilot-scale, including the removal of actinides by the TRPO process, the removal of Sr by the CESE process, and the removal of Cs by ion exchange using KTiFC was carried out in 2005. [13,54] In the pilot test of the total partitioning process, 30% TRPO-kerosene was used to extract Nd, Zr and U from the simulated HLW (see Table 11) in the first pulsed column, then the loaded 30% TRPO-kerosene after scrubbing in the second pulsed column was successively stripped with 5.5 mol/L HNO_3 and 0.1 mol/L HNO_3 to recover Nd and to scrub the organic phase (1B), with 0.6 mol/L $H_2C_2O_4$ to recover Zr (1C), with 5.5 mol/L HNO_3 and 0.01 mol/L HNO_3 to scrub $H_2C_2O_4$ loaded in the organic phase (1CA1 and 1CA2), with 5% $(NH_4)_2CO_3$ to recover U (1D), and with 0.1 mol/L HNO_3 to scrub the 30% TRPO-kerosene (1E) in the 24 stages ϕ70 mm centrifugal contactors. The stripping section of TRPO process for actinides removal is shown in Figure 18. The required flow rate of every effluent is shown in Table 12. Based on their similar extraction behaviors in the TRPO process, Nd and Zr were used to simulate Am and Pu, respectively. In the meantime, Sr was extracted with DCH18C6/ n-octanol from the simulated HLW after extraction with the 30% TRPO-kerosene in the third pulsed column, and the loaded crown ether / n-octanol after scrubbing in the fourth pulsed column entered the stripping section (the fifth pulsed column), lastly, the simulated HLW after actinides and strontium removal entered the ion-exchange column loaded with KTiFC.

Table 11. The component of the simulated HLW

Element	Concentration, mg/L	Element	Concentration, mg/L
Na	1.83×10^4	Sr	222
Al	5.68×10^3	Cs	715
Fe	5.98×10^3	Zr	78
Cr	715	Ba	27
Mo	309	U	1.63×10^3
K	160	HNO_3	1.0 mol/L
Ni	2.93×10^3	SO_4^{2-}	0.034
Nd	1.50×10^3		

A series of hydraulic tests of 24-stage ϕ70 mm centrifugal contactors under the required flow rate of every effluent were firstly carried out. After 24-stage ϕ70 mm centrifugal contactors were operated for 6 h, the temperature rise of every stage was different, and all was lower than 55°C. The operation of the whole cascade was stable, moreover, both vibration and noise of every stage were very low. Both the rotor speed (ω) and the diameter of heavy phase weir (D) in every stage centrifugal contactor determined are shown in Table 13. Under these operation factors and the required flow rate (see Table 12), all of the end stream entrainment in either phase was below 0.5%.

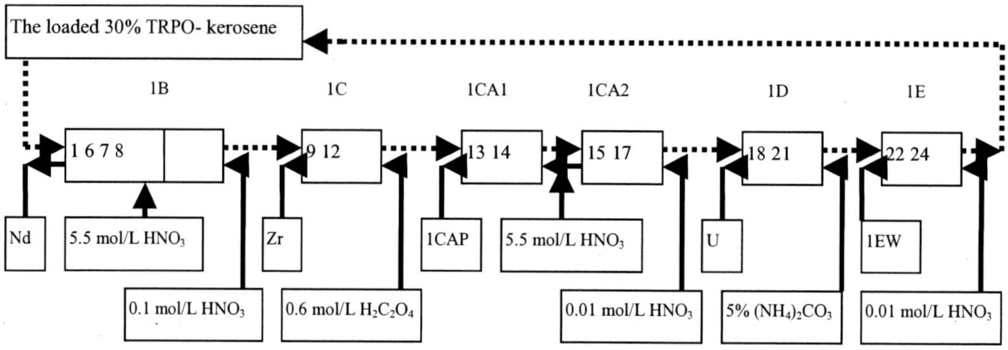

Figure 18. The stripping section of TRPO process.

Table 12. The required flow rate of every effluent

Effluent	***30% TRPO-kerosene***	***5.5 mol/L HNO_3***	***0.1mol/L HNO_3***	***0.6 mol/L $H_2C_2O_4$***	***5.5 mol/L HNO_3***	***0.01mol/L HNO_3***	***5% $(NH_4)_2CO_3$***	***0.01 mol/L HNO_3***
Flow rate (L/h)	55	44	11	55	20	35	55	55

Table 13. The operation factors of centrifugal contactors

Stage	**ω (r/min)**	**D (mm)**
1-17	2700-3000	38-39
18	2400	38
19-21	2700-3000	40
22-24	2700-3000	38-39

A 72 h pilot test of the total partitioning with the simulated HLW was carried out to demonstrate the feasibility and reliability of the process and the used separation equipments. Moreover, after the 66th hour of the pilot test, the simulated HLW containing U was used to replace the simulated HLW without U. In the pilot test of the total partitioning process, TRPO extracted Nd, Zr and U from the simulated HLW in the first pulsed column, and the loaded TRPO was scrubbed in the second pulsed column and directly flew into the stripping section which consisted of 24-stage ϕ70 mm centrifugal contactors. Figure 19 shows the evolution of the metal (Nd, Zr and U) concentration in the exit aqueous phase. The stripping in centrifugal contactors reached steady state within 3 hours because it needed 3 h to reach steady state of the extraction in the first pulsed extraction column although it only needed 30 min to reach steady state of mass-transfer in centrifugal contactors. It was similar to the mass-transfer test, namely, 99.9% of Nd and Zr was stripped by 5.5 mol/L HNO_3 and 0.6 mol/L $H_2C_2O_4$ respectively, and 99% of Fe and Mo was stripped in the whole stripping section respectively. Moreover, 99.9% of U was stripped by 5% $(NH_4)_2CO_3$.

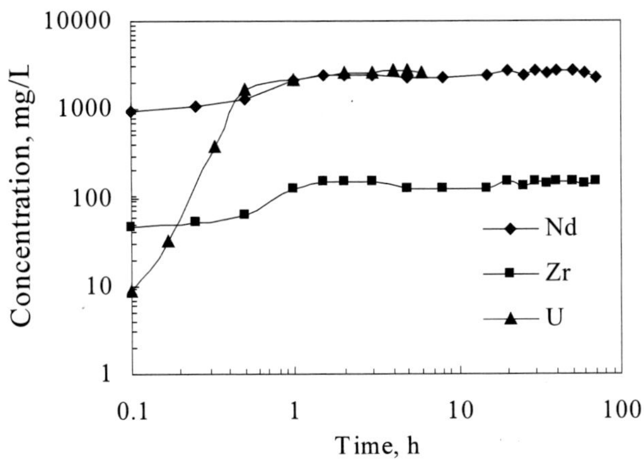

Figure 19. The evolution of Nd, Zr and U concentration in the exit aqueous phase of stage 1, 9 and 18 respectively.

During the 72 h pilot test of the total partitioning process, 24-stage ϕ70 mm centrifugal contactors worked stable, and no stage broke down thereby stopped operation, which demonstrated that these centrifugal contactors were feasible and reliable to be used in the stripping section of the TRPO process.

The decontamination factor was used to describe the removal efficiency of Nd, Zr, U, Sr and Cs. From Table 14, it is found that all decontamination factors in this pilot test are more than the required ones which meet the removal requirement to change the HLW into a non-α and intermediate level waste.

Table 6. The test results [13]

Element	DF_{ex}	DF_{re}*
U	>3000	
Nd	>500	100
Zr	>1000	100
Sr	~160	50
Cs	~94	50

* The requirement is to change HLW into a non-α and intermediate level waste.

4.5. The Cold Test for Removal of Cs by iPr-C[4]C-6

Although KTiFC ion exchanger can meet requirement of removal of Cs from HLW, operation of the ion exchange equipment is complicated, moreover, both drying and curing of the used KTiFC are uneasy to control. It is necessary to develop new method for removal of Cs from HLW.

Recently, iPr-C[4]C-6 (see Figure 20)/n-octanol was used to extract Cs, and has a good extraction ability and selectivity to Cs. [55-58] The complex of iPr-C[4]C-6 with Cs is iPr-C[4]C-6:Cs, and final species is [iPr-C[4]C-6:Cs]NO_3 according to the electric neutral rule.

D_{Cs} is increased with increase of HNO_3 concentration when the HNO_3 concentration is between 0.001 to 3 mol/L, but is decreased with increase of the HNO_3 concentration when the HNO_3 concentration is more than 3 mol/L. 0.025 mol/L iPr-C[4]C-6/n-octanol has a little extraction abilities to K, Mo. D_M(M=K, Mo) was lower than 0.2, while under the same conditions, D_{Cs} is about 9.78, and much larger than those of other elements. The cold test for removing Cs from the simulated HLW was carried out with 15 stages ϕ10 mm centrifugal contactors at INET. The main composition of the simulated HLW is listed in Table 15. The process for removing Cs is shown in Figure 21. The result of the cold test (see Figure 22) showed 99.5% Cs in the simulated HLW was removed, and 99.2% Cs loading in iPr-C[4]C-6/ n-octanol was stripped, therefor the process with iPr-C[4]C-6/ n-octanol was an effective process for removing Cs from HLW,

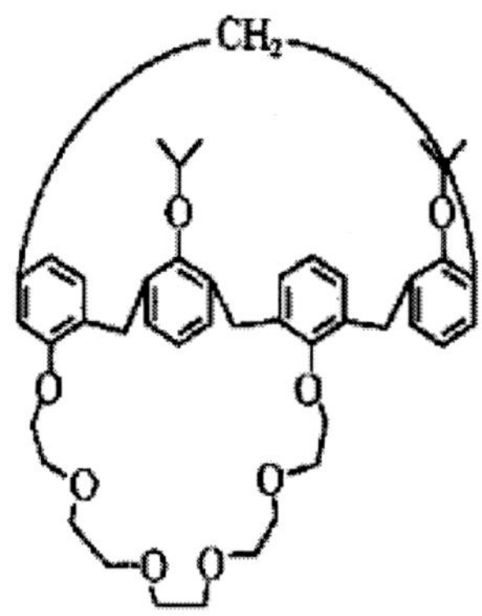

Figure 20. iPr-C[4]C-6.

Table 15. The main composition of the simulated HLW [58]

Element	Cs	Na	K	Sr	Al	Fe	Cr	Ni	Mo	Nd
Concentration, g/L	0.93	16.5	0.15	0.24	5.21	6.45	0.62	3.43	0.32	1.74

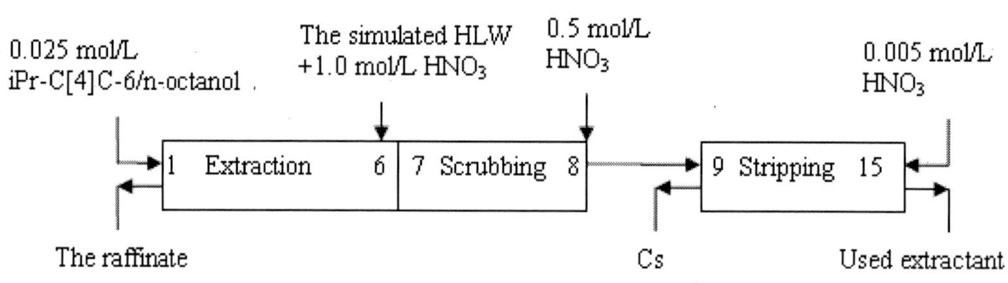

Figure 21. The flowsheet of the cold test for removing Cs from the simulated HLW.

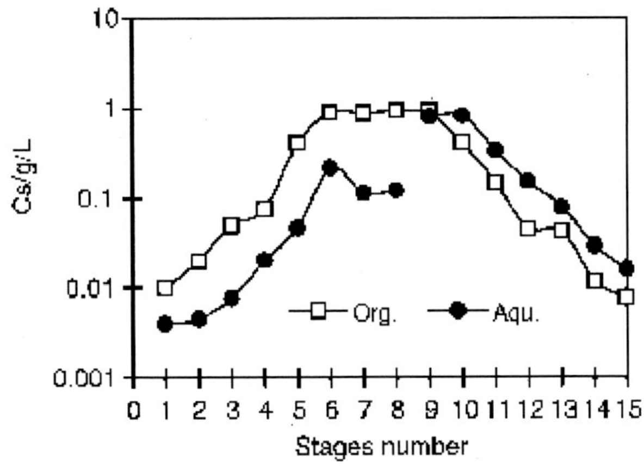

Figure 22. The distribution of Cs in aqueous and organic phase at each stage.

4.6. The Hot Test for Separation of Lanthanides and Actinides by the Purified Cyanex 301

The fission product lanthanides together with trivalent actinides are recovered in various partitioning processes of HLW such as TRPO process. [4-8] However, it is necessary to separate efficiently trivalent lanthanides from actinides due to some lanthanides' high thermal neutron absorption cross sections and the high lanthanides/actinides (Ln/An) mass ratio which is in the range of 20 hence interface with reactor (transmutation) efficiency. [59,60]

Separation of An(III) and Ln(III) is a much difficult task due to similarity of their chemistries. Different complexants/extractants containing soft-donor (N, S, Cl) have been used to accomplish this separation for laboratory-scale, and some extraction processes for this separation have been developed. [61-66]

In 1990s the extraction behavior of Am and fission product Lns were studied with various extractants. [67-70] Cyanex 301 had been proven to be an effective extractant for the separation of trivalent Am from Lns and is promising for practical use. [71] The main composition of Cyanex 301 is bis(2,4,4 trimethylpentyl) dithiophospninic acid (HBTMPDTP). The content of R_2PS_2H is 75-83%, R_3PS 5-8%, R_2PSOH 3-6% and unknown compound ~2%. Some impurities in commercial Cyanex 301 can extract both Am(III) and Lns(III) with strong bonding at low pH. It would effect on the separation of Am and Lns in tracer amount of Lns. So Cyanex 301 must be extensively purified before being used. A purified HBTMPDTP (>99%) could be obtained by purification of Cyanex 301.[3,71]

The purified Cyanex 301 used for extraction of Am(III) and Eu(III) from nitrate medium gives an SF (separate factor) of 5.9×10^3 in favour of Am. The SF for other lanthanides was found to be Am/La ~3500, Am/Ce, Pr ~1000, Am/Nd ~1900, Am/Sm ~4500, and the average Am/Ln SF was greater than 2300. [3,71]

The processes for separation of Ln/An with the purified Cyanex 301 has been developed and verified by the hot test in hot cell with 10 stages ϕ10 mm centrifugal contactors at INET.[72-74] The feed solution was the Am/Lns fraction obtained from the hot test of the total partition process for the Chinese HLW. It contains ~4.7 mol/L HNO_3, FPLns, micro amount of Cm and Tc, small amount of Am, Fe and Mo, Ru, Pd and so on. After denitration and removing impurity, the feed solution was adjusted to pH 3.5. The flowsheet of the hot test is shown in Figure 23.

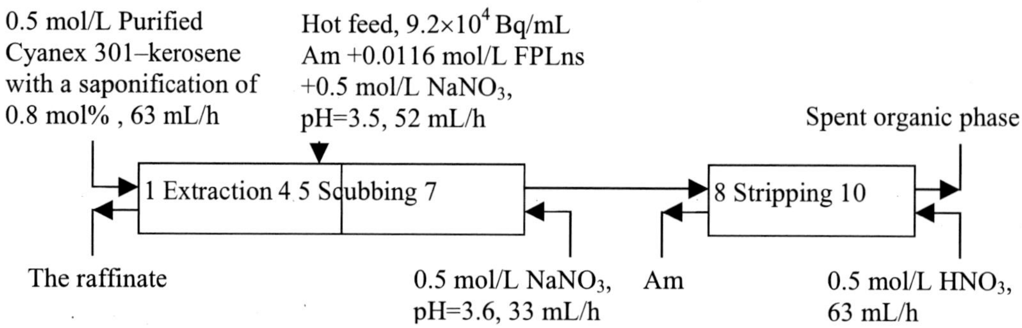

Figure 23. The flowsheet for Am/Lns separation by the purified Cyanex 301.

More than 99.93% of Am was extracted into the organic phase. The Am concentration in the raffinate was less than 60 Bq/mL. Only ~1.01% Lns was extracted by the purified Cyanex 301. The separation of Am from FPLns was more than 1000 and the separation of FPLns from am is about 100. In the hot test, the material balance was found to be 93.8% for Am and 99.9% for Eu. The hot test results proved that the separation process was effective.

4.7. The Cold Test for Partitioning of HLW by DHDECMP-TBP

Partitioning of HLW by DHDECMP-TBP/kerosene has been studying at China Institute of Atomic Energy since 1990s. [75-78] DHDECMP is dihexyl N,N-diethylcarbamoylmethyl-phosphonate. When 22% DHDECMP-42% TBP/kerosene contacts with 3 mol/L HNO_3 solution or the simulated HLW (see Table 16), a second organic phase does not form above 15 °C. The distribution ratios (see Table 17) of U(VI), Np(IV), Np(VI), and Pu(IV) are higher than those of Am(III) and Np(V).

Table 16. The main composition of the four-fold concentrated simulated HLW [78]

Element	Concentration, g/L	Element	Concentration, g/L	Element	Concentration, g/L
Na	9.0	Nd	2.34	Mo	2.01
Cr	1.05	Eu	0.09	Ag	1×10^{-3}
Fe	1.12	In	9.8×10^{-3}	I	0.17
Ni	1.13	Cs	0.60	PO_4^{3-}	0.1 mol/L
Gd	5.31	Ba	1.20	NO_3^-	3.5 mol/L
Sr	0.45	Sb	6.1×10^{-3}	H^+	3.0 mol/L
Sn	0.01	Ru	0.7	U	~0.7
Zr	1.84	Y	0.26	Np(IV,V,VI)	Trace
Pd	0.73	Pr	0.37	Pu(IV)	Trace
Rb	1×10^{-3}	Sm	0.52	Am(III)	Trace
La	0.64	Ce	1.0		

Table 17. Distribution ratios of U, Np, Pu, Am in the simulated HLW (25 °C) [78]

Element	U(VI)	Np(IV)	Np(V)	Np(VI)	Pu(IV)	Am(III)
HLW	53	53	6	24	225	3.2
Concentrated HLW	51	123	8	58	312	3.3

O/A=1, Organic phase: 22% DHDECMP-42% TBP/kerosene.

The cold test for removal of actinides by 22% DHDECMP-42% TBP/kerosene has been carried out using 14 stages ϕ10 mm centrifugal contactors. The flowsheet of the test is shown in Figure 24. The results showed that the removal efficiency of actinides including U(VI), Np(V), Pu(IV) and Am(III) and RE (except for Y) was greater than 99.9%, while Rb, Cs, Sr and Ba were nearly not extracted, and the D.F for Zr, Mo, Ru, Cs, Sr and Ba was 3.2, 18, $>10^3$, $>2\times10^2$, $>4\times10^2$ and $>4\times10^2$,respectively. The stripping efficiency of U(VI), Np(V), Pu(IV) and Am(III) from the loaded organic phase by 0.05 mol/L HNO_3 was 2%, 39%, 2%, and 99.9%, respectively. Moreover, using DHDECMP to remove actinides from HLW, both prior denitration and neutralization (dilution) are unrequired.

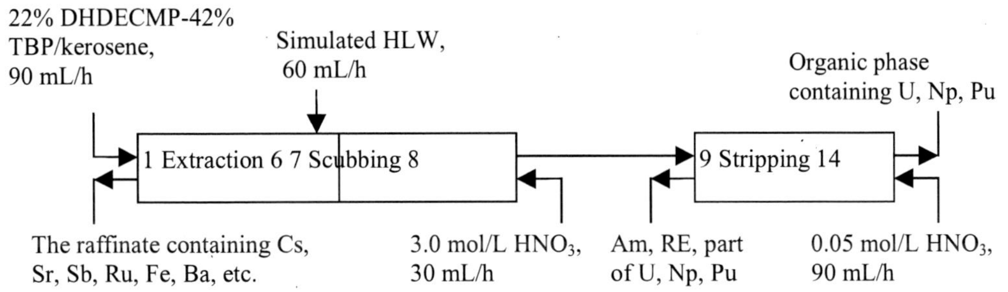

(From Ref. 78, courtesy of Ye, Y. X..).

Figure 24. Flowsheet of removal of actinides from simulated HLW by DHDECMP-TBP/kerosene.

5. Conclusion

Centrifugal contactors are efficient liquid-liquid extraction equipments. Compared with conventional contactors such as mixer-settlers and pulsed columns, centrifugal contactors offer some advantages, therefore, are favored for reprocessing of the spent nuclear fuel and partitioning of HLW. In order to meet requirements of tests for partitioning of HLW in China, several types of centrifugal contactors for both laboratory-scale and industry-scale have been developed at INET. A series of cold and hot tests of both the separate processes and the total partitioning process for partitioning of HLW were carried out with centrifugal contactors. Moreover, satisfactory results were obtained from these tests.

References

[1] Mathur, J. N., Murali, M. S., Nash, K. L. *Solvent Extr. Ion Exch.*, 2001, 19(3), 357-390.
[2] Paiva, A. P., Malik, P. *J. Radioanal. Nucl. Chem.*, 2004, 261(2), 485-496.

[3] Song, C. L. *Proceedings of the International Symposium "Energy Future in the Asia/Pacific Region'2000",* Beijing, China, Mar. 27-29, 2000, 89-99.

[4] Modolo, G., Vijgen, H., Serrano-Purroy, D., Christiansen, B., Maimbeck, R., Sorel, C., Baron, P. *Sep. Sci. Technol.,* 2007, 42(3), 439-452.

[5] Kubota, M., Morita, Y. *Global'97,* Pacifico Yokohama, Yokohama, Oct. 5-10, 1997, 458-462.

[6] Cecille, L., Casarci, M., Pietrelli, L. *New Separation Chemistry Techniques for Radioactive Waste and Other Specific Application.* Elsevier: London, UK, 1991, 21-29.

[7] Herbst, R. S., Law, J. D., Todd, T. A., Romanovskiy, V. N., Smirnov, I. V., Babain, V. A., Esimantovskiy, V. N., Zaitsev, B. N. *Sep. Sci. Technol.,* 2003, 38 (12&13), 2685-2708.

[8] Zhu, Y. J., Jiao, R. Z. *Nucl. Technol.,* 1994, 108(3), 361-369.

[9] Glatz, J. P., Song, C. L., Koch, L., Bokelund, H., He, X. M. *Global'95,* Palais Des Congres Versailles, France, Sept. 11-14, 1995, 548-555.

[10] Song, C. L., Glatz, J. P., He, X. M., Bokelund, H., Koch, L. *RECOD'94,* London, UK, Apr. 24-28, 1994, VII.

[11] Song, C. L., Wang, J. C., Lian, J. F. *Global'97,* Pacifico Yokohama, Yokohama, Oct. 5-10, 1997, 475-480.

[12] Song, C. L., Wang, J. C., Jiao, R. Z. *Global'99,* Jackson Hole, Wyoming, U.S.A, Aug.29-Sept.3, 1999, 235-240.

[13] Chen, J., Wang, J. C., Jing, S. *Global'07,* Boise, Idaho, U.S.A, Sept. 9-13, 2007, 1831-1835.

[14] Leonard, R. A. *Sep. Sci. Technol.,* 1988, 23(12&13), 1473-1487.

[15] Taylor, R. J., May, I., Denniss, I. S. *RECOD'98,* Nice Acroplic, France, Oct. 25-28, 1998, 417-424.

[16] Bernstein, G. J., Grosvenor, D. E., Lenc, J. F., Levitz, N. M. Argonne National Laboratory: Argonne, IL, *ANL-7968,* 1973.

[17] Bernstein, G. J., Grosvenor, D. E., Lenc, J. F., Levitz, N. M. Argonne National Laboratory: Argonne, IL, *ANL-7969,* 1973.

[18] Leonard, R. A., Bernstein, G. J., Ziegler, A. A., Pelto, R. H. *Sep. Sci. Technol.,* 1980, 15(4), 925-943.

[19] Leonard, R. A., Chamberlain, D. B., Conner, C. *Sep. Sci. Technol.,* 1997, 32 (1-4), 193-210.

[20] Herbst, R. S., Law, J. D., Todd, T. A., Romanovskiy, V. N., Smirnov, I. V., Babain, V. A., Esimantovskiy, V. N., Zaitsev, B. N. *Sep. Sci. Technol.,* 2003, 38 (12&13), 2685-2708.

[21] Walker, D. D., Norato, M. A., Campbell, S. G., Crowder, M. L., Fink, S. D., Fondeur, F. F., Geeting, M. W., Kessinger, G. F., Pierce, R. A. *Sep. Sci. Technol.,* 2005, 40(1), 297-309.

[22] Kuznetsov, G. I., Pushkov, A. A., Kosogorov, A.V. *ISEC'2002,* Cape Town, South African, March 17-21, 2002, 1174-1179.

[23] Rivalier, P., Lanoe, J. Y. *ATALANTE 2000*, Palais, France, October 24-26, 2000, P2-28.

[24] Taylor, R. J., May, I. *Sep. Sci. Technol.,* 2001, 36(5&6), 1225-1240.

[25] Koganti, S. B., Periasamy, K., Rajagopalan, C. V., Sreedharan, V., Balasubramanian. *RECOD'94,* London, UK, April 24-28, 1994, V1, 3B-4.

[26] Natarajan, R., Raj, B. *J Nucl. Sci. Technol.,*2007, 44(3), 393-397.

[27] Washiya, T., Takeuchii, M., Ogino, H., Aose, S. *GLOBAL'05,* Tsukuba, Japan, Oct 9-13, 2005, Paper No. 305.
[28] Nakahara, M., Sano, Y., Koma, Y., Kamiya, M., Shibata, A., Koizumi, T., Koyama, T. *J Nucl. Sci. Technol.,* 2007, 44(3), 373-381.
[29] Zhou, J. Z. *Chinese Chem. Eng.,* 1984, 12(6), 25-29. (in Chinese)
[30] Duan, W. H., Wang, J. C., Chen, J., Zhou, J. Z., Zhou, X. Z. *J. Radioanal. Nucl. Chem.,* 2007, 273(1), 103-107.
[31] Duan, W. H., Zhang, C. X., Zhou, X. Z., Chen, J., Zhou, J. Z. *Chinese J. Nucl. Sci. Eng.,* 2006, 26(2), 97-102. (in Chinese)
[32] Duan, W. H., Song, C. L., Wu, Q. L., Zhou, X. Z., Zhou, J. Z. *Sep. Sci. Technol.,* 2005, 40(9), 1871-1883.
[33] Duan, W. H., Yu, W. D., Zhou, X. Z., Li, F. Q., Zhou, J. Z. *Atom. Energy Sci.Technol.,* 2001, 35(1), 51-55. (in Chinese)
[34] Zhou, J. Z. *Chinese Chem. Eng.,* 1982, 10(5), 50-56. (in Chinese)
[35] Lu, Q., Li, K. H., Zhang, M., Shi, Y. J. *Sep. Sci. Technol.,* 1998, 33(15), 2397-2409.
[36] Leonard, R. A. *Sep. Sci. Technol.,* 1995, 30(7-9), 1103-1122.
[37] Thornton, J. D. *Science and Practice of Liquid-liquid Extraction.* Clarendon Press: Oxford, UK, 1992, pp. 362-473.
[38] Li, Z., Li, Y. G., Fei, W. Y. *The Processes and Facilities of Liquid-liquid Extraction.* Nuclear Energy Press: Beijing, China, 1985, pp.85-537. (in Chinese)
[39] He, X. M., Yan, Y. S., Zhang, Q. R., Liu, B. R. *Nucl. Sci. Technol.,* 1998, 9(3), 157-162.
[40] He, X. M., Liu, B. R. *Chinese J. Nucl. Sci. Eng* 1997, 17(4), 341-345. (In Chinese)
[41] Jubin, R. T. US Patent 4,925,441, May 15, 1990.
[42] Song, C. L., Jin, G. Y., Wang, J. C., Zhu, Y. J. *Tsinghua Sci. Technol.,* 1996, 1(1), 81-85.
[43] Morss, L. R., Fuger, J. *Transuranium Elements: A Half Century,* American Chemical Society: Washington D.C. U.S.A., 1992, 318-330.
[44] Wang, J. C., Tian, G. X., Chen, J., Song, C. L. *Sep. Sci. Technol.,* 2001, 36(13), 2983-2996.
[45] Wang, J. C., Liu, B. R., Song, C. L., Tang, W. C. *J. Nucl. Radiochem.,* 1996, 18(2), 94-99.
[46] He, L. H., Weng, X. M., Yang, D. Z., Jiao, R. Z., Song, C. L. *Chinese J. Nucl. Sci. Eng.,* 1994, 14(4), 350-356. (in Chinese)
[47] He, L. H., Weng, X. M., Yang, D. Z., Jiao, R. Z., Song, C. L. *Chinese J. Nucl. Sci. Eng.,* 1995, 15(3), 259-264. (in Chinese)
[48] He, L. H., Weng, X. M., Yang, D. Z., Song, C. L. *J. Nucl. Radiochem.,* 1994, 16(1), 18-22. (in Chinese)
[49] Wang, Q. P., Wang, J. C., Song, C. L., Tang, W. C. *J. Nucl. Radiochem.,*1996, 18(2):89-93. (in Chinese)
[50] Feng, X. G., Jing, S., Wu, Q. L., Chen, J., Song, C. L. *Chinese J. Chem. Eng.,* 2007, 15(2), 184-189.
[51] Liu, B. R., Jiao, R. Z., He, X. M., Han, S. Y., Meng, Z. K., Song, C. L. *Chinese J. Nucl. Sci. Eng.,* 1997, 17(4), 372-376. (in Chinese)
[52] Wang, J. C., Song, C. L. *Radiochim. Acta,* 2001, 89(3), 151-154.
[53] Wang, J. C., Song, C. L. *Solvent Extrt. Ion Exch.,* 2001, 19(2), 231-242.
[54] Xu, C., Feng, X. G., Jing, S. *Chinese J. Chem. Eng.,* 2007, 15(5), 654-660.

[55] Zhu, X. W., Wang, J. C., Tong, L. B., Song, C. L. *Atom. Energy Sci.Technol.,* 2003, 37(5), 428-433. (in Chinese)
[56] Zhu, X. W., Wang, J. C., Song, C. L. *Atom. Energy Sci.Technol.,* 2004, 38(1), 21-24. (in Chinese)
[57] Zhu, X. W., Wang, J. C., Song, C. L. *Atom. Energy Sci.Technol.,* 2004, 38(4), 301-305. (in Chinese)
[58] Wang, J. C., Zhu, X. W., Song, C. L. *Sep. Sci. Technol.,* 2005, 40(16), 3381-3392.
[59] Mathur, J. N., Murali, M. S., Nash, K. L. *Solvent. Extr. Ion Exch.*, 2001, 19(3), 357-390.
[60] Zhu, Y. J., Chen, J. *Tsinghua Sci. Technol.*, 2001, 6(2), 151-155.
[61] Vandegrift, G. F., Regalbuto, M. C., Aase, S., Bakel, A., Battisti,T. J., Bowers, D., Byrnes, J. P., Clark, M. A., Emery, J. W., Falkenberg, J. R., Gelis, A. V., Pereira, C., Hafenrichter, L., Tsai, Y., Quigley, K. J., Vander Pol, M. H. *ATALANTE 2004,* Nices, France, Jun 21-24, 2004, 012-01.
[62] Chen, J., Zhu, Y. J., Jiao, R. Z. *Sep. Sci. Technol.,* 1996, 31(19), 2723-2731.
[63] Nakahara, M., Sano, Y., Koma, Y., Kamiya, M., Shibata, A., Koizumi, T., Koyama, T. *J. Nucl. Sci. Technol.,* 2007, 44(3), 373-381.
[64] Facchini, A., Amato, L., Modolo, G., Nannicini, R., Madic, C., Baron, P. *Sep. Sci. Technol.,* 2000, 35(7), 1055-1068.
[65] Bilancia, G., Facchini, A., Ferrando, M., Giola, M., Maluta, F., Mariani, M., Mazzuccato, M., Madic, C., Sabbioneda, S. *Solvent. Extr. Ion Exch.,* 2005, 23(6), 773-780.
[66] Modolo, G., Vijgen, H., Serrano-purroy, D., Christiansen, B., Malmbeck, R., Sorel, C., Baron, P. *Sep. Sci. Technol.,* 2007, 42(3), 439-452.
[67] Zhu, Y. J., Song, C. L., Jiao, R. Z. *Global'95,* Palais Des Congres Versailles, France, Sept. 11-14, 1995, 571-577.
[68] Zhu, Y. J. *Solvent. Extr. Ion Exch.*, 1996, 14(4), 543-553.
[69] Zhu, Y. J., Xu, J. M., Chen, J., Chen, Y. Z. *J. Alloy. Compd.,* 1998, 271-273, 742-745.
[70] Zhu, Y. J., Jiao, R. Z. *Radiochim. Acta,* 1995, 69(4), 191-193.
[71] Chen, J., Zhu, Y. J., Jiao, R. Z. *Sep. Sci. Technol.,* 1996, 31(9), 2723-2731.
[72] Chen, J., Jiao, R. Z., Zhu, Y. J. *Radiochim. Acta,* 1997, 76(3), 129-130.
[73] Zhu, Y. J., Chen, J., Jiao, R. Z. *Global'97,* Pacifico Yokohama, Yokohama, Oct. 5-10, 1997, 581-584.
[74] Chen, J., Tian, G. X., Jiao, R. Z., Zhu, Y. J. *J. Nucl. Sci. Technol.,* 2002, S3, 325-327.
[75] Ye, Y. X., Wu, G. M., Zhu, W. B. *Atom. Energy Sci.Technol.,*2001, 35(S), 70-75. (in Chinese)
[76] Ye, Y. X., Wu, G. M., Zhu, W. B. *Atom. Energy Sci.Technol.,*2003, 37(3), 242-245. (in Chinese)
[77] Ye, Y. X., Tang, H. B., Shi, W. Q., Wang, X. R., Sheng, D., Dong, Y., Luo, F. X., Zhao, Y. Z. *J. Nucl. Radiochem.,* 2003, 25(2), 96-101. (in Chinese)
[78] Ye, Y. X., Hu, J. X., Huang, Z. L., Shi, W. Q., Wu, G. M., Tang, H. B., Zhu, W. B., Wang, X. R., Zhao, H. G., Dong, Y., Sheng, D., Luo, F. X., Zhao, Y. Z. *J. Radioanal. Nucl. Chem.,* 2004, 261(2), 365-374.

In: Nuclear Chemistry: New Research
Editor: Axel N. Koskinen, pp. 241-251

ISBN: 978-1-60456-957-5

Chapter 8

EFFECTIVENESS ASSESSMENT OF CORROSION INHIBITORS OF BRONZE SURFACES, AN APPLICATION OF THE THIN LAYER ACTIVATION TECHNIQUE TO MATERIALS OF INTEREST IN THE FIELD OF CULTURAL HERITAGE

G. Laguzzi**[*]*****, L. Luvidi and E. Gregori***
Istituto di Metodologie Chimiche del Consiglio Nazionale delle Ricerche (CNR),
Area della Ricerca di Roma 1,
Via Salaria km. 29,300 – 00016 Monterotondo Stazione (Roma), Italy.

Abstract

In the frame of a European project (Eureka Σ!2210 "Eurocare-Bronzart"), the effectiveness of several corrosion inhibitors for bronze surfaces: benzotriazole derivatives and 3-mercapto-propyl-trimethoxy-silane, was comparatively evaluated by the Thin Layer Activation method. Bronze surfaces were irradiated with a 11.5 MeV proton beam, generated by a MC-40 energy variable cyclotron. The tracer radio nuclide ^{65}Zn ($t_{1/2}$ = 244 days) was produced by the following nuclear reaction: $^{65}Cu(p,n)^{65}Zn$. Activated surfaces of different bronze alloys, whose composition is suitable for production of outdoor sculptures, were coated with the inhibitors and exposed to natural and artificial corrosive environments. After determination of the activity versus depth functions, the thickness loss of the activated samples, due to the effects of corrosion, was calculated by measurements of the ^{65}Zn γ-activity at 1115 KeV (I_γ = 0.50), carried out with a multi channel analyser equipped with a Ge detector. In evaluating the sample thickness loss two different procedures were applied. The first one consisted in the quantification of the activity of ^{65}Zn on the solid sample, prior to exposure and after stripping. The second one consisted in a step by step electrolytic dissolution of the solid sample, after stripping the corrosion layer. The ^{65}Zn activity was successively measured in the pickling solution and in each solution arising from the electrolytic dissolution of the solid sample. The last method, because of its high sensitivity,

[*] E-mail address: giuseppe.laguzzi@imc.cnr.it (G. Laguzzi). Corresponding author: Tel. +39 06 90672515; fax +39 06 90672519

was applied to evaluate the thickness loss at sub micrometer scale, as in the case of a mild corrosion.

1. Introduction

Outdoor bronze statues are mostly located in polluted urban areas and consequently exposed to a hostile environment that can readily cause modifications of the original appearance and long term damages of the metallic surface.

In the tempt to prevent corrosive attack from atmospheric reagents and anthropogenic pollutants, the bronze surfaces are usually treated with protective coatings which include wax and organic corrosion inhibitors [1, 2]. For artworks applications the organic film, forming the coating layer, needs to be transparent, removable and possibly with a long lifetime in order to minimize the number of conservation treatments. One of the most efficient corrosion inhibitor is 1,2,3-benzotriazole (BTA) [3, 4, 5] which is included in Incralac, an acrylic polymer based coating, commonly utilised for outdoor sculptures. In the past few years the protective effectiveness of BTA has been investigated [6] respect to its alkyl derivatives, pointing out that (5-esyl) 1,2,3-benzotriazole exhibits a better capability to hinder corrosion on copper. Silane coatings have also been tested for inhibition of copper corrosion and 3-mercapto-propyl-trimethoxy-silane (MPTS), in particular, showed a very good capability [7, 8, 9].

In the frame of a European project (Eureka Σ!2210 "Eurocare-Bronzart"), the present work is addressed toward two main objectives. The first one consists in the evaluation of the effectiveness of MPTS, that could be proposed for future applications on artworks, respect to a mixture of BTA and (5-esyl)1,2,3-benzotriazole (BTA_{mix}). For these experiments bronze alloys No.1 and No.2 (Tab.1), suitable for production of outdoor sculptures, were coated with the inhibitors under investigation. The second objective refers to the comparative evaluation of the performances of BTA respect to several BTA alkyl derivatives as: (5-esyl)benzotriazole (C6-BTA), (5-dodecyl)1,2,3-benzotriazole (C12-BTA) and [5(1-undecyl)dodecyl]1,2,3benzotriazole-(C23-BTA). Considering that C6-BTA showed a better performance respect to BTA as corrosion inhibitor of bronze, in this test is investigated the influence of the chain length of BTA alkyl derivatives on corrosion inhibition of bronzes. A bronze alloy, whose composition is similar to materials produced for outdoor artefacts, was used in these experiments.

Before exposure to artificial and natural corrosive atmospheres, the surface of bronze coupons, made by the different bronze alloys and coated with the inhibitors, were investigated by the Atomic Force Microscopy (AFM) technique in order to determine the homogeneity of the covering and the thickness of the protective film. Artificial ageing experiments were carried out exposing coated specimens to corrosive atmospheres in a salt spray cabinet, in a climatic chamber, here in the presence of a controlled concentration of SO_2. Outdoor exposures were carried out in a polluted urban area (Rome).

In the present work the effectiveness assessment of corrosion inhibitors of bronze surfaces was evaluated by the Thin Layer Activation (TLA) technique [10, 11]. Even if normally applied to on line wear and corrosion monitoring of industrial components, the TLA method, due to its high sensitivity, has been successfully used for metal decay measurements in the field of cultural heritage [11]. Bronze surfaces were irradiated by a high energy proton beam generating, along a defined depth, the tracer radio-nuclide (^{65}Zn). The extent of corrosion on the coated bronze alloys was calculated by the activity loss of materials after removal of degradation products from the surface, by using a pickling solution. Activity loss and thickness loss were related by a calibration curve.

2. Experimental Procedures

2.1. Materials

Bronze alloys No.1 and No.2 were purchased by the Italian artistic foundry "Venturi Arte" (Cadriano di Granarolo-Bologna). Bronze No.3, a pure single phase alloy, was provided by Goodfellow. The materials composition is reported in Tab.1,

Table 1. Quantitative analysis of the composition of bronze alloys obtained by x-ray fluorescence.

Alloy No	%Cu	%Sn	%Zn	%Pb	%Si	%Ni	%Bi	%Fe
1	85.95	8	0.25	3	2.7	0.06	--	0.04
2	84.47	7.8	2.1	0.25	2.9	0.03	2.4	0.05
3	94.0	6.0	--	--	--	--	--	--

The elemental composition was determined by the x-ray fluorescence technique, using a Panalytical Magix PRO XRF Spectrometer.

Coupons (10 x 10 x 2 mm) of the bronze alloys were cut from plates purchased by the "Venturi Arte" foundry. The surfaces were polished up to 0.1μm diamond paste.

2.2. Bronze Surfaces Coating

Solutions respectively $5 \cdot 10^{-3}$M of BTA, $1 \cdot 10^{-3}$M *of* C6BTA in chloroform (BTA_{mix}) and $2 \cdot 10^{-3}$M of MPTS (90/6/4 v/v ethyl alcohol/water/silane) were prepared for the treatment of the alloys No.1 and No.2. Solutions $1 \cdot 10^{-3}$M in chloroform of BTA, C6BTA, C12BTA and C23BTA were used for treatment of the alloy No.3.

BTA and MPTS are commercially available (Aldrich) while the C6BTA was synthesized at Dept. of Chemistry of the University of Ferrara. C12-BTA and C23 BTA were synthesized at Institute of Chemical Methodologies of CNR.

Specimens were dipped into the above mentioned solutions for five minutes at room temperature. After drying, the procedure was repeated a second time.

2.3. The TLA Method

The tracer radio nuclide ^{65}Zn ($t_{1/2}$ = 244 days) was obtained irradiating 10 x 10 x 2 mm bronze samples with a 11.5 MeV proton beam, generated by a Scanditronix MC-40 energy variable cyclotron. The proton beam, having an intensity of 170 nA, was focused on target with an irradiation angle of 30 degrees. The irradiation time was 15 minutes. The relevant nuclear reaction is: ^{65}Cu (p,n)^{65}Zn. Prior to calibration and corrosion experiments the samples remained isolated for at least 2 weeks until the activity dose decreased, due to the decay of short lived radioisotopes. These were produced, by side nuclear reactions, from the different elements composing the bronze alloys (Tab.1).

In the irradiated samples the activity depth and the distribution of activity versus depth were both theoretically determined and experimentally evaluated for each bronze alloy. The activity-depth function was theoretically determined from the energy degradation in the samples that was calculated using the polynomial approximation of H.H. Andersen and J.F. Ziegler [9] and by the experimental cross section curve of the nuclear reaction as function of the energy.

The distribution of activity versus depth function was experimentally obtained by a serious of steps which involve an electrodissolution of activated specimens followed by mechanical polishing. At the end of each step the thickness and the remaining ^{65}Zn activity of the samples were measured. This combined procedure was used to minimize the roughness of the surface and consequently to allow reliable measurements of thickness using a mechanical micrometer.

Accordingly, the error of the activity measurement and depth determination remained less than 0.5% and 1 μm respectively.

The ^{65}Zn γ- activity was quantitatively measured from its emission peak at 1115 KeV (I_γ = 0.50) by a Canberra multi channel analyser equipped with a Ge detector.

The thickness loss of bronze samples, due to corrosion processes, is calculated by γ-activity measurements before and after the exposure of samples at corrosive environments. After exposure and before measurements the bronze surfaces were treated with a (1:1 HCl/H_2O) pickling solution to remove the layer of corrosion products. In order to relate activity loss with thickness loss, activity versus depth functions were used.

2.4. AFM Measurements

A PSIA XE-100 atomic force microscope was used to analyse the surface of 10x10 mm bronze specimens coated with BTA, alkyl derivatives of BTA and MPTS. Measurements were done in air, using a silicon nitride tip (PSIA 910M-NSC36). To image the surfaces a scan size of 50x50 μm was used. Several images were taken in different positions of the coated surfaces in order to control the homogeneity of the covering. The thickness of the inhibitors on the surfaces was then determined digging, by the tip, a series of 2x2 μm square areas into the organic films, until the metal was reached. An average value of the gap between the organic coating and the metallic surfaces was then measured.

2.5. Artificial Ageing Experiments

2.5.1. Salt Spray Cabinet Test

Activated bronze specimens, coated with the inhibitors under investigation and fixed on a plane with an angle of 75° from horizontal, were exposed in a salt spray cabinet operating with a 5% NaCl solution (pH = 6.5), in accordance with the ASTM B117-03 standard. Samples were withdrawn after periods of 4, 10, 17 and 31 days.

A salt spray cabinet, "Angelantoni- DCTC 600P", was used for the experiments.

2.5.2. Climatic Chamber Tests

Activated bronze specimens, coated with the inhibitors, were exposed in a climatic chamber operating by four steps cycles at the following experimental conditions:

1) 8 hrs. at 35°C, 65% R. H. in air flow containing 1 ppm of SO_2
2) 4 hrs in a controlled ramp to reach temperature and R.H. of step 3
3) 8 hrs. at 5°C, 85% R. H.
4) 4 hrs. in a controlled ramp to reach temperature and R.H. of step 1.

Samples were withdrawn after periods of 10, 20 and 30 days.

A climatic chamber, "Angelantoni-Challenge 340", was used for the experiments.

2.6. Outdoor Exposures

Activated coupons, once coated with the inhibitors, were exposed, facing south at an angle of 30° to the ground [12], on the roof of a building, located in the campus of the University of Rome "La Sapienza". The aggressiveness rate of the site referred to copper, according to the ISO 9224 Standard, is C4-C5. Samples were withdrawn after periods of 120, 330 and 540 days.

3. Results and Discussion

3.1. Coated Surface Characterization

Preliminarily to the exposure of specimens at corrosive atmospheres, an investigation of the characteristics of the organic films, coating the bronze surfaces, was carried out by Atomic Force Microscopy. In particular the thickness of the inhibitors on the metallic surfaces and the homogeneity of the covering was controlled. To this purpose, a series of solutions of BTA, alkyl derivatives of BTA and MPTS were prepared by different solvents for the treatment of the surfaces and for the successive AFM analysis.

Among different solvents, Chloroform, used for BTA, alkyl derivatives of BTA and the hydro-alcoholic solution used for MPTS, gave rise to a very homogeneous covering. The thickness of BTA and alkyl derivatives of BTA was ranging between 37-42 nm while MPTS was about 60 nm. These measurements were carried out with the method indicated in the experimental section. A lower difference of thickness between the layers of inhibitors on the surfaces was observed using other solvents. Water solutions of BTA and alkyl derivatives of BTA gave rise, for instance, to a layer of about 50 nm. In this case, however, the bronze surface shows large areas where the coating was absent. On the ground of these results, to ensure an homogeneous covering, chloroform was chosen as solvent for BTA and alkyl derivatives of BTA. To verify possible damages produced by the solvents on the bronze surfaces, experiments were carried out dipping bronze specimens respectively into chloroform and hydro-alcoholic solutions. Roughness measurements of the bronze surfaces, before and after the treatment, showed a little effect on the specimen dipped in chloroform while the sample dipped in ethyl alcohol/water solution did not show any detectable difference.

3.2. Distribution of the Activity versus Depth Functions in Bronze Alloys

In order to obtain thickness loss data by the TLA method, it is essential to determine the distribution of the activity along the activated layer of the irradiated materials. The function of relative activity of ^{65}Zn versus material depth (calibration curve) was determined for each alloy under investigation. The calibration curve of the bronze alloy no. 1 is reported in Fig. 1.

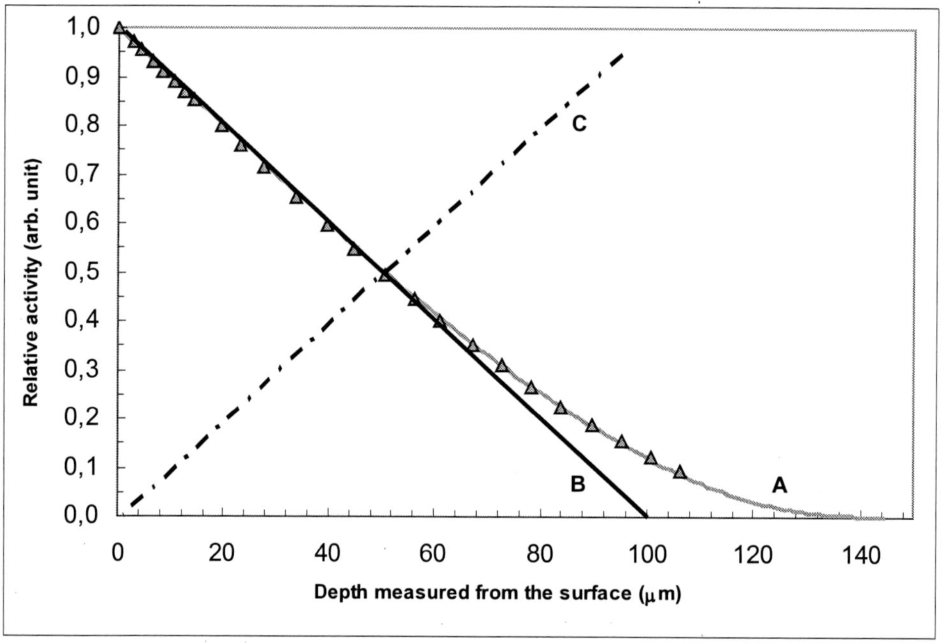

Figure 1. Calibration curve of the bronze alloy No.1.

The curve A represents the calculated function while triangles along the curve represent measurements obtained by the experimental procedure (see experimental section). The straight line B, arising from the linear fitting of measurements, intercepts the abscissas axis around 100 μm. For all the investigated alloys, the linear fitting of experimental data gave rise to a linear function indicating a homogeneous activated layer ranging between 100 and 120 μm. According to the experience of TLA measurements, the sensitivity of thickness loss determination, obtained by activity measurements of solid samples, normally corresponds to about 1/100 of the homogeneously activated layer. For this reason thickness loss lower than 1-1.2 μm cannot be correctly evaluated by activity measurements of solid samples. For determination of the thickness loss lower than these values, the method of the electrolytic dissolution of solid samples and the relevant activity measurements in solution was used.

In a previous paper [11] where both methods of measurements were employed two calibration curves were experimentally determined. For measurements of solid samples an experimental calibration curve whose linear fitting was similar to that showed in Fig.1 (B), was obtained. For measurements in solution, the calibration curve was experimentally carried out through a step by step electrolytic dissolution of the solid sample and a sequence of corresponding measurements of the activity of the electrolytic solutions. In that experiment, the linear fitting of data gave rise to a linear function having an opposite angular coefficient with respect to the calibration curve used for measurements of solid samples.

On the ground of these considerations, in the present work, the dashed line C (Fig. 1) used for activity measurements in solution was derived from the corresponding B.

3.3. Activated Specimens Exposures

3.3.1. Corrosion Tests of Bronze Specimens Coated with MPTS and BTA_{mix}

After treatment with inhibitors, i.e. MPTS and a mixture of 1,2,3-benzotriazole and (5-esyl) 1,2,3-benzotriazole (BTA_{mix}), activated specimens were exposed to natural and artificial corrosive atmospheres. The thickness loss of specimen, obtained by removal of corrosion products with a picking solution, was evaluated by the TLA method. On the ground of the characteristics of the bronze alloys, tested in a previous work [15], the alloys No.1 and No.2 were chosen for these experiments.

Table 2. Thickness loss of bronze alloys from artificial ageing test in a salt spray cabinet after 4, 10, 17 and 31 exposure days.

	Average thickness loss (μm)			
No. Alloy (inhibitors)	**I Step 4 d.**	**II Step 10 d.**	**III Step 17 d.**	**IV Step 31 d.**
1	1.1 ± 0.1	1.8 ± 0.2	2.1 ± 0.2	2.3 ± 0.3
1 (MPTS)	< 1.1	< 1.1	< 1.1	1.2 ± 0.2
1 (BTA mix)	< 1.1	< 1.1	< 1.1	< 1.1
2	2.1 ± 0.3	2.6 ± 0.8	3.6 ± 0.9	3.6 ± 0.7
2 (MPTS)	< 1.1	< 1.1	< 1.1	< 1.1
2 (BTA mix)	1.3 ± 0.2	1.4 ± 0.3	1.9 ± 0.5	1.9 ± 0.4

Artificial ageing experiments were carried out in a salt spray cabinet and in a climatic chamber here in the presence of a controlled concentration of SO2 (1ppm). Thickness loss results are respectively reported in Tab. 2 and in Tab. 3. Each data is an average value obtained from the measurements of five specimens.

Thickness loss results, related to the experiments carried out in the salt spray cabinet were calculated measuring the γ-activity of specimens before and after ageing: The sensitivity of these measurements, as a general rule of the TLA method, is 1/100 of the activated layer of specimens [16]. Because of the experimental conditions used in the activation stage the activated layer is about 110 μm, therefore, the sensitivity corresponds to 1.1 μm. Data in Tab.2 show that, in the case of the alloy No.1, it is not possible to observe any difference of thickness loss between specimens coated with BTA_{mix} and MPTS. In the case of the alloy No.2 we can notice that MPTS offers a better resistance toward corrosion, respect to the BTA_{mix}. In the experiments carried out in a climatic chamber in the presence of a controlled concentration of SO_2 (Tab. 3), MPTS confirms a better capability respect to BTA_{mix} to inhibit corrosion on the surfaces of the two alloys under investigation.

In this experiment the thickness loss calculations refer to measurements of activated specimens that were electrochemically dissolved. The γ-activity was measured in the rising solution and then related to the γ-activity of the corresponding picking solution. This method exalts the sensitivity of measurement since the small activity lost in the corrosion processes, which corresponds to the activity detected in the pickling solution, is measured directly and not by difference as in the previous case. We have to take into account that an activity whatever small can be correctly measured if only shielded from the background activity. The method is however laborious, since every activated specimen has to be electrochemically dissolved, therefore, is only applied in those cases where a very small thickness loss is expected. For this reason we applied this procedure, besides to experiments carried out in the climatic chamber also to the analyses of outdoor exposed samples.

Table 3. Thickness loss of bronze alloys from artificial ageing test in a climatic chamber in the presence of controlled concentration of SO_2 after 10, 20 and 30 exposure days.

	Average thickness loss (μm)		
No. Alloy **(inhibitors)**	I step **10 d.**	**II Step 20 d.**	**III Step 30 d.**
1	0.48 ± 0.08	0.56 ± 0.04	0.70 ± 0.09
1 (MPTS)	0.20 ± 0.03	0.24 ± 0.03	0.30 ± 0.03
1 (BTA mix)	0.26 ± 0.04	0.29 ± 0.03	0.41 ± 0.05
2	0.67 ± 0.09	0.79 ± 0.08	0.86 ± 0.09
2 (MPTS)	0.29 ± 0.04	0.31 ± 0.04	0.36 ± 0.04
2 (BTA mix)	0.33 ± 0.04	0.39 ± 0.04	0.43 ± 0.05

Once verified in artificial atmospheres the higher capability of MPTS to inhibit corrosion, respect to the mixture of benzotriazoles, its effectiveness was tested in a natural environment. To this purpose, samples of bronze alloys No.1 and No.2, part of them coated with the inhibitor, were exposed outdoor on the top of a building located in the campus of the University of Rome for a period of eighteen months. Data related to outdoor exposure in polluted urban areas are particularly significant to verify the performance of a bronze corrosion inhibitor, specially considering that most of bronze statues are located in these atmospheres. The results are reported in Tab. 4. Each data is an average value obtained from the measurements of five specimens.

Table 4. Thickness loss of bronze alloys exposed outdoor for 120, 330 and 540 days.

	Average thickness loss (μm)		
No. Alloy **(inhibitors)**	I step **120d.**	II Step **330d.**	III Step **540d.**
1	0.70 ± 0.11	1.21 ± 0.08	2.20 ± 0.28
1 (MPTS)	0.42 ± 0.06	0.71 ± 0.03	3.71 ± 0.49
2	0.75 ± 0.07	2.48 ± 0.03	3.30 ± 0.36
2 (MPTS)	0.57 ± 0.06	0.73 ± 0.18	3.54 ± 0.42

From the inspection of table 4 we can observe that MPTS exhibits a good capability to inhibit corrosion, at least for a period as long as 330 days. In the period ranging between 330 and 540 days, in the case of the alloy No.2, the values of thickness loss of the specimens coated with the inhibitor are very similar to those relative to uncoated specimens. In the case of the alloy No.1, the thickness loss of the surface coated with MPTS is even higher than the uncoated surfaces. A reasonable hypothesis we can advance is that the organic film, originally covering the metal, flaked off after 330 days, leaving a surface that could be, to some extent, deteriorated and therefore more vulnerable toward corrosion.

The chemical interaction between MPTS and the surface, that is the formation of a strong thiolate bond [17] between mercapto group of MPTS and copper could be responsible of the surface modification.

3.3.2. Corrosion Tests of Bronze Specimens Coated with BTA and BTA Alkyl Derivatives

A comparative evaluation of the performances of 1,2,3-benzotriazole respect to several BTA alkyl derivatives (C6-BTA, C11-BTA and C23-BTA) was carried out exposing activated specimens in a climatic chamber in the presence of a controlled concentration of SO_2. Thickness loss results were obtained by the same method used in the experiments relevant to the specimens coated with MPTS and BTA_{mix} and exposed to the same corrosive environment (§ 3.3.1). Results reported in Tab.5, that refer to an exposure period of 30 days, indicate that alkyl derivatives of BTA provide a better capability to hinder corrosion respect to BTA.

Table 5. Thickness loss of bronze alloy No.3 from artificial ageing test in a climatic chamber in the presence of controlled concentration of SO_2 after an exposure period of 30 days.

	Uncoated	BTA	C6-BTA	C12-BTA	C23-BTA
Average thickness loss (μm)	0.75± 0.11	0.58± 0.08	0.40± 0.06	0.31± 0.04	0.29± 0.04

In particular, C_6-BTA acts as a better corrosion inhibitor of bronze, respect to BTA. This confirms a previous investigation [6], carried out by electrochemical techniques, where the effectiveness of the same corrosion inhibitors was evaluated on copper surfaces. Moreover, from the inspection of Tab.5 it is possible to observe that the alkyl chain length of BTA derivatives has an influence on their properties as corrosion inhibitors of bronzes. Molecules with alkyl chain having 11 and 23 carbon atoms show a better effectiveness to hinder corrosion respect to C6-BTA. This is probably due to the increased hydrophobic capability of these BTA alkyl derivatives and to their relevant arrangement on bronze surface. However, this will be matter of a future investigation that will be carried out by using theoretical calculations.

Acknowledgements

The present work was carried out under the European Project Eureka Σ!2210 "Eurocare-Bronzart" and was financially supported by the Italian "Ministero dell'Istruzione, dell'Università e della Ricerca Scientifica."

The authors wish to thank Dr. Gabriella Venturi Pagani Cesa who was the co-ordinator of the project. Acknowledgement is also due to Prof. F. Zucchi for his interest in this work and for providing BTA_{mix} and finally to Dr. A. de Stefanis, as valued adviser in the use of the AFM technique.

References

[1] T. Drayman-Wasser, in *Proc, Dialogue/89 – The conservation of Bronze Sculptures in the Outdoor Environment*, National Association of Corrosion Engineers, Houston, Tx, 1992.

[2] L. B. Brostoff, E. René de la Rie in Proc. Metal 95, James & James (Science Publishers), London, 1997, pp. 242-244.

[3] I. Dugdale and J.B. Cotton, *Corrosion Science* **3**, 69 (1963).

[4] G.W. Poling, *Corrosion Science* **10**, 359 (1970).

[5] P.G. Fox, G. Lewis and P.J. Boden, *Corrosion Science* **19**, 457 (1979).

[6] L. Tommesani, G. Brunoro, A. Frignani, C. Ponticelli and M. Dal Colle, *Corrosion Science* **39**, 1221 (1997).

[7] F. Zucchi, V. Grassi, A. Frignani and G. Trabanelli *Corrosion Science*, **46**, 2853 (2004).

[8] F. Zucchi, A. Frignani, V. Grassi, G. Trabanelli and M. Dal Colle, *Corrosion Science,* **49**, 1570 (2007).

[9] F. Sinapi, J. Delhalle and Z. Mekhalif, *Materials Science and Engineering C,* **22**, 345 (2002).

[10] M.F. Stroosnijder, in : P. Misaelides (Ed.), *Application of Particle and Laser Beams in Materials Technology*, Kluwer press, Dordrecht, p.399, (1995).

[11] G. Laguzzi, L. Tommesani, L. Luvidi, R. Bucci and G. Brunoro, *Corrosion Science*, **41**, 197, (1999).

[12] G. Laguzzi, L. Luvidi and G. Brunoro, *Corrosion Science*, **43**, 747, (2001).

[13] M. Stern and A.L. Geary, *J.Electrochem. Soc.* **104**, 56 (1957).

[14] Mansfeld, *Advances in Corrosion Science and Technology*, **6** 163-262, (1976).

[15] F. Gallese, G. Laguzzi, L. Luvidi, V. Ferrari, S. Takacs and G. Pagani Cesa, *Corrosion Science*, **50**, 954 (2008).
[16] P.M. Racolta, *Applied Radiation Isotopes*, **46**, 663, (1995).
[17] W.R. Thomson, Mei Cai, Mankit Ho and J.E. Pemberton, *Langmuir*, **13**, 2291-2302, (1997).

In: Nuclear Chemistry: New Research
Editor: Axel N. Koskinen, pp. 253-266 ISBN 978-1-60456-957-5

Chapter 9

BINDING AND STRUCTURE PROPERTIES IN MONTE CARLO SIMULATION FOR NON-ISOSYMMETRIC NUCLEI

M. Ángeles Pérez-García[*]
Department of Fundamental Physics and IUFFyM
Universidad de Salamanca, Plaza de la Merced s/n 37008 Salamanca

Abstract

We study the binding energies and radii of spin saturated light nuclei with nucleon number in the range $8 \leq A \leq 30$ in the framework of semiclassical Monte Carlo many-body simulations. Using a model Hamiltonian consisting of the usual kinetic term, hadronic nucleon nucleon (NN) interaction potential, Coulomb and an effective Pauli potential, which depends on density, results are obtained by the simulation of a set of isotopic chains for selected nuclei. In particular, especial attention is devoted to nuclei in the vicinity of proton and neutron drip lines. In this latter case proton and neutron skins develop. This work suggests a pragmatic procedure for modelling a set of non-isosymmetric nuclei focusing on the description of the empirical binding energies and different (matter, neutron, charge) radii, deduced from experimentally available data, starting from a given effective nuclear interaction potential picture and statistical treatment.

PACS numbers: 07.05.Tp,21.10.D,21.65.+f

1. Introduction

The properties of finite nuclear systems have been long studied since they are not only important by itselves but they provide a terrestrial insight into properties of nuclear matter in the more extreme regimes as, for example, matter in the inside of compact astronomical objects such as neutron stars [1] or in supernova matter.

Many-body techniques have proved to be a powerful tool in the description of this type of complex systems. In this sense mean field theories [2] have provided a valuable method

[*]E-mail address: mperezga@usal.es

to reproduce nuclear observables such as nuclear binding energies, charge distributions, excitation properties, etc, available from experiment [3, 4, 5]. In particular, Random Phase Approximation (RPA) and Hartree-Fock (HF) calculations are just some examples of techniques that have been used to gain knowledge on the behaviour of finite size nuclear systems and then to extract information in order to be able to discard between the myriad of possible equations of state (EOS) of nuclear matter existing in the literature in the astrophysical context [6].

Molecular Dynamics techniques seem specially suitable for studying nuclear many body systems since they allow to study nuclear correlations beyond first order. In systems corresponding to the NVT ensamble, and, in the case of interaction potentials dependent on positions and momenta, the dynamics of each particle in the simulation is determined by solving the equations of motion. In order to fix temperature constant it is necesary to set a thermostat using additional auxiliar hamiltonian variables to keep the system in equilibrium with a thermal bath at temperature T [7]. With Monte Carlo calculations, however, the system is placed in the configurational ground state at a given temperature T by construction [8].

In an exploratory work [9], we studied the nuclear binding energies for isosymmetric ($N = Z$) spin saturated medium mass nuclei with nucleon number $8 \leq A \leq 44$ using semiclassical simulations via Monte Carlo many-body techniques. It was demonstrated that the empirical binding energies for these nuclei can be reproduced satisfactorily by the use of a short range nuclear interaction plus a density dependent effective Pauli potential, where the density dependence is parameterized by one variable, the Fermi momentum of the constituent nucleons, asuming a simplified Fermi gas model [10].

This obtained procedure is robust since the reproduction of the nuclear binding energies does not depend on the details of the NN interaction potential. One can always expect to find a proper counter-balancing density dependent Pauli potential to reproduce the empirical binding energies.

When extending this treatment [11] to a wider range of nuclei with $8 \leq A \leq 100$ for spin-isospin saturated nuclei the density dependence of the Pauli potential can be parametrized well by the nucleon number A instead of the Fermi momentum. This reflects the fact that the average Fermi momenta in nuclei with $A > 50$ saturate to a value $\simeq 260$ MeV/c (e.g., 265 MeV/c for a ^{208}Pb nucleus) [10].

Although the simulations presented in this work are not based on the fundamental physics of strong interaction, QCD, it is a effective guidance for treating complicated many-nucleon systems in practical manner. In this same line some efforts have been made [12]. In the nuclear matter low density regime, where densities do not exceed 10% of the nuclear saturation density, $\rho_0 = 0.17\, fm^{-3}$, this technique may be very helpful to study many-nucleon systems such as clustered configurations in neutron rich matter arising in the outer crust of neutron stars and known as *pasta phase* [13, 14, 15, 16].

Other nuclear finite systems in the vicinity of the proton and neutron drip lines are also interesting since there is an increasing effort in the experimentation with radioactive beams such as, among others, ISOLDE [17], those at RIKEN [18] or the proposed RIA [19] and FAIR [20]. Special attention is devoted to the new available (proton) neutron rich nuclei where observables such as the (proton) neutron skin radius can be measured, and other phenomenology like halo nuclei [21] is observed. Other experiments, like P-ReX at

Hall A in JLAB, will measure the neutron radius in ^{208}Pb using parity violating electron scattering [22].

2. Nuclear Simulations

In this picture nucleons are treated as classical structureless particles that interact typically by a Hamiltonian consisting of nucleon kinetic energy, NN interaction (V_{NN}), Coulomb (V_{Coul}) and Pauli (V_{Pauli}) potentials. The Pauli potential is a phenomenological potential depending both on positions and momenta of the interacting particles and it is included to mimic the Pauli exclusion principle in the present treatment. A Gaussian form introduced originally by Dorso *et al.* [23] is adopted in this work. We will allow, however, in the present treatment a density dependence in the Pauli potential.

The model Hamiltonian can be written as,

$$H = \sum_{i=1}^{A} \frac{\mathbf{p}_i^2}{2m_N} + \sum_{i=1,j>i}^{A} \left[V_{NN}(r_{ij}) + V_{Coul}(r_{ij}) + V_{Pauli}(r_{ij}, p_{ij}) \right], \tag{1}$$

where $\mathbf{p}_i$ is the 3-momentum of i-th nucleon and $r_{ij} = |\mathbf{r}_i - \mathbf{r}_j|$ ($p_{ij} = |\mathbf{p}_i - \mathbf{p}_j|$) the relative distance (momentum) of the i-th and j-th nucleons.

Since we are interested in the properties of finite nuclei under the influence of a phenomenological potential we will use a simplified NN interaction potential keeping only S-wave interactions [9]. We will consider firstly a square well potential with a repulsive core of strength V_{Core} of width a and an attractive well of strength V_0 and width b. We will refer to it, generically, as SW model. We will take values $a = 2\,fm$, $b = 4\,fm$, $V_{Core} = 100\,MeV$ and $V_0 = 3\,MeV$ (SW1) or $V_0 = 5\,MeV$ (SW2).

Also we will consider a second NN potential consisting of two Gaussians whose explicit expressions are as follows:

$$V_{NN}(r_{ij}) = ae^{-r_{ij}^2/\lambda_1^2} + [b + c\tau_i\tau_j]e^{-r_{ij}^2/\lambda_2^2} \tag{2}$$

where τ_i (τ_j) is the isospin third-component of i-th (j-th) nucleon ($+1/2$ for protons, $-1/2$ for neutrons). We will label this interaction potential as GM model. The parameter values are $a = 50\,MeV$, $b = 3\,MeV$, $c = -4\,MeV$, $\lambda_1 = 0.8\,fm$ and $\lambda_2 = 4\,fm$.

Due to the fact that protons are electrically charged particles the Coulomb potential must be introduced. It can be written as,

$$V_{Coul}(r_{ij}) = \frac{e^2}{4\pi r_{ij}}(1/2 + \tau_i)(1/2 + \tau_j), \tag{3}$$

where e is the proton electric charge. Finally the Pauli potential reads

$$V_{Pauli}(r_{ij}, p_{ij}) = V_P \, \exp\left(-\frac{r_{ij}^2}{2q_0^2} - \frac{p_{ij}^2}{2p_0^2} \right) \delta_{\tau_i\tau_j}\delta_{\sigma_i\sigma_j}, \tag{4}$$

where $\delta_{\tau_i\tau_j}$ ($\delta_{\sigma_i\sigma_j}$) is the Kronecker's delta for the isospin (spin) third-component. The Pauli potential mimics the Pauli exclusion principle to prevent nucleons from occupying the same phase space volume when they have the same quantum numbers. Although this effective procedure allows to somewhat take care of the genuine antisymmetrization of the fermion sector, however, this treatment is not absent of some particular issues regarding a more selective analysis [24].

As demonstrated in Ref. [9], it is crucial to allow a density dependence for this Pauli potential if one focuses on the reproduction of the empirical binding energies for a set of nuclei. Since the Fermi momenta for nuclei in the middle mass range saturate to a value of $\simeq$ 260 MeV/c [10] then one can expect the nucleon number A may be an alternative variable to characterize the density dependence of the Pauli potential.

The parameters q_0 and p_0 in the Pauli potential, are density dependent and can be related to the typical interparticle distance in a nucleus where the nucleon sphere of radius r is given by,

$$r = \left(\frac{3}{4\pi\rho}\right)^{1/3}, \tag{5}$$

where ρ is the nucleon density. As done in [10] this could be related to the model where nucleons are considered in the Fermi gas approximation. In this case then $\rho = 2p_F^3/3\pi^2$ and p_F is the nucleon Fermi momentum. Then, the average inter-nucleon distance $2r$ may be estimated as $(2r/\sqrt{2}q_0) \simeq 1$, where q_0 is an "effective range" of the Pauli potential. With the uncertainty principle, $q_0\ p_0 \simeq \hbar$, this leads to:

$$q_0 \simeq \frac{(9\pi)^{1/3}\hbar}{\sqrt{2}p_F}, \tag{6}$$

$$p_0 \simeq \frac{\hbar}{q_0} = \frac{\sqrt{2}}{(9\pi)^{1/3}}p_F. \tag{7}$$

We are mainly interested in this work in showing that Monte Carlo techniques can be used to describe binding energies and radii in finite nuclear systems that may have some asymmetry in the isospin component. We will consider systems in the light mass range $8 \leq A \leq 30$. In this range the p_F asociated to nuclei [10] only changes by 10% so that we will average the density dependence in the q_0 and p_0 parameters and set them to be constant with a value of $(q_0, p_0) = (1.8\, fm, 110\, MeV/c)$.

In order to calibrate the model we choose a set of mirror nuclei where the neutron number, N, and the proton number, Z, can be interchanged as, for example, in ^{14}C and ^{14}O and fit the experimentally available root mean square (RMS) matter and charge radius. Proton and charge radius can currently be measured with high relative precision using electromagnetic probes by the cross section measurement of elastic scattering of electrons [25], with muonic atoms [26], etc. However, for matter and neutron radii the accuracy is not so good due to the fact that there is uncertainty in the nuclear regions probed by the various processes and due to the effective nuclear interactions between nucleons.

It is worth mentioning that, for simplicity, in this work we have considered only S-wave interactions so, in principle, only spherical non-deformed nuclei could be meaningfully simulated. However, if a proper account on the angular momentum is taken, one could simulate nuclear configurations where dipole or quadrupole deformations are non-negligible [27].

3. Nuclear Binding and Structure Results

In the present study, all simulations are performed in the NVT ensamble with a fixed temperature $T = 1$ MeV. The different isotopic nuclear chains are simulated by means of a finite size system with A nucleons. Initially, nucleons are uniformly distributed inside a sphere of radius R_0 of range 2-3 fm within a cubic box of volume $V = L^3$ and impose $L >> R_0$. We consider only even number of protons and neutrons, since in our simplified treatment all isospin populations are paired off in spins. Then, using the Metropolis algorithm [8], the ground state configuration forming a compact nucleus, is searched by thermal relaxation. After this, we sample the configurations in order to calculate the statistical average for binding energies and RMS radii. The Pauli potential strength, V_P, in Eq.(4) has been shown to be A dependent in order to reproduce the empirical binding energies for the nuclei as can be seen in Fig. 1 for (even Z and N) spin-isospin saturated nuclei using the SW1 potential. As expected, the strength V_P of the Pauli potential increases as A (or nuclear density for pre-saturation mass range) increases since correlations tend to be more important. This behaviour may be analogous to that of the vector potential in Hartree approximation in relativistic mean field models [2].

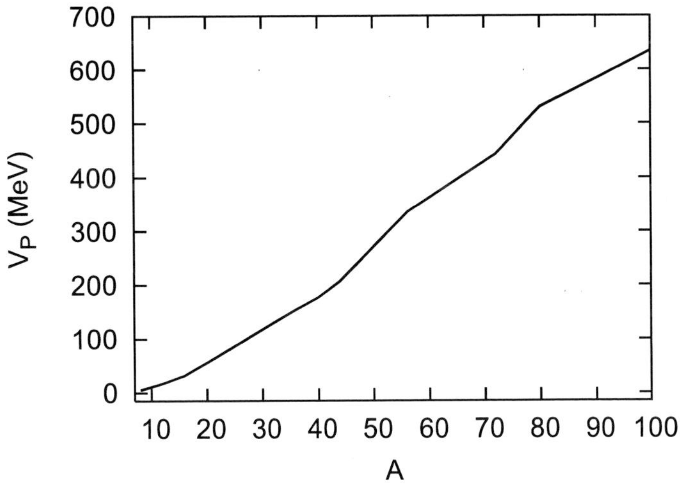

Figure 1. Pauli potential strength V_P vs. nucleon number A for (even Z and N) spin-isospin saturated nuclei using the SW1 model.

In Fig. 2 we plot binding energy per nucleon vs. the nucleon number, A, for even-even nuclei up to the $A = 100$ value using the SW1 model. Statistical error bars are within the 5% experimental values. We can see that there is a satisfactory reproduction of the binding energy curve.

Kinetic and potential energetic contributions to the binding energy per nucleon for even-even nuclei can be seen in Fig. 3. We use the SW model with two different values of the potential well strength. Solid (dashed) lines denote calculations using set SW1 (SW2). Total binding energy is depicted by the dotted line. The total binding energy results as a delicate counterbalance of the positive kinetic energy contribution plus Pauli and Coulomb

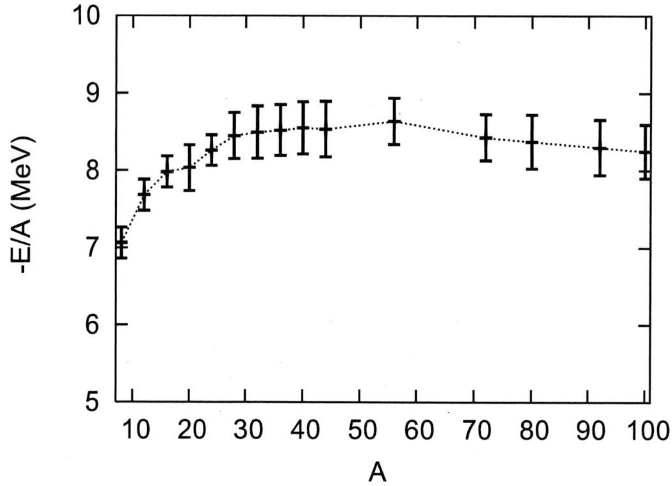

Figure 2. Binding energy per nucleon vs. nucleon number A for even-even nuclei using the SW1 model.

correlations and the negative atractive NN potential interaction to give the approximate $\approx -(7-8)$ MeV/nucleon in the selected mass range.

When simulating a set of nuclei we will be interested not only in reproducing the nuclear binding energies, but also the structure properties such as matter, proton and neutron RMS radii [11], charge distributions or phenomenology like the presence of skins or halos for some non-isosymmetric nuclei.

In our approach nucleons are considered as point-like particles and the matter RMS radius is computed by averaging the number density operator over the box volume $V = L^3$

$$\rho_{matter}(\mathbf{r}) = \frac{\mathbf{1}}{\mathbf{L^3}} \sum_{\mathbf{i}} \delta(\mathbf{r} - \mathbf{r_i}) \tag{8}$$

In this way the matter RMS radius can be obtained as

$$< r_m^2 >^{1/2} = [\int d^3\mathbf{r} \rho_{\mathbf{matter}}(\mathbf{r})\, \mathbf{r}^2]^{1/2} \tag{9}$$

The proton number density operator is written as

$$\rho_{proton}(\mathbf{r}) = \frac{\mathbf{1}}{\mathbf{L^3}} \sum_{\mathbf{i}} (\mathbf{1/2} + \tau_{\mathbf{i}}) \delta(\mathbf{r} - \mathbf{r_i}) \tag{10}$$

and in order to compare with the nuclear charge radius, r_c, that can be experimentally measured, it is usual [28] to assume a proton finite size, s_p, and typically $s_p \approx 0.8$ fm [29]. The expresion that will be taken for the charge RMS radius is

$$< r_c^2 >^{1/2} = [< r_p^2 > + s_p^2]^{1/2} \tag{11}$$

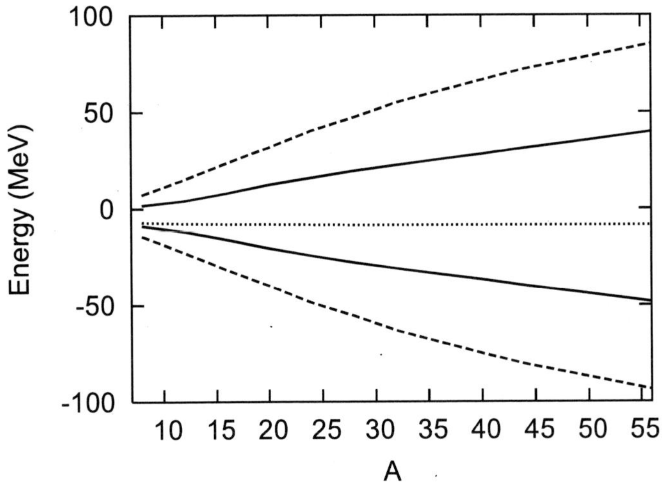

Figure 3. Kinetic and potential energetic contributions to the total binding energy per nucleon vs. nucleon number A for even-even nuclei using the SW model. See details in the text.

where $< r_p^2 >^{1/2}$ is computed in similar fashion to Eq.(9). Then the neutron number density operator is

$$\rho_{neutron}(\mathbf{r}) = \frac{\mathbf{1}}{\mathbf{L^3}} \sum_{\mathbf{i}} (\mathbf{1/2} - \tau_{\mathbf{i}}) \delta(\mathbf{r} - \mathbf{r_i}) \tag{12}$$

With these definitions we can proceed and obtain the neutron skin radius that is supposed to be available experimentally as

$$r_{NS} = < r_n^2 >^{1/2} - < r_p^2 >^{1/2} \tag{13}$$

This observable is related [30] to the shift in the surface to volume ratio of symmetry energy, $r_{S/V} = a_S/a_V$. For heavy nuclei the neutron skin is related to the neutron excess by the hydrodynamical formula to the lowest expansion [31] in $1/A$ as

$$\frac{\delta r^2}{< r^2 >} \approx \frac{N-Z}{A}(1 + \frac{2}{3}\frac{r_{S/V}}{A^{1/3}} - ...) \tag{14}$$

where $< r^2 >= \frac{3}{5}R^2 A$ and $R = R_o A^{1/3}$ with $R_o \approx 1.1$ fm. Skyrme-Hartree-Fock models [32] predict a ratio $r_{S/V} = 1.6$ in accordance with the mass formula ratio. Other calculations yield values in the range 2.2 ± 0.6 [30, 33, 31].

In Fig. 4 we show the shift in binding energy per nucleon ($-E/A$) from the isosymmetric species ($N = Z$) versus nucleon number A for Carbon (C), Oxygen (O), Neon (Ne) and Magnesium (Mg) isotopes. We can see that for Ne and Mg the minimum is shifted to the neutron rich side.

In Fig. 5 we plot matter (boxes), neutron (circles) and charge (triangles) RMS radii for the Carbon isotopic chain versus the nucleon number A. We can see that for the symmetric species with $Z = 6, N = 6$ there is no skin present. For the proton-rich isotopes a proton

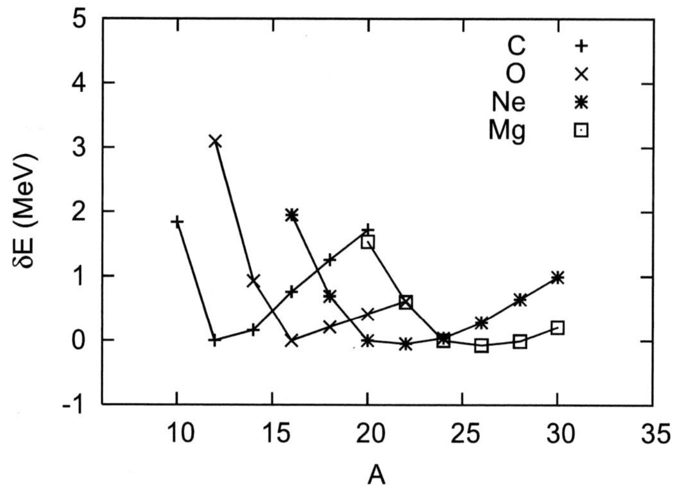

Figure 4. Energy shift per nucleon for the isotopic chains of C, O, Ne and Mg nuclei.

skin develops while for nuclei on the neutron rich side a neutron skin forms. As A increases this neutron skin grows since the charge radius remains approximately constant.

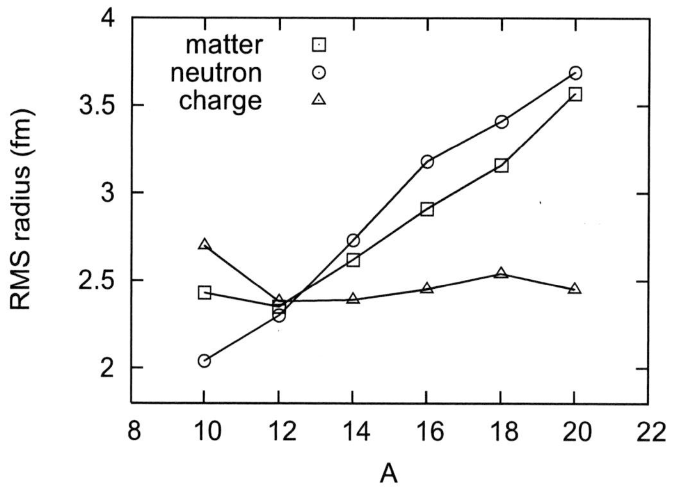

Figure 5. From top to bottom: neutron, matter and charge RMS radii for the isotopic chain of Carbon.

We can see more in detail in Fig. 6 matter (boxes) RMS radii for the Carbon isotopes along with experimentally deduced values from Ref. [34]. Data are extracted from hadronic reactions involving break up of nuclei. For example, in the reaction

$$a + A \rightarrow b + X \tag{15}$$

the interaction cross section depends on the mass number of the projectile (A_a) and of the fragment (A_b) but it is found to be essentially independent of the target mass and beam energy. It can be expressed using the Goldhaber model [35] as

$$\sigma = \sqrt{A_b(A_a - A_b)/(A_a - 1)}\sigma_0 \quad (16)$$

with this prescription and in the approximation where each isotopic component is a Fermi gas with Fermi momentum p_F, then the nucleon density is $\rho = \frac{2p_F^3}{3\pi^2}$ and $\sigma_0 = p_F/\sqrt{5}$.

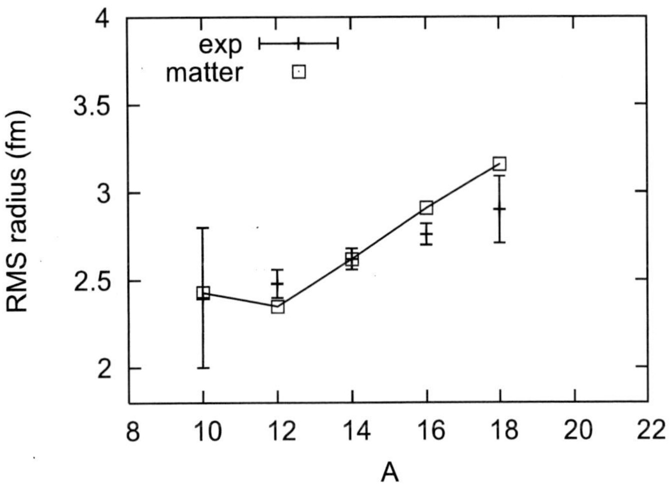

Figure 6. Matter RMS radii for the isotopic chain of Carbon as obtained with the GM model. Empirical data are shown along with error bars taken from Ref. [34].

In Figs. 7 and 8 we plot neutron (circles) and charge (triangles) RMS radii, respectively, for the Carbon isotopes along with experimental data. We can see that while the neutron RMS radius is a rising function of A the nuclear charge radius stabilizes from the symmetric isotope up to values of $T_z = 3$, where the isospin in the z-component, T_z, is defined as $T_z = (N - Z)/2$.

In Fig. 9 we show a comparative of the matter RMS (boxes), neutron (circles) and charge (triangles) for the Oxygen isotopes as obtained with the GM model. We can see that as isospin asymmetry grows a neutron (proton) skin develops on the neutron (proton) rich side.

In Fig. 10 we can see the neutron skin radius, r_{NS}, as defined in Eq.(13), versus nucleon number A for the Neon isotope chain as obtained in more sofisticated calculations as those of the relativistic mean field approximation with the constant gap approximation in axially deformed oscillator basis (DEF) model from Ref.[36] and compared to our semiclassical simulation with the GM model. We can see that the divergence of both models is bigger as the isospin asymmetry ($I = |N - Z|$) grows. Note that our model assumes no spatial deformation from sphericity whatsoever.

In Fig. 11 we can see the energetic contributions to the Mg isotope chain [37]. Solid, dotted and dashed lines show the potential, kinetic and total binding energy, respectively. Positive and negative energetic contributions allow to fit the experimental isotopic binding energy values.

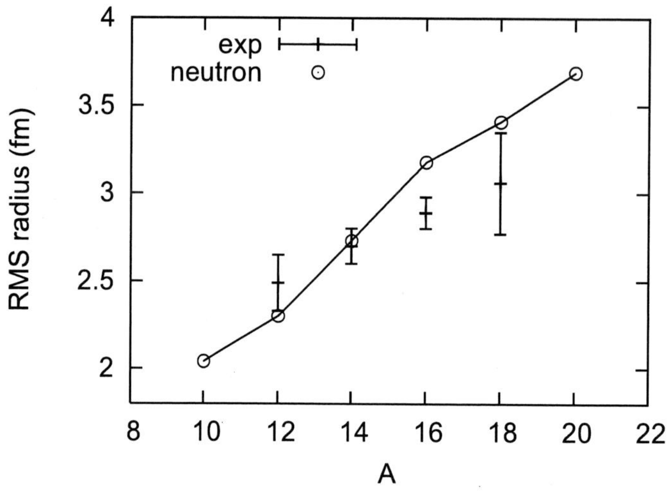

Figure 7. Neutron RMS radii for the isotopic chain of Carbon as obtained with the GM model. Empirical data are shown along with error bars taken from Ref. [34].

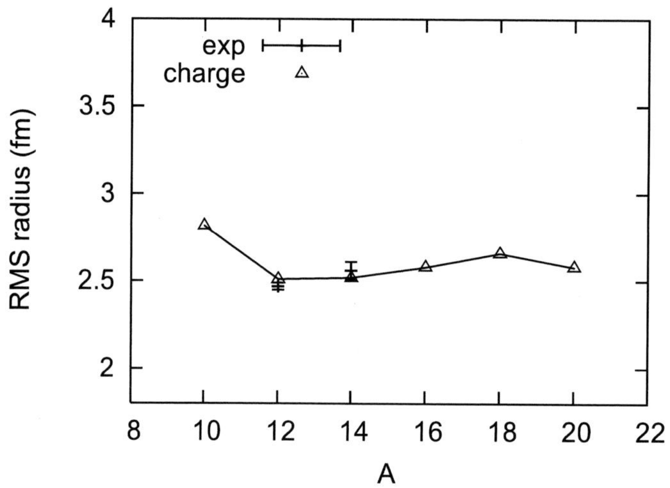

Figure 8. Same as Fig. 7 for the charge RMS radii.

In Fig.12 we show matter and charge radii vs. the neutron excess for the Mg isotopic chain. Experimental values for matter radii are depicted for comparison as taken from Ref. [34]. Neutron excess (fault) in the isotope results in a developing neutron (proton) skin.

4. Conclusion

Using many-body semiclassical simulations with Monte Carlo techniques, we have studied the nuclear binding energies for spin saturated nuclei in the light mass range with $8 \leq A \leq 30$. We use a Hamiltonian formulation with a simplified model where apart from the kinetic

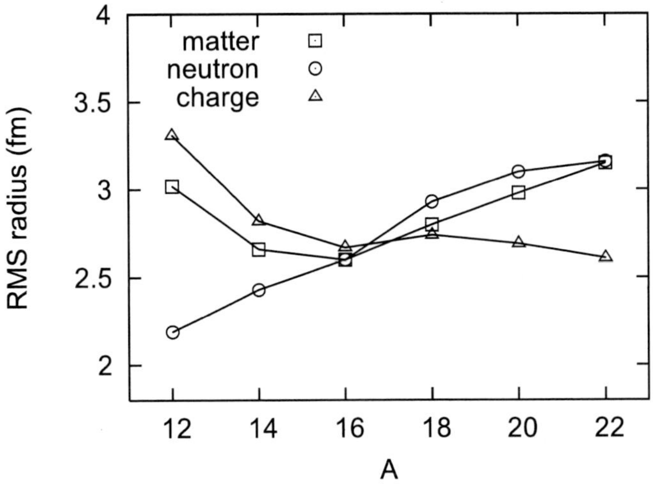

Figure 9. matter RMS (boxes), neutron (circles) and charge (triangles) radii for Oxygen isotopes as obtained with the GM model.

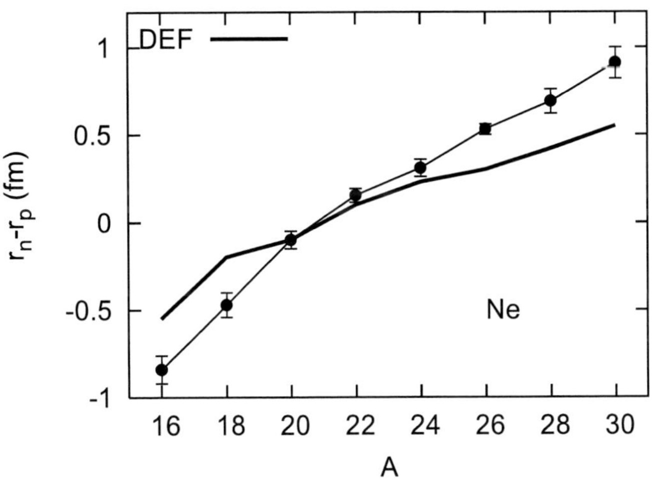

Figure 10. Neutron skin radius versus nucleon number A for the Neon isotope chain as deduced from the DEF model [36] and compared to our calculation with the GM model. We can see that the divergence of both models is more important as the isospin asymmetry grows.

term the S-wave only NN interaction potential consists of a hadronic interaction, Coulomb and Pauli terms. We have shown that for a set of nuclear isotopic chains (C, O, Ne, Mg) binding energies and nuclear structure properties such as matter, neutron and charge radii can be reproduced in an approximate way. As isospin asymmetry grows the departure from the experimental values as obtained with this simplified model is larger.

The present study confirms our previous conclusion that to allow a density dependence for the Pauli potential is crucial to reproduce the empirical binding energies and radii in

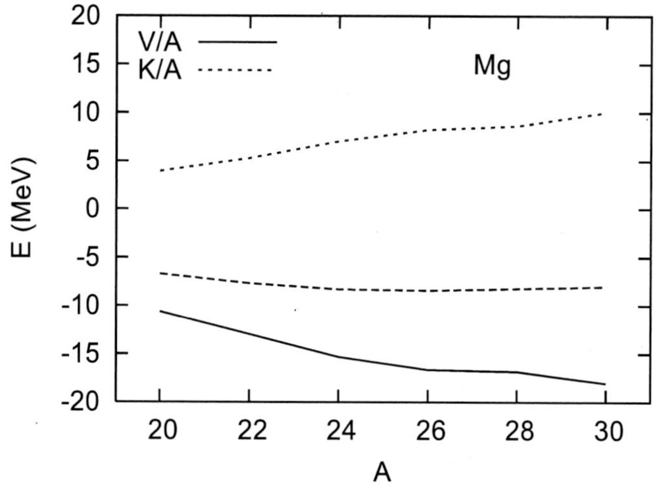

Figure 11. Potential (solid line), kinetic (dotted line) and total (dashed line) binding energy per nucleon for the isotopic chain of Mg.

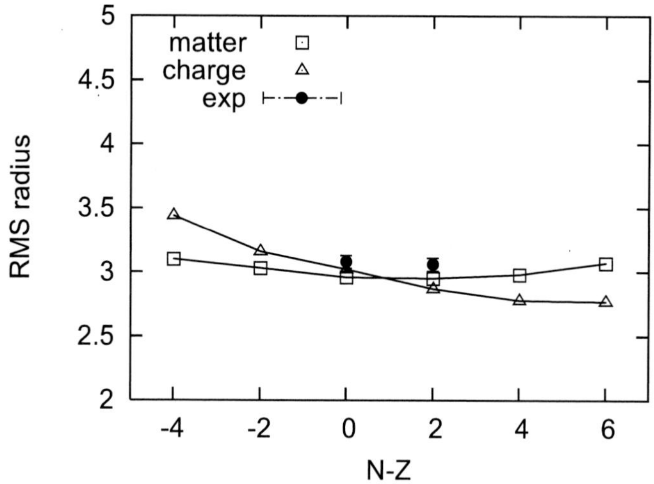

Figure 12. Matter and charge radii vs. the neutron excess for the Mg isotopic chain. Experimental values are depicted for comparison as taken from Ref. [34].

semiclassical simulations.

We find that a neutron (proton) excess in principle results in a neutron (proton) skin forming in the exterior shell of the nuclei and computed as the difference of both RMS radii. For isosymmetric nuclei there is a vanishing skin in our model, in accordance with the liquid drop formula. Charge radii are computed from point-like particles in our simulation by adding the proton finite size ≈ 0.8 fm. Our computed values are within experimental errors. Charge radii stabilizes as A increases giving the possibility of a skin formation. However, in particular, the Neon isotope anomaly [36] can not be reproduced within our simplified spherical configuration. For this type of phenomenology the nuclear deformations play an importatn role. Although the procedure presented in this work is not based on the fundamental theory of strong interaction QCD, we would like to emphasize that, it suggests a pragmatic procedure for modelling a set of nuclei.

Acknowledgements

We would like to thank K. Tsushima and A. Valcarce for suggesting us this calculation. This work has been partially funded by the Spanish Ministry of Education and Science project DGI-FIS2006-05319.

References

[1] Glendenning, N. K. *Compact Stars*, 2nd edition, Springer-Verlag, New York (2000)

[2] Serot, B. D. and Walecka, J. D. *Adv. Nucl. Phys.* **16** (1986) 1

[3] Nadjakov et al, *At. Data Nucl. Data Tables* **56** (1994) 133

[4] Audi, G., Wapstra, A. H. and Thibault, C. *Nucl. Phys. A* **729** (2003) 337.

[5] Anagnostatos, G. S. *Int. J. Mod. Phys.* **E5** (1996) 557

[6] Horowitz C. J. and Piekarewicz, *J. Phys. Rev. C* **64** (2001) 062802

[7] Perez-Garcia, M. A. *J. Math. Chem.* **40**, 1 (2006) 63

[8] Metropolis, M. et al., *J. Chem. Phys.* **21** (1953) 1087

[9] Perez-Garcia, M. A., Tsushima, K.,Valcarce, A., *Int. Jour. Mod. Phys. E,* in press.

[10] Moniz E. J. et al., *Phys. Rev. Lett.* **26** (1971) 445

[11] Perez-Garcia, M. A. Tsushima, K. Valcarce, A. *Physics Letters B* **660** (2008) 600

[12] Akimura, Y. Maruyama, T. Yoshinaga, N. Chiba, S. *European Physical Journal A* **25** (2005) 405

[13] Ravenhall, D. G. Pethick, C. J. and Wilson, J. R. *Phys. Rev. Lett.* **50** (1983) 2066; Hashimoto, M. Seki, H. and Yamada, M. *Prog. Theor. Phys.* **71** (1984) 320

[14] Watanabe G. et al., *Phys. Rev. Lett.* **94** (2005) 031101; Watanabe G. and Sonoda, H. arXiv:cond-mat.soft/0502515

[15] Maruyama T. et al., *Phys. Rev. C* **72** (2005) 015802

[16] Horowitz, C.J. Pérez-García, M.A. and Piekarewicz, J. *Phys. Rev. C* **69** (2004) 045804;

[17] http://isolde.web.cern.ch/isolde/

[18] Ishihara, M. *Nuclear Physics A*, Volume 538 (1992) 309

[19] "The Science of the Rare Isotope Accelerator (RIA)", Internal Report, 2005, http://www.orau.org/ria/pdf/RIAFINAL.pdf

[20] Rosner, G. Proceedings of the 7th International Conference on Hyperons, Charm and Beauty Hadrons, *Nuc. Phys. B* **167** (2007) 77

[21] For a review see, e.g., Tanihata, I. *J. Phys. G.* **22** (1996) 157

[22] Horowitz, C. J. Pollock, S. J. Souder, P. A. Michaels, R. *Phys. Rev. C* **63** (2001) 025501

[23] Dorso, C. Duarte, S. and Randrup, J. *Phys. Lett. B* **188** (1987) 289 ibid. B 215 (1988) 611

[24] Taruna, J. Piekarewicz J. and Pérez-García, M. A. J. *Phys. A: Math. Theor.* **41** (2008) 035308

[25] Singhal, R. P. Moreira, J. R.Caplan, H. S. *Phys. Rev. Lett.* **24** (1970) 73

[26] Fricke G. et al., *Phys. Rev. C* **45** (1992) 80

[27] Gambhir, Y. K. Ring P. and Thimet, A. *Ann. Phys. (NY)* **198** (1990) 132

[28] Buchinger, F. Pearson, J. M. and Goriely, S. *Phys. Rev. C* **64** (2001) 067303

[29] Particle Data Book, Yao, W.-M. et al., *Journal of Physics,* **G 33**, 1 (2006)

[30] Moller, P. et al, *At. Data Nucl. Data Tables* **59** (1995) 185

[31] Lipparini, E. and Stringari, S. *Phys. Lett. B* **112** (1982) 421

[32] Satula, W. and Wyss, R. *Phys. Lett. B* **572** (2003) 152

[33] Danielewicz, P. *Nuc. Phys. A* **727** (2003) 233

[34] Ozawa, A. *Riken Review* N.**39**: Focused on Physics at Drip Lines (2001)

[35] Goldhaber, A. S. *Phys. Lett.* **53 B** (1974) 306

[36] Bhagwat, A.and Gambhir, Y. K. *Phys. Rev. C* **68** (2003) 044301

[37] Ren, Z. Zhu, Z. Y. Cai, Y. H. and Xu, G. *Phys. Lett. B* **380** (1996) 241

INDEX

A

D

E

F

G

H

I

J

K

L

M

N

O

P

Q

R

S

T

U

V

W

X

Y

Z